Advanced Research in VLSI

Proceedings of Previous Conferences

Proceedings of Caltech Conference on Very Large Scale Integration, 1979, Charles L. Seitz, Editor, Computer Science Department, California Institute of Technology, 256–80, Pasadena, CA 91125.

Proceedings of Second Caltech Conference on Very Large Scale Integration, 1981, Charles L. Seitz, Editor, Computer Science Department, California Institute of Technology, 256–80, Pasadena, CA 91125.

Proceedings, Conference on Advanced Research in VLSI, MIT, January 1982, Paul Penfield, Jr., Editor, Artech House, Inc., 610 Washington St., Dedham, MA 02026.

Third Caltech Conference on Very Large Scale Integration, 1983, Randal Bryant, Editor, Computer Science Press, Inc., 1803 Research Boulevard, Rockville, MD 20850.

Proceedings, Conference on Advanced Research in VLSI 1984, Paul Penfield, Jr., Editor, Artech House, Inc., 610 Washington St., Dedham, MA 02026.

1985 Chapel Hill Conference on Very Large Scale Integration, Henry Fuchs, Editor, Computer Science Press, Inc., 1803 Research Boulevard, Rockville, MD 20850.

Advanced Research in VLSI: Proceedings of the 4th MIT Conference, Charles Leiserson, Editor, MIT Press, 55 Hayward Street, Cambridge, MA 02142.

Advanced Research in VLSI: Proceedings of the 1987 Stanford Conference, Paul Losleben, Editor; The MIT Press, 55 Hayward St., Cambridge, MA 02142.

All of the above volumes are available directly from the publishers.

Advanced Research in VLSI

Proceedings of the Fifth MIT Conference, March 1988

edited by
Jonathan Allen and
F. Thomson Leighton

The MIT Press
Cambridge, Massachusetts
London, England

PUBLISHER'S NOTE

This format is intended to reduce the cost of publishing certain works in book form and to shorten the gap between editorial preparation and final publication. Detailed editing and composition have been avoided by photographing the text of this book directly from the authors' prepared copy.

This book was printed and bound in the United States of America

Library of Congress Cataloging-in-Publication Data

Advanced Research in VLSI.

   Papers presented at the Fifth MIT Conference on Advanced Research in VLSI, organized by the Microsystems Research Center of the Massachusetts Institute of Technology, held in Cambridge, Mass., Mar. 1988.
   Includes bibliographies and index.
   1.   Integrated circuits--Very large scale integration--Congresses. I. Allen, Jonathan, 1934-   .
II. Leighton, Frank Thomson. III. Massachusetts Institute of Technology. Microsystems Research Center.
IV. MIT Conference on Advanced Research in VLSI (5th: 1988: Cambridge, Mass.)
TK7874.A3352 1988 621.395   88-621
ISBN 0-262-01100-X

Contents

Preface

As the size and complexity of VLSI circuits has increased over the last decade, the emphasis among the many VLSI disciplines has changed from its intial focus on circuits and technology to a growing concern with design issues. Additional levels of abstraction have encouraged computer scientists to contribute in a complementary way with electrical engineers. While computer scientists were concerned with logic, architecture, and complexity issues, electrical engineers addressed technology and circuit issues. Early in the decade, there was a tendency for the two disciplines to produce designs in different representational domains. The meeting ground was at the artwork level, where both groups shared a concern for the details of layout and interconnection. In recent years, however, there has been increased integration as local optimizations within the individual representational domains were recognized as inadequate in terms of providing the best overall system. Performance for specific applications is acknowledged as the key motivator, and it has necessitated a greater awareness of the interactions of various design representations, such as layout, circuit, and architecture.

In this conference, the broad range of important contributions to the design of high-performance circuits (often targeted at novel applications) is demonstrated. New architectures aimed at improved system performance are discussed, and the use of novel analog circuits in connectionist architectures is described. Given the importance of speed, it is not surprising to find several papers on clock design methodologies, and the use of self-timed circuits. Fault tolerance, reconfiguration, testing, and verification are topics of growing concern due to the huge complexity of existing chips and systems which require fresh approaches to design and CAD tools. Design optimization continues to be a major theme, focusing on logic, circuit size, and speed. The specification of high-performance layout (both in theory and practice) is also addressed, and the integration of these design tools into easy-to-use and extendable CAD systems deals with the need to coherenly interrelate programs oriented to many facets of the overall design. We believe this broad range of interest illustrates the vitality in all aspects of contemporary VLSI research.

We gratefully acknowledge the strong contributions of those on the program committee, who collectively and carefully read the eighty submitted paper, and selected nineteen from an outstanding field. The desire to avoid parallel sessions, and the need to cover a substantial range of topics, prevented us from accepting many excellent papers. We also thank the authors of the nineteen contributed papers, and the nine invited speakers, whose expertise characterizes the ongoing progress in the many diverse aspects of VLSI as well as Robert Lucky who is the Conference banquet speaker. Thanks to Bernard Chern and Robert Grafton of the National Science Foundation for their support of the conference. Barbara Lory, Barbara Tilson, and the staff of the MIT Microsystems Research Center have contributed to all aspects of the Conference's organization and planning. Paul Penfield is the guiding hand who has set the standard for this continuing series of MIT conferences. Carver Mead and Chuck Seitz started these university-based conferences in 1979, and MIT is pleased to carry on the tradition during the even-numbered years. We hope you enjoy these papers and the view they provide of today's VLSI research.

Jonathan Allen
F. Thomson Leighton

Program Committee

Advanced Research in VLSI

A Pulse-Width Modulation Design Approach and Path-Programmable Logic for Artificial Neural Networks

Neil E. Cotter, Kent Smith, and Martin Gaspar

Electrical Engineering Department
University of Utah
Salt Lake City, UT 84112

Simple circuits for performing mathematical operations on analog signals are described. Multiplication, addition, and nonlinear scaling of voltages are accomplished by a pulse-width modulation scheme that utilizes comparators and switched capacitor filters implemented in CMOS. A set of standard cells containing these circuit elements may be used in a Path-Programmable Logic scheme that provides for extremely rapid artificial neural network integrated circuit design. Design of a Kohonen self-organizing feature map is described.

1. Introduction

Biological neural networks rely on analog signal processing and massive connectivity to achieve low-power, high-speed computation. Artificial Neural Networks (ANNs) mimic this ability by performing computations electronically or optically. A new approach for designing ANN integrated circuits based on Pulse-Width Modulation (PWM) circuits is described here. The PWM approach has several advantages over other proposed approaches [1,2,5,7-11,19,20]: (1) it provides a reliable inter-chip communication scheme, (2) it provides a circuit for multiplication that has wide dynamic range and tolerates variations in component values, (3) it provides a nonlinear scaling circuit with a precisely controllable transfer function, (4) it provides a small number of cell types that can be used to implement almost any neural network algorithm, (5) it provides convenient digital PWM test signals, and (6) it provides design cells that require small numbers of components.

The Path-Programmable Logic (PPL) system, developed by Smith [21,26-29] at the University of of Utah, may be used to layout PWM circuits for ANNs. In PPL, layout starts with the definition of signal lines that are laid down in a grid on the integrated circuit. Circuit cells are then placed at appropriate locations on the grid. The primary advantage of PPL is that complete integrated circuits can be designed in as little as a few hours, but PPL has additional advantages for the design of ANNs: (1) circuits for simulation are automatically extracted from the layout, (2) analog and digital circuits are easily combined, and (3) mixed types of ANNs can be implemented on a single chip. In the last section of the paper the simplicity of the PPL approach is demonstrated for a somewhat complicated algorithm: Kohonen's self-organizing feature map [17].

2. Circuit building blocks for ANNs

CMOS circuit building blocks used in the PPL design of ANNs are shown in Figure 1. The first building block is a comparator with complementary outputs. As shown below, the comparator is useful in at least four roles: (1) in converting analog voltage levels to PWM signals, (2) in multiplying voltage levels, (3) in finding the smallest of a set of voltages, and (4) in scaling voltages by nonlinear functions.

The second building block is a summation and low pass filter circuit designed with switched capacitors. This circuit is useful as a stand-alone low pass filter or as a summation circuit. It accepts either PWM waves or analog levels as input, and it outputs a slowly varying signal that is proportional to the duty cycle of the sum of the input voltages. A complementary analog output (denoted by a solid circle) is provided for use in multiplication circuits. All resistances in the circuit have the same value and are simulated by switched capacitors that require a clock signal ($\phi 1$).

The third building block is a capacitive storage circuit for synaptic weights. Values stored as voltages on synaptic weight capacitors act as the memory elements in an ANN. The function of these synapses is to control the strength of connections between neurons, exactly paralleling the function of synaptic connections between biological neurons.

Changes in synaptic weights are effected by a voltage connected to a capacitor through a CMOS switch and a resistance R. If the switch is closed for a short time t_Δ, the capacitor starts charging from its initial voltage V_0 towards the input voltage V_I. The net change ΔV in voltage on the capacitor is proportional to t_Δ provided t_Δ is small:

$$\Delta V \approx \frac{t_\Delta}{RC}(V_I - V_0)$$

This form of synaptic weight change is useful in proportional increment training [4,22,25], backward error propagation [23], self-organizing feature maps [17], and adaptive resonance [6].

If $VI \gg V0$ then the dependence on V_0 is eliminated from the equation for ΔV, and the synaptic weight can be stepped by fixed increments proportional to t_Δ and V_I:

$$\Delta V \approx \frac{t_\Delta}{RC}V_I$$

This form of synaptic weight change is useful in outer product rule learning for Hopfield networks [13-16,30].

The synaptic weight can also be set to a desired voltage level by leaving the CMOS switch closed until the capacitor charges to the value of V_I. This is useful for initializing synaptic weights.

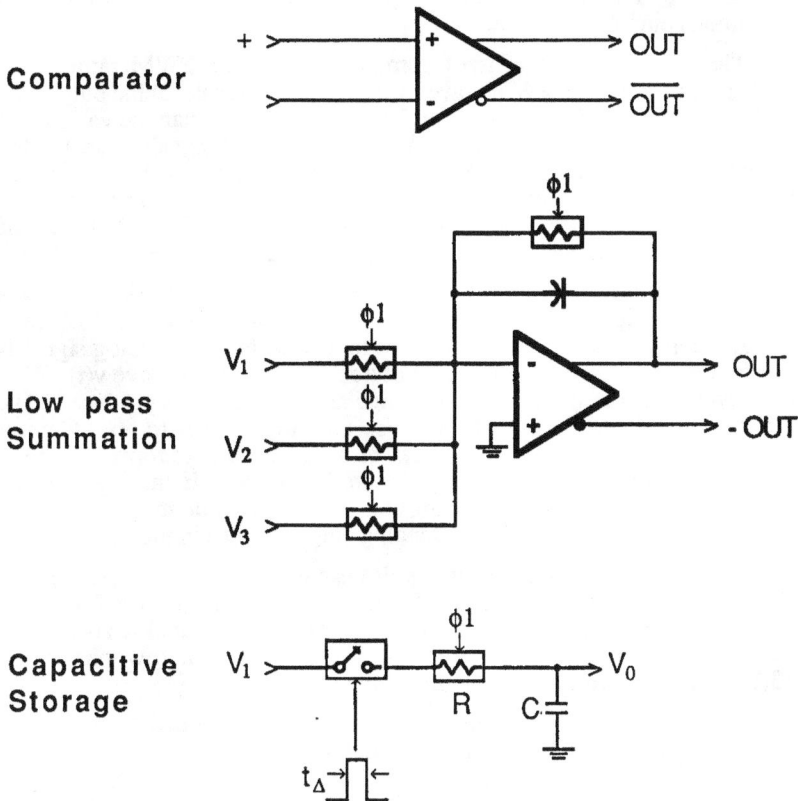

Figure 1 – Circuit building blocks for ANNs.

3. PWM interchip communication scheme

This and the following sections describe how the ANN circuit building blocks can be combined to provide useful functions. The first such function is the communication of signals between ANN integrated circuits, a central issue in the design of expandable ANNs.

Biological neurons use pulses called action potentials to communicate with one another. To a first approximation, pulse rates encode the activity of neurons. Because all action potentials are the same shape and size, the duty cycle of neural output is proportional to pulse rate. Hence, neurons convey information by altering the duty cycle of a pulse train. Equivalently, neurons may be said to be communicating sample values of an analog signal.

A PWM wave accomplishes the same task; the duty cycle of a PWM wave conveys information about an analog signal. By low pass filtering the PWM signal, the underlying analog signal is recovered. This scheme is used in computer sound generators that sample a desired output waveform, convert it to PWM, and then low pass the PWM signal to produce a continuous output waveform.

There are three benefits of communicating with PWM rather than analog signals: (1) the PWM waveform is a digital signal, (2) the duty cycle of the PWM waveform can be controlled more precisely than the value of an analog signal, and (3) the information content of PWM signals is unaffected by moderate attenuation. These benefits permit communication of analog information to be accomplished with a minimum of analog circuitry. Consequently, circuit complexity is reduced and the reliability of communication is increased.

Figure 2 illustrates the PWM interchip communication scheme. In a transmitting chip, analog output signal V_0 is converted into a PWM wave i(t) by a comparator whose output goes high whenever the analog signal is greater than the instantaneous height of a reference triangle wave v(t). With each cycle of the triangle wave a rectangular current pulse is produced. The width and DC component of the pulse are proportional to V_0. Current pulses are used so that output lines can be wired together to form the sum of signal strengths, a convenient operation for ANNs. If the input to the downstream ANN is a virtual ground, then a resistor may be used to convert voltages to current in the sending chip as shown in the figure.

The width (or duty cycle) of a pulse is equivalent to a sample value of the original analog signal. If the frequency of the triangle wave is much higher than the frequency of the analog signal, then the analog signal may be recovered by passing the PWM wave through a low pass filter [3,12,24], as shown in the receiving chip in Figure 2.

It is assumed that the reference triangle wave is provided by an external signal source. Although this requires additional circuitry, the same triangle wave can be used by every chip in a system, and triangle waves are easily generated. Furthermore, amplification or other analog processing of the triangle wave is unnecessary.

4. Multiplication

A common element of ANN algorithms is the need to compute weighted sums of voltages, where the weights are synaptic strengths and the voltages represent incoming activity from sensors or other neurons. Perceptrons provide a typical example:

$$V_i = \text{sgn}(\sum_{j=0}^{N} w_{ij}V_j)$$

$$\text{where} \quad V_i \equiv \text{output of } i\text{th neuron}$$

$$w_{ij} \equiv \text{synaptic weight for } j\text{th input}$$

$$V_j \equiv j\text{th input}$$

Since synaptic strengths and input signals are both represented by voltages, a voltage multiplier is needed. Figure 3 shows the block diagram of a PWM multiplier. Multiplication is accomplished by using the

Transmitting chip

Receiving chip

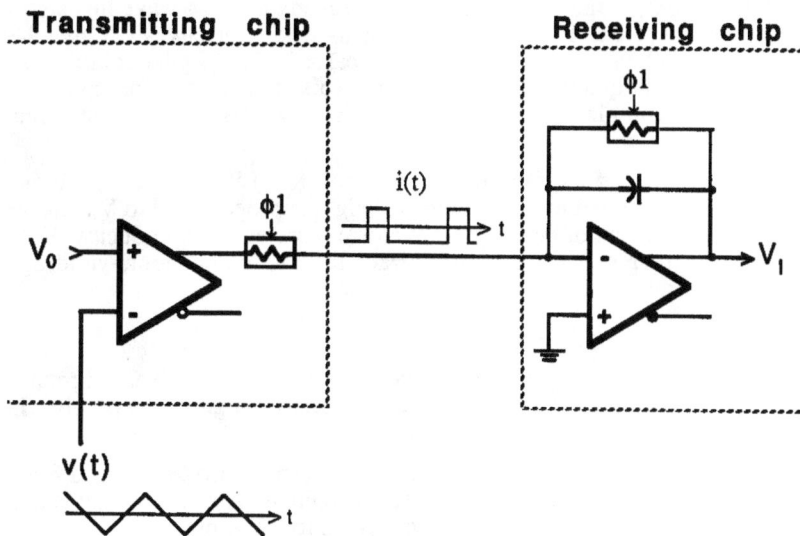

Figure 2 – PWM interchip communication scheme for ANNs.

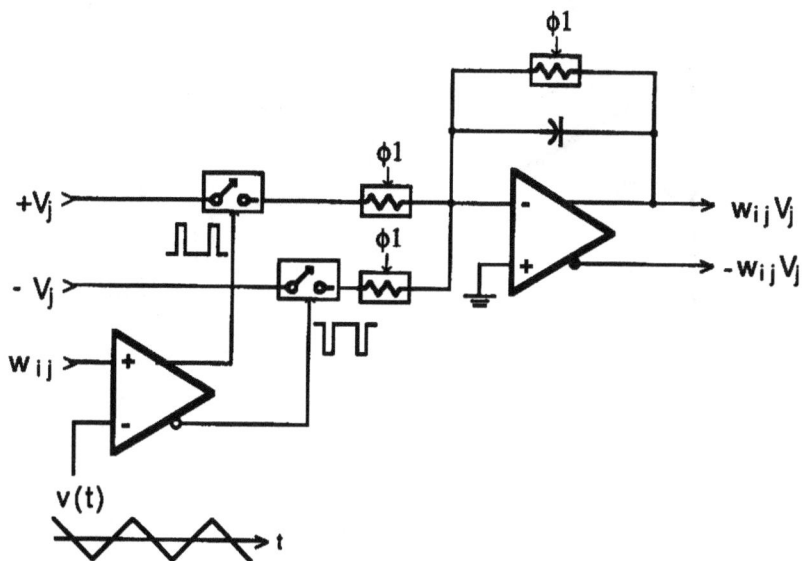

Figure 3 – Schematic diagram of PWM multiplier.

comparator described in the previous section to convert voltage w_{ij} into a PWM wave . (If the incoming signal is already a PWM wave this step is unnecessary.) The PWM wave is then used to control CMOS switches through which voltages $+V_j$ and $-V_j$ are passed. Dual signal polarities are needed for four quadrant multiplication. This is the reason for having op-amps with complementary outputs in the low pass summation circuit discussed earlier.

The output of the multiplication circuit is a PWM wave whose duty cycle is proportional to w_{ij} and whose height is proportional to V_j. Hence, the DC component of the output waveform represents the product $w_{ij}V_j$. A low pass filter can be used to extract the DC component, yielding a continuous analog signal.

5. Nonlinear scaling function

In most ANN algorithms a sharp threshold or nonlinear sigmoidal function is required. If the inputs to a comparator are ground and a signal, a sharp threshold function of the signal is obtained.

A nonlinear sigmoidal transfer function $g()$ may also be obtained from a comparator. The scheme is illustrated in Figure 4. A signal V_i that is to be compressed by the $g()$ function is connected to the + input of a comparator, and a reference signal $g_p^{(-1)}(t)$ is connected to the - input of the comparator. The output of the comparator is a PWM wave with DC component V_o equal to $g(V_i)$. A low pass filter extracts the DC level.

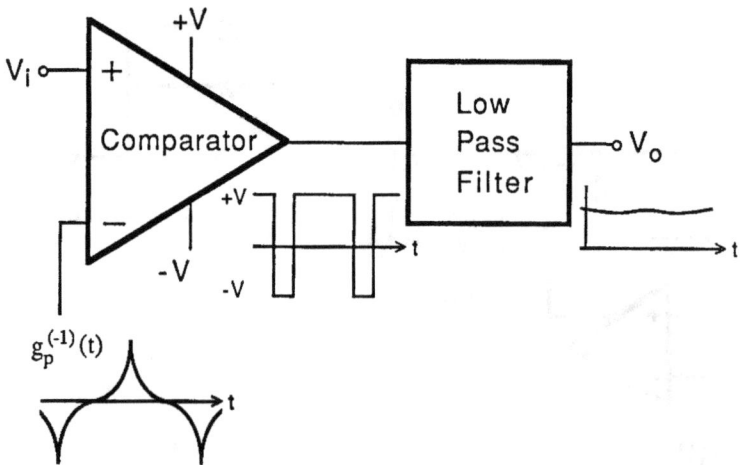

Figure 4 – Nonlinearity Circuit with arbitrary transfer function.

To obtain a given transfer function $g()$, the needed reference signal $g_p^{(-1)}(t)$ is calculated in four steps illustrated in Figure 5:

1. Find the function inverse $g^{(-1)}()$ of $g()$. (The function inverse $g^{(-1)}()$ satisfies $g^{(-1)}(g(V)) = V$ and should not be confused with the multiplicative inverse $\frac{1}{g(V)}$.) For $g^{(-1)}()$ to exist, $g()$ must be invertible. This requirement is met by all strictly increasing or decreasing functions, a class of functions that includes sigmoids used in ANNs.

2. Scale the horizontal axis of $g^{(-1)}()$ so that its domain corresponds to one time interval T, where T is much longer than the clock $\phi 1$ used for switched capacitors. For this scaling to be possible, $g()$ must have a restricted range. This requirement is met by sigmoids that are bounded above and below.

3. Create a periodic signal by replicating $g^{(-1)}()$ along the horizontal time axis.

4. To reduce the bandwidth required for $g_p^{(-1)}(t)$, flip every other copy of $g^{(-1)}()$ right-to-left, and round off the infinitely high points on the signal. Start rounding off at voltages just above the comparator supply voltages. The resulting waveform is the desired reference signal $g_p^{(-1)}(t)$.

The DC level V_0 of the comparator output is equal to the percentage of time the comparator output is high $(V_i > g_p^{(-1)}(t))$ minus the percentage of time the comparator output is low $(V_i \leq g_p^{(-1)}(t))$. Voltage V_0 may be computed from one interval of $g_p^{(1)}()$ as illustrated in Figure 6. The mathematical equation relating V_0 and V_i involves integrals corresponding to the time intervals where $V_i > g_p^{(-1)}(t)$ and $V_i \leq g_p^{(-1)}(t)$:

$$V_0 = \frac{1}{2}\left(\int_{-1}^{t_i} V\, dt - \int_{t_i}^{-1} V\, dt \right) = g(V_i)$$

where t_i is implicitly defined by $g_p^{(-1)}(t_i) = V_i$

and $\pm V$ are the comparator supply voltages

A single external circuit may be used to generate and distribute the $g_p^{(-1)}(t)$ signal to any number of ANN integrated circuits. A stairstep waveform passed through a smoothing filter is a suitable way to generate $g_p^{(-1)}(t)$, and the shape of the waveform may be slowly varied over time if desired. This is useful in Hopfield networks where increasingly steep sigmoidal functions are used in solving optimization problems.

8

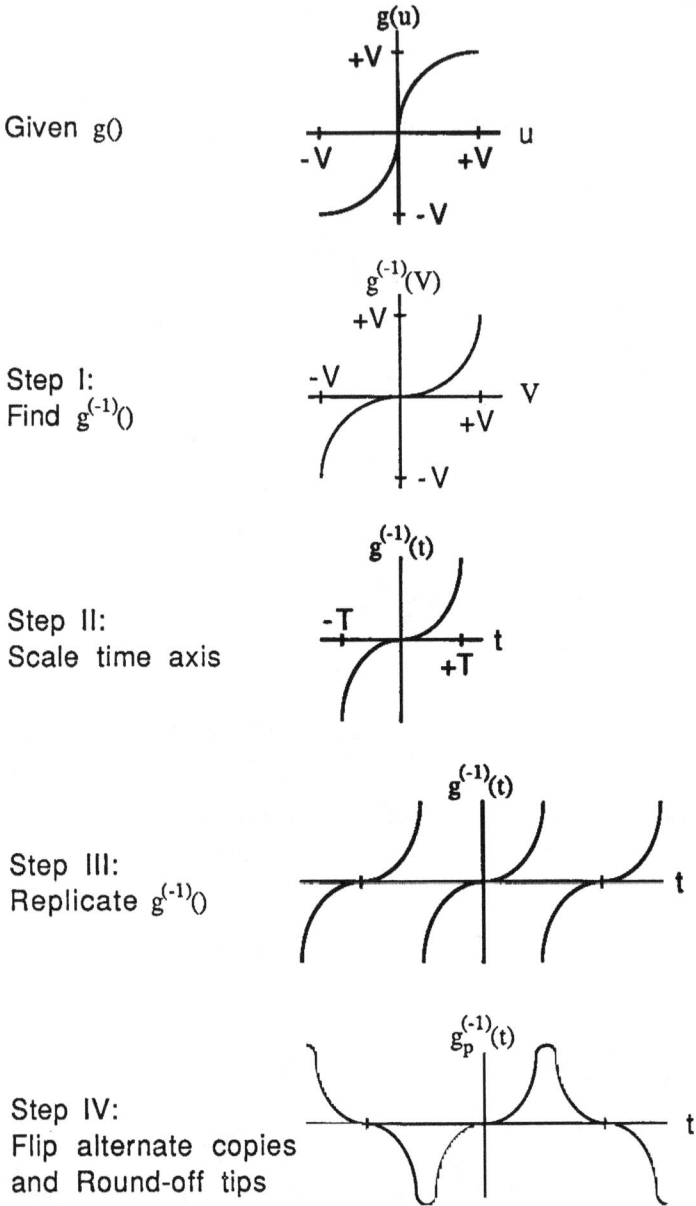

Given g()

Step I:
Find $g^{(-1)}()$

Step II:
Scale time axis

Step III:
Replicate $g^{(-1)}()$

Step IV:
Flip alternate copies
and Round-off tips

Figure 5 – Construction of reference waveform for nonlinear scaling circuit.

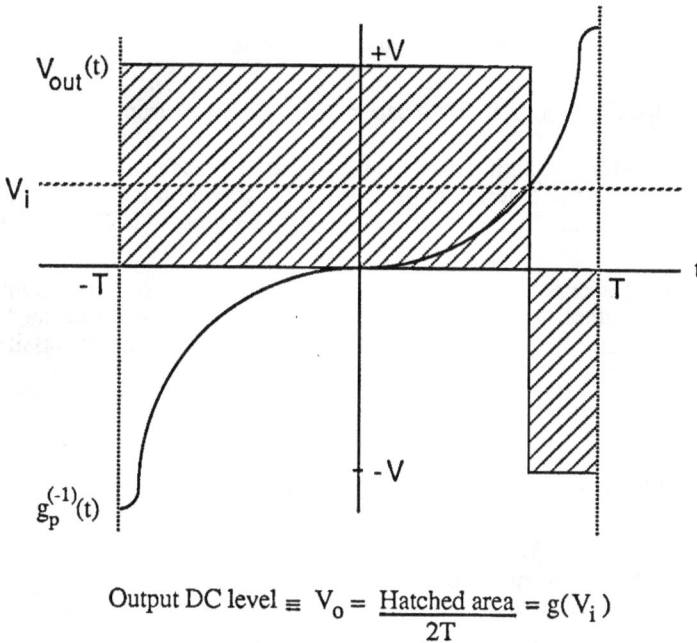

Output DC level $\equiv V_o = \dfrac{\text{Hatched area}}{2T} = g(V_i)$

Figure 6 – Calculation of DC component of nonlinear scaling circuit.output.

6. Implementation of a self-organizing feature map

The interconnection of ANN building blocks is illustrated here for Kohonen's algorithm for self-organizing feature maps [17,18]. This algorithm is more complex than most ANNs and demonstrates the flexibility of the PWM approach to network design.

Kohonen's algorithm is an adaptive vector quantization scheme in which a vector $X \equiv (x_1(t),...,x_M(t))$ of sample values is classified as belonging to one of N possible sets $y_1,...,y_N$. In practice, the x_i's are inputs to the neural network, and one of the N neurons in the network fires to indicate which class of pattern the input belongs to.

A minimal self-organizing feature map has two inputs and two outputs. Input patterns are ordered pairs (x_1,x_2) that may be identified with points on a unit square. The two neurons in the circuit may also be identified with points on the unit square. Positions of the neurons are identified by synaptic weights: (w_{11},w_{21}) for neuron 1, and (w_{12},w_{22}) for neuron 2. Kohonen's algorithm distributes neuron positions according to the statistics of input patterns, with more neurons being positioned where input patterns are more likely to occur. When an input pattern is presented to the neural network, the neuron with position closest to the input pattern is activated. Which neuron is closest to the input pattern is determined by computing Euclidean distance:

$$d_j = \sum_{i=1}^{N} (x_i - w_{ij})^2$$

where $d_j \equiv$ distance from input pattern (x_1, x_2) to jth neuron

Circuitry for computing distance is replicated in each neuron in analog form, allowing rapid determination of the closest neuron by a comparator circuit that finds the smallest neural output. This output is latched, and the pattern classification is complete.

Learning occurs after each classification and consists of positioning some neurons closer to the point being classified. This action is defined by an equation applied to neurons lying within a distance $r(t)$ of the classified point:

$$\Delta w_{ij} = w_{ij} + \eta(t)(x_i - w_{ij})$$

where $0 < \eta(t) < 1$ is a gain term that decreases in time

The values of $r(t)$ and $\eta(t)$ decrease as training proceeds, causing smaller changes to be made to fewer neurons. This is necessary for proper convergence of the algorithm, but it leads to stubbornness in old age.

A block diagram for the minimal version of Kohonen's network is shown in Figure 7. Starting at the top of the diagram, input signals x_i enter a summation and low-pass filter, and the difference signal $x_i - w_{ij}$ is computed. This x_i signal is also fed to the synaptic weight for the learning part of the algorithm, where the synaptic weight is incremented by $\eta(t)(x_i - w_{ij})$ with $\eta(t)$ encoded by the width of a pulse controlling a CMOS switch in the synapse. For the illustrated circuit, only the one of the two neurons is updated by the learning algorithm, (implying $r(t) = 0$), but more complicated $r(t)$ functions could easily be implemented in larger networks.

Proceeding downward, the Euclidean distance is computed. A bank of comparators with a triangle ramped reference input $m(t)$ then determines which of the neurons is producing the smallest output. The first comparator to trip locks out the other comparators, latches the network output $(y1, y2)$, and pulls the End Of Conversion (EOC) signal low. Upon receipt of a control signal R from the outside world, the bank of comparators is re-enabled and the network is ready to classify a new pattern. A timing diagram for the circuit is shown in Figure 8.

Cell types for laying out ANNs in PPL are cataloged in Figures 9 and 10, and a hypothetical layout for a Kohonen self-organizing feature map is illustrated in Figure 11. The PPL layout of the circuit begins with the definition of signal lines running horizontally and vertically. Each PPL cell occupies some number of squares of fixed size, and horizontal and vertical wires in the cell are automatically connected to those of neighboring cells. Routing problems are minimized by this uniformity of cell size and wiring.

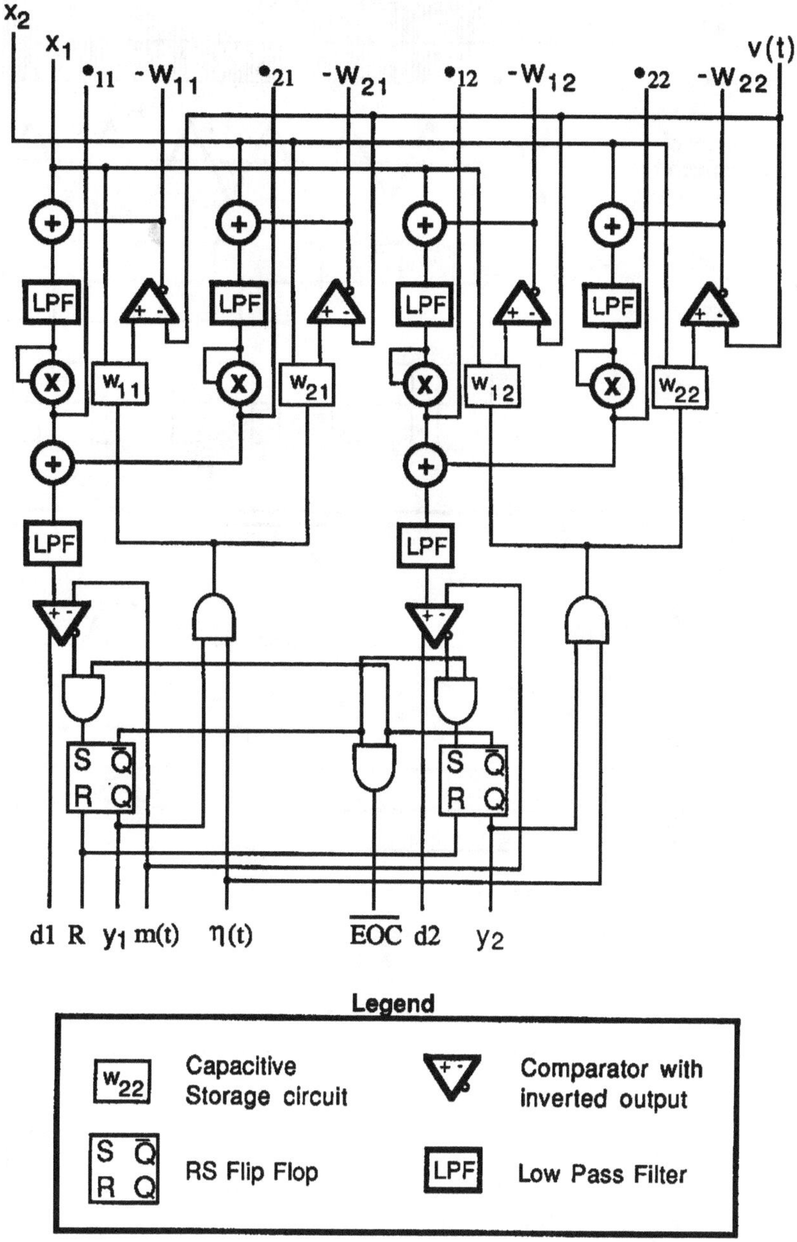

Figure 7 – Block diagram for a minimal self-organizing feature map.

12

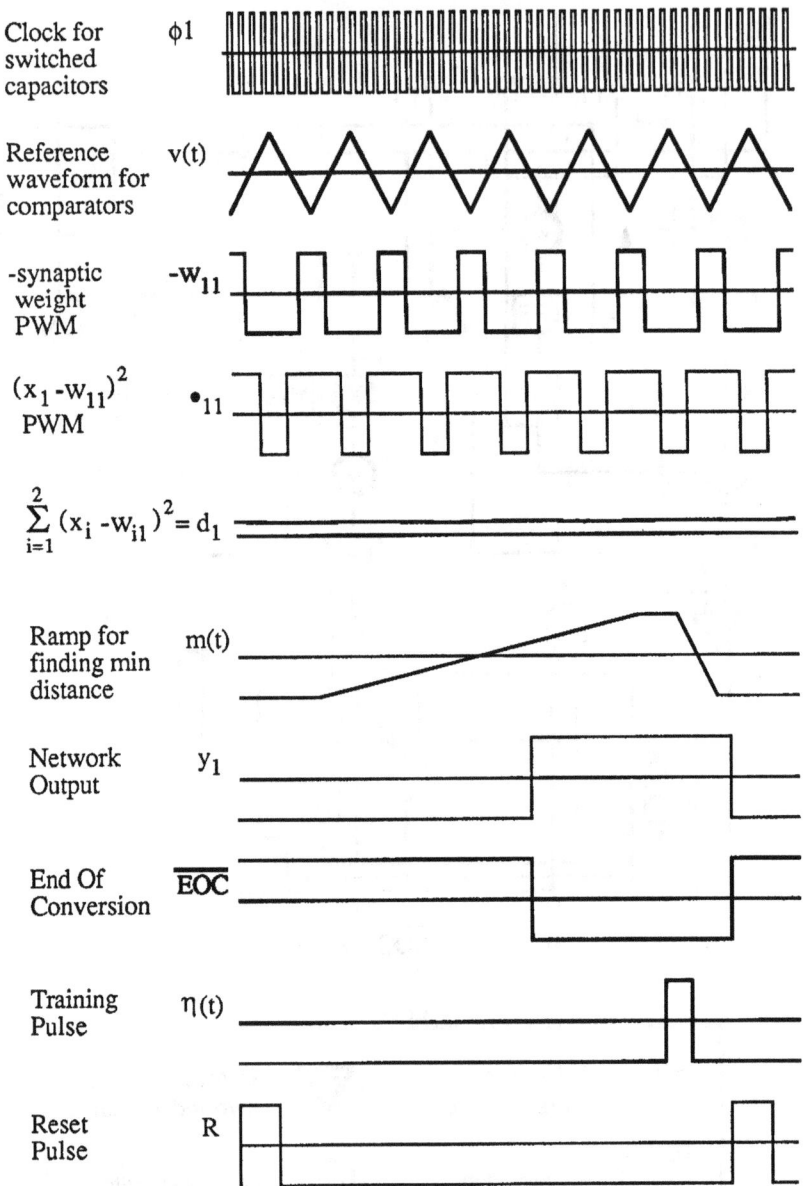

Clock for switched capacitors	$\phi 1$	
Reference waveform for comparators	$v(t)$	
-synaptic weight PWM	$-w_{11}$	
$(x_1 - w_{11})^2$ PWM	$\bullet_{11}$	
$\sum\limits_{i=1}^{2} (x_i - w_{i1})^2 = d_1$		
Ramp for finding min distance	$m(t)$	
Network Output	y_1	
End Of Conversion	$\overline{EOC}$	
Training Pulse	$\eta(t)$	
Reset Pulse	R	

Figure 8 – Timing diagram for self-organizing feature map.

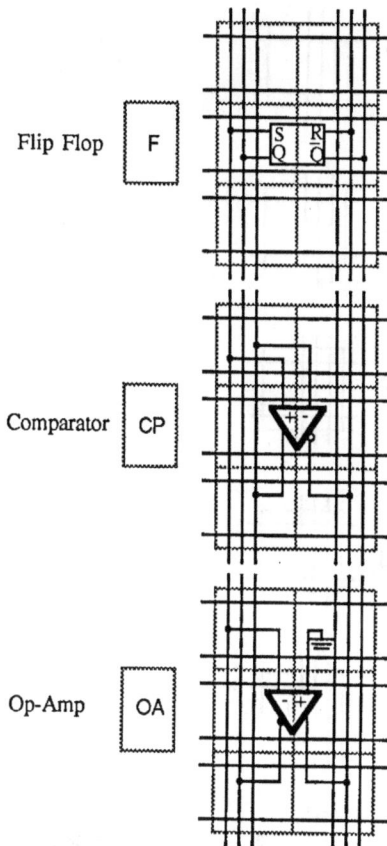

Figure 9 – PPL cell catalog.

Numerous digital integrated circuits have been designed and fabricated with existing PPL cells: (#,*) for wiring connections, (0,1,+) for logic cells, and (FF) for flip-flops. With the addition of the new cell types (@,G,-,R,C,CP,OA) ANNs may be designed. The layout shown in Figure 11 translates the block diagram of Figure 7 into an integrated circuit. Placement of components in the two figures is similar, but there are several considerations that guide the layout of the ANN. First, resistors sharing a common $\phi 1$ clock appear in the same column. The $\phi 1$ clock runs the length of the column and is therefore automatically connected to all the resistors. Second, comparators that share reference signal v(t) are placed in the same column by similar reasoning. Third, bringing test signals off the chip requires some planning of where to cut signal wires, where cuts are indicated by heavy black lines. However, less time need be spent on such planning if compactness is not particularly important.

Connect left wire to horizontal wire	#️		Resistor	R
Connect middle wire to horizontal wire	•		Capacitor	C
Connect right wire to horizontal wire	@		Cut horizontal wires	☐
Connect horizontal wire to gnd	G		Cut left vertical wire	#️
Vertical S OR'ed with horizontal S	+		Cut middle vertical wire	#️
Vertical A AND'ed with horizontal S	1		Cut left and middle vertical wires	#️
Switch	S		Cut right vertical wire	-

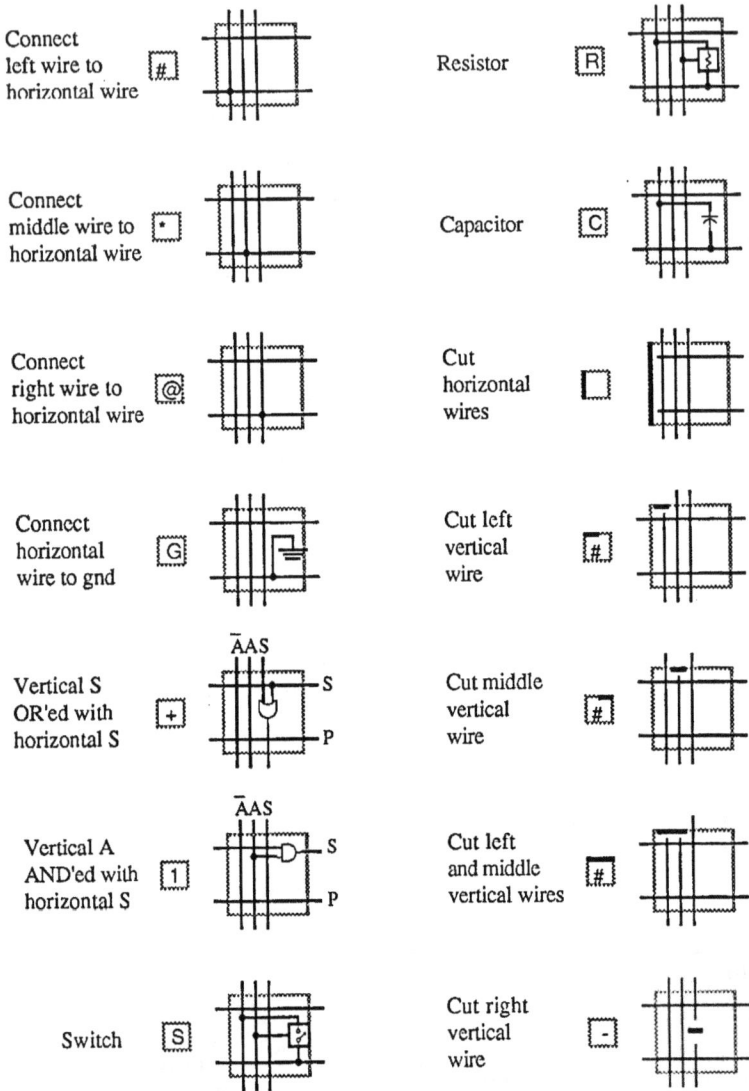

Figure 10 – PPL cell catalog continued.

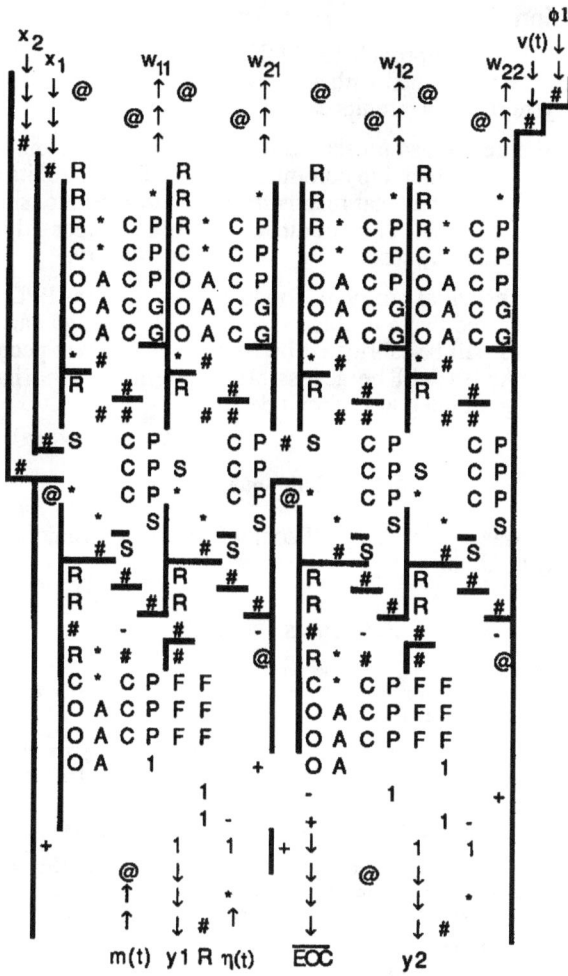

Figure 11 – PPL layout for self-organizing feature map.

7. Conclusion

With the PWM approach to ANN design, complex mathematical operations can be performed with a minimal amount of circuitry. With the PPL layout system, ANN circuits can be designed very rapidly.

Among the new ideas presented here are circuits for multiplication, nonlinear scaling, and interchip communications. These circuits exploit the simplicity of PWM signals that results from the close relationship between PWM and digital signals. The circuits are easily implemented in standard CMOS and are minimally susceptible to process variability.

With the substantial software foundation for digital PPL already in place, it is anticipated that a usable system for ANN design can be created quickly. Designs will be fabricated by MOSIS, and it is expected that the PPL layout of ANNs will be accessible to engineers who have only a limited knowledge of integrated circuit design.

Acknowledgements

This work was supported by a Faculty Research Committee Grant from the University of Utah.

References

[1] A. Agranat and A. Yariv, "A new architecture for a microelectronic implementation of neural network models," *Proc. IEEE 1st Conf. Neural Networks*, San Diego, 1987.

[2] J. Alspector and R. B. Allen, "A neuromorphic VLSI learning system," *Proc. Conf. Advanced Research VLSI*, Stanford University, 1987, pp. 313 - 349.

[3] H. S. Black, *Modulation Theory*. Toronto: D. Van Nostrand, 1953

[4] H. D. Block and S. A. Levin, "On the boundedness of an iterative procedure for solving a system of linear inequalities," *Proceedings of the American Mathematical Society*, vol. 26, pp. 229-235, 1970.

[5] P. B. Brown, R. Millecchia and M. Stinely, "Analog memory for continuous-voltage, discrete-time implementation of neural networks," *Proc. IEEE 1st Conf. Neural Networks*, San Diego, 1987.

[6] G. A. Carpenter and S. Grossberg, "Neural dynamics of category learning and recognition: Attention, memory consolidation, and amnesia," in J. Davis, R. Newburgh, and E. Wegman, Eds. *Brain Structure, Learning, and Memory*, AAAS Symposium Series, 1986.

[7] L. T. Clark and R. O. Grondin, "Comparison of pipelined 'best match' content addressable memory with neural networks," *Proc. IEEE 1st Conf. Neural Networks*, San Diego, 1987.

[8] J. G. Cleary, "A simple VLSI connectionist architecture," *Proc. IEEE 1st Conf. Neural Networks*, San Diego, 1987.

[9] N. El-Leithy, R. W. Newcomb, and M. Zaghloul, "A basic MOS neural-type junction: a perspective on neural-type microsystems," *Proc. IEEE 1st Intl. Conf. Neural Networks*, San Diego, 1987.

[10] S. C. J. Garth, "A chipset for high speed simulation of neural network systems," *Proc. IEEE 1st Intl. Conf. Neural Networks*, San Diego, 1987.

[11] H. P. Graf, "A CMOS associative memory chip," *Proc. IEEE 1st Intl. Conf. Neural Networks*, San Diego, 1987.

[12] S. Haykin, *Communication Systems*. New York: John Wiley, 1983.

[13] J. J. Hopfield, "Neural networks and physical systems with emergent collective computational abilities, *Proc. Natl. Acad. Sci. USA*, vol. 79, pp. 2554-2558, 1982.

[14] J. J. Hopfield, "Neurons with graded response have collective computational properties like those of two-state neurons," *Proc. Natl. Acad. Sci. USA*, vol. 81, pp. 3088-3092, 1984.

[15] J. J. Hopfield and D. W. Tank, "Computing with neural circuits: a model," *Science*, vol. 233, pp. 625-633, August 8, 1986.

[16] J. J. Hopfield, D. I Feinstein, and R. G. Palmer, "'Unlearning' has a stabilizing effect in collective memories," *Nature*, vol. 304, pp. 158-159.

[17] T. Kohonen, *Self-Organization and Associative Memory*, Berlin: Springer-Verlag, 1984.

[18] R.P. Lippmann, "An introduction to computing with neural nets," *IEEE ASSP Mag.*, pp. 4-22, April 1987.

[19] J. Moody, "Simplified circuits for neural nets," *Proc. IEEE 1st Intl. Conf. Neural Networks*, San Diego, 1987.

[20] A. Moopenn, A. P. Thakoor, T. Duong, and S. K. Khanna, "A neurocomputer based on an analog-digital hybrid architecture," *Proc. IEEE 1st Intl. Conf. Neural Networks*, San Diego, 1987.

[21] B. E. Nelson, K. F. Smith, "Computer-aided design of integrated circuits using an IBM PC," *Proc. ICCD Conf.*, New York, 1984, pp. 640 - 644.

[22] R. Rosenblatt, *Principles of Neurodynamics*. New York: Spartan Books, 1959.

[23] D. E. Rumelhart, J. L. McClelland, and the PDP Research Group, *Parallel Distributed Processing Vol. I,II*. Cambridge, MA: MIT Press, 1986.

[24] M. Schwartz, W..R. Bennett, and S. Stein, *Communication Systems and Techniques*. New York: McGraw-Hill, 1966.

[25] J. Sklansky and G. N. Wassel, *Pattern Classifiers and Trainable Machines*. New York: Springer-Verlag, 1981.

[26] K. F. Smith, T. M. Carter, C. E. Hunt, "Structured logic design of integrated circuits using the stored logic array," *IEEE J. Solid-State Circuits*, vol. SC-17 (2), pp. 395-406, 1982.

[27] K. F. Smith, T. M. Carter, D. A. Fisher, B. E. Nelson and A. B. Hayes, "The path programmable logic (PPL) user's manual," *Technical Report of University of Utah VLSI Research Group*, 1982.

[28] K. F. Smith, B. E. Nelson, T. M. Carter and A. B. Hayes, "Computer-aided design of integrated circuits using path programmable logic," *IEEE Electro '83, Prof. Program Session Record 22*, New York, 1983.

[29] K. F. Smith, "Design of regular arrays using CMOS in PPL," *Proc. ICCD Conf.*, New York, 1983.

[30] D. W. Tank and J. J. Hopfield, "Simple 'Neural' optimization networks: An A/D converter, signal decision circuit, and a linear program circuit," *IEEE Trans.Circuits and Systems*, vol. CAS-33 (5), pp. 533-541, 1986.

Automatic Determination of Optimal Clocking Parameters in Synchronous MOS VLSI Circuits

Michel R. Dagenais*, Nicholas C. Rumin**

* Department of Electrical Engineering, Ecole Polytechnique
  P.O. box 6079 Station A, Montréal, Québec, Canada, H3C 3A7
** Department of Electrical Engineering, McGill University
  3480 University Street, Montréal, Québec, Canada, H3A 2A7

Abstract

A new algorithm is presented which automatically determines the optimal clocking parameters in synchronous circuits that contain level-sensitive latches, and use arbitrarily complex clocking schemes. The borrowing of time between clock phases, enabled by level-sensitive latches, comes at the cost of additional complexity for the timing analysis algorithms. An iterative algorithm is proposed here which provides a lower bound on the shortest possible cycle time at each iteration. This algorithm has been implemented in a circuit-level timing analysis program for MOS VLSI circuits.

INTRODUCTION

Among the more important problems that are being faced by the designers of MOS VLSI circuits is the one of specifying optimal clocking parameters for synchronous systems. This is a problem because they rely on level-sensitive latches for their clocked elements which, in turn, necessitates the use of multi-phase clocking methodologies. These can be quite complex in high performance circuits. In contrast, it is much easier to analyze the timing of circuits which are based on MOS MSI components or which exploit the bipolar technology. The reason lies in the fact that such designs are based on edge-triggered latches.

What simplifies the timing analysis of designs based on edge-triggered latches is the fact that the timing equations are decoupled at the latches, since the input and output signals do not depend directly on each other. This has permitted the development of powerful timing analysis tools, of which probably the best known one is SCALD [1].

In a level-sensitive latch, on the other hand, a signal can propagate from its input to its output at any time during the entire period when the latch is on. The resulting unknown that is added to the timing equations is what leads to the complexity of computing the optimal clocking parameters [2]. One of the best known attempts to solve this problem is that of Ousterhout [3]. Although his timing analyzer CRYSTAL represents important pioneering work, its weakness lies in the fact that it cannot handle properly complex clocking schemes. In particular it can not handle signals which cross phases, which is a very common situation in MOS circuits.

Jouppi's TV [4] is in many respects similar to CRYSTAL and again it is hampered by the fact that it is tied to a very specific clocking scheme. Szymanski [5] uses an alternative approach wherein the procedure verifies if the circuit will function correctly with a specified set of clocking parameters. Thus, to find the optimal set of parameters, the designer might have to go through many tries. This is somewhat alleviated by the fact that the step of verifying the validity of a specified set of parameters is extremely fast.

The main objective here is to present a procedure which determines automatically the optimal clocking parameters for circuits using level-sensitive latches. A simplified version of this procedure has been implemented and tested in the transistor-level timing analysis program TAMIA [6].

In this paper, the notion of timing intervals is first introduced. Then, the timing requirements associated with level-sensitive latches are described and formalized. The problem of determining the clocking parameters for circuits using level-sensitive latches is discussed. An iterative algorithm to find the optimal clocking parameters is then proposed. Finally, results obtained with a simplified implementation are discussed.

TIMING INTERVALS AND SYNCHRONIZING EVENTS

The approach to the computation of optimal clocking parameters which is presented below is based on the idea of dividing the clock cycle into timing intervals. The clock transitions are the synchronizing events, and these can be either dependent or independent. Every timing interval may contain only one independent event. A clocking methodology is specified by enumerating the timing intervals and the synchronizing events which each interval contains. The clocking parameters are then determined by computing the durations of the timing intervals.

In the two-phase, non-overlapping clocking scheme shown in Figure 1 all four clock transitions are independent events and, therefore, there are four timing intervals. The duration of the *phi1 rise* and *phi2 rise* intervals are determined by the set-up times of the *phi1* and *phi2* latches. On the other hand, clock skew on the two clock phases determines the durations of the other two intervals, *phi1 fall* and *phi2 fall*. These non-overlap intervals are needed to ensure that the the *phi1* and *phi2* latches are never *on* at the same time.

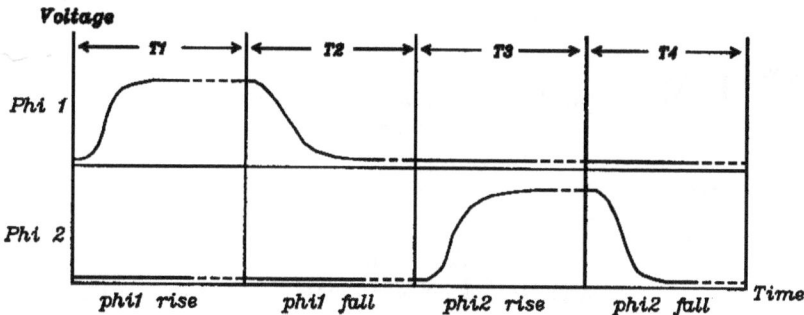

Figure 1 Two phase non-overlapping clocking scheme decomposed in four timing intervals.

The three-phase, overlapping scheme shown in Figure 2 illustrates the idea of dependent events. Here the third clock phase ensures that no more than two out of three latches in a feedback loop can conduct simultaneously. Thus only three timing intervals are needed and, for example, the *phi2 rise* and *phi1 fall* events can be dependent, and are therefore put in the same interval.

TIMING REQUIREMENTS

The timing requirements of level-sensitive latches are examined in this section. The input to a timing analysis procedure consists of a circuit description, a definition of the signals at the primary inputs, and a specification of the clocking methodology. Each input signal is defined in time coordinates relative to the beginning of the interval in which that signal originates, i.e. the interval which contains the synchronizing event which caused a transition of that signal. Indeed, the absolute time in the clock cycle at which the signal changes is not known since the duration of the timing intervals is yet to be determined.

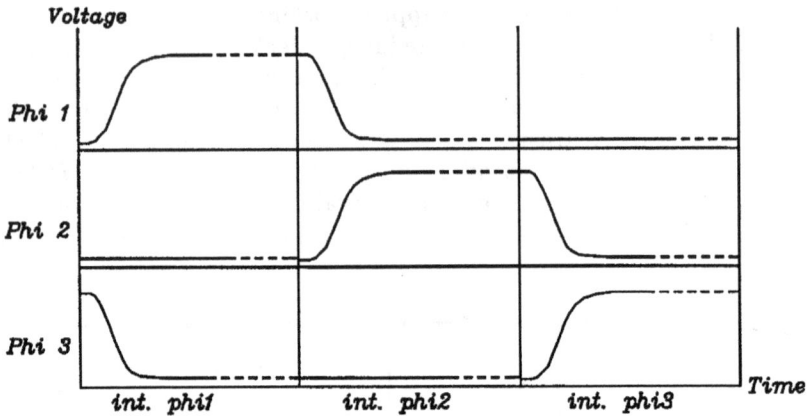

Figure 2   Three phase overlapping clocking scheme decomposed in three timing intervals.

The circuit elements are defined here at a high level, namely combinational blocks and latches. Nevertheless, the theory which is presented here can be applied equally well to timing analysis procedures which function at the gate, switch or transistor level. Figure 3 shows the circuit elements, which are assumed to be unidirectional with a minimum and maximum delay associated with each input-output pair.

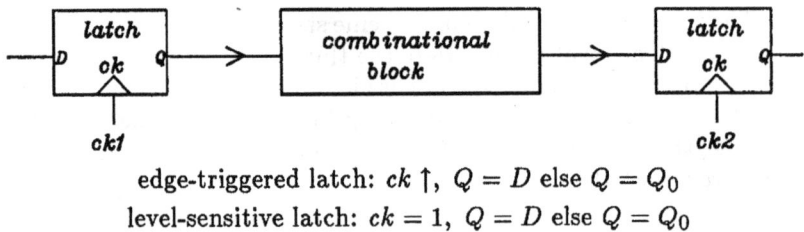

edge-triggered latch: $ck \uparrow$, $Q = D$ else $Q = Q_0$

level-sensitive latch: $ck = 1$, $Q = D$ else $Q = Q_0$

Figure 3   Circuit decomposed in combinational blocks and latches.

Definitions

To express the timing requirements as inequalities, a few definitions are in order. Different time values are defined here. Unless otherwise stated, these time values are relative to the beginning of the clock cycle. For the sake of brevity, it is assumed that the clocks are active *high* and have sharp transitions.

Input and Output of the latch

in_stable : latest time at which the latch input stabilizes
in_change : earliest time at which the latch input is no longer stable
out_stable : latest time at which the latch output stabilizes
$retard$: Signal retardation at the input to a level-sensitive latch

Delay through the latch

$delay_{et}$: Maximum propagation delay from rising clock to output
$delay_{ls}$: Maximum propagation delay from latch input to output

Interval boundaries

$start_a, end_a$: Start and end time of interval a
$start_b, end_b$: Start and end time of interval b $(end_a = start_b)$
end : End time of the complete clock cycle

Position of clock transitions relative to timing intervals

$rise\_min$: Minimum delay between $start_a$ and clock rise
$rise\_max$: Maximum delay between $start_a$ and clock rise
$fall\_min$: Minimum delay between $start_b$ and clock fall
$fall\_max$: Maximum delay between $start_b$ and clock fall

Level-Sensitive Latches

In the present section, the operation of a level-sensitive latch is examined. The latch studied turns *on* with the rising of the clock in interval a and turns *off* with the falling of the clock in interval b. The time at which the output of the latch stabilizes depends on both the clock and the latch input signal. Indeed, if the input of the latch is stable before the latch turns *on*, the out_stable value depends on the clock rising transition $start_a + rise\_max$. On the other hand, if the input of the latch stabilizes after the latch turns *on*, the out_stable value depends on the input signal in_stable. This double dependency can be expressed by introducing the notion of retardation. The retardation is null when the input signal is stable before the latch turns *on* with the rising clock, in which case it does not depend on the value in_stable. When the input signal settles after the latch turned *on*, the retardation indicates by how much the input signal arrives later than the clock rising transition. Thus, it is computed as follows:

$$retard = max(0, in\_stable - (rise\_max + start_a)) \qquad (1)$$

The out_stable value is then easily computed using the retardation value $retard$, the clock rising transition $start_a + rise\_max$ and the combinational delay through the latch $delay_{ls}$.

$$out\_stable = start_a + rise\_max + retard + delay_{ls} \qquad (2)$$

To insure proper latching, the output of the level-sensitive latch *out_stable* must be stable before the latch starts to turn *off* $start_b + fall\_min$. Moreover, the input of the latch *in_change* should remain stable until the latch is completely *off* $start_b + fall\_max$. These are respectively the *setup* and *hold time* for a level-sensitive latch. They correspond to the following inequalities:

$$setup\ time:\ out\_stable \leq fall\_min + start_b \qquad (3)$$

$$hold\ time:\ in\_change \geq fall\_max + start_b \qquad (4)$$

It is generally assumed that a signal originating in a subsequent interval cannot corrupt the content of a latch turning *off* in the current interval. However, in the case of a very severe clock skew, it could happen that the latch takes a long time to turn *off* allowing its content to be corrupted by a transition which originates in a subsequent interval. The *hold time* already insures that the content of the latch cannot be corrupted by a signal originating in the current interval. It is completed by the *min pulse width* requirement which insures that the latch is completely *off* at a time $start_b + fall\_max$ before the next interval starts, end_b. Thus, the latch is completely turned *off* before any transition originating in a subsequent interval can reach its input. This requirement is expressed by the following inequality:

$$min\ pulse\ width:\ fall\_max + start_b \leq end_b \qquad (5)$$

The above inequalities insure correct operation of the circuit. They are used to determine the minimum acceptable duration for each timing interval. For level-sensitive latches, the timing requirements arise when the falling transition of the clock turns them *off*. The case of the latches turning *on* in interval a and turning *off* during interval b is examined here.

When the analysis reaches interval b, all the latches turning *off* during that interval are examined. When the signal at the input of one such latch originates in a preceding interval, its *in_stable* value is known at that point. Thus, the *out_stable* value is obtained with equations (1) and (2). The *setup* time requirement (3) can be satisfied by adjusting the beginning of interval b such that the clock does not fall at $start_b + fall\_min$, to turn the latch *off*, before its content is stable at *out_stable*. To compute the minimum time at which interval

b can start, inequality (3) is used as an equality and the maximum value over all the latches turning *off* in interval b is retained:

$$start_b = max_{i=1,n}(out\_stable_i - fall\_min_i) \qquad (6)$$

There is a problem, however, when a signal which arises in a subsequent interval is connected to the input of a latch active during interval a. This is the case when a signal is generated at the end of a clock cycle, and is latched at the beginning of the next cycle. Such a signal is said to cross the cycle boundary. If such a signal is connected to the input of a latch active during interval a, the *in_stable* value for that latch is not known. In such a case, the *setup* requirement can be satisfied by adjusting the duration of the clock cycle. Thus, the minimum duration for the clock cycle which satisfies the *setup* time at such latches can be computed as follows:

$$end = max_{i=1,n}(out\_stable_i - fall\_min_i - start_b) \qquad (6a)$$

Nevertheless, there remains a problem which cannot easily be solved. Indeed, when the value *in_stable* is unknown because the signal originates in a subsequent interval, the value *out_stable* cannot be computed from equations (1) and (2). Effectively, these two values are not decoupled as they would be in the case of edge-triggered latches. This prevents the analysis of the signals which are connected to the output of this latch. In fact, at each feedback loop there is a set of simultaneous equations which relates the time at which each signal in the loop becomes stable. Moreover, these simultaneous equations cannot necessarily be solved easily since they are not linear.

In order to pursue the analysis, a bound on the allowable retardation at the examined latch can be set. Thus, if the maximum retardation assumed for latch i is max_retard_i, the maximum time at which the output of the latch becomes stable can be computed using equation (2). The resulting *out_stable* value will be pessimistic. Thus, the interval durations computed using this value will still insure correct operation of the circuit but may lead to pessimistic values. Indeed, the pessimistic durations computed may not represent the shortest allowable interval durations.

At the end of the analysis, the duration of the clock cycle must be adjusted to satisfy the bounds imposed on the retardation at such latches. At those latches where the input signal crosses the cycle boundary, the value *in_stable* represents the time at which the signal settles one clock cycle later. Thus, the value for the current clock cycle is obtained by subtracting the cycle duration *end*. In that case, equation (1) is replaced by:

$$retard = max(0, (in\_stable - end) - (start_a + rise\_max)) \quad (7a)$$

This retardation value must be smaller than the maximum retardation allowed for that latch. Thus, the following inequality is derived from (7a):

$$max\_retard \geq max(0, (in\_stable - end) - (start_a + rise\_max)) \quad (7b)$$

$$max\_retard \geq (in\_stable - end) - (start_a + rise\_max) \quad (7c)$$

This inequality insures the validity of the analysis performed on the circuit using the pessimistic values of retardation. Such inequalities as (7c) can be satisfied by adjusting the cycle duration *end*. Thus the minimum duration of the clock cycle which guarantees the validity of the analysis can be obtained by taking the maximum value of *end* which barely satisfies (7c) at all the relevant latches:

$$end = max_{i=1,n}(in\_stable_i - start_a - rise\_max_i - max\_retard_i) \quad (8)$$

In addition to these requirements, the *min pulse width* inequality (5) must be satisfied. It determines the minimum time at which all the latches turning *off* are indeed completely *off* and the interval can end. This insures that no signal originating in a subsequent interval can corrupt the content of the latch while it turns *off*. The minimum time at which interval *b* can end, end_b, is obtained by taking the maximum time at which any of the latches turn completely *off*:

$$end_b = max_{i=1,n}(start_b + fall\_max_i) \quad (9)$$

The procedure that uses these equations to compute the minimum interval durations in a synchronous circuit, (and thus the maximum clock frequency), is described in the next sections.

SOLVING FOR THE OPTIMUM CLOCKING PARAMETERS

The problem in finding the optimal clocking parameters resides in avoiding having to solve simultaneous equations. This can generally be achieved by ordering the equations properly , so that only one unknown remains in each equation.

In the circuit, the ordering is such that an element is examined only when all the signals on which it depends have been processed. Also, within each interval, the latches turning *off* must be analyzed before the latches turning *on*. Indeed, the start time of an interval is determined by the latches turning *off* using equation (6). All the values used in this computation are relative to the beginning of the interval , and can be obtained without prior knowledge of the start time. Once the start time of the interval is determined all the time values can be converted to time relative to the beginning of the clock cycle. Then, the latches turning *on* can be processed and the retardation at their input can be computed with equation (1), which requires the start time of the interval.

There is, however, an essential difference between level-sensitive latches and edge-triggered latches. The time at which the output of a level-sensitive latch settles depends on its input, as can be seen from (2). Consequently, circular dependencies exist around feedback loops which cannot be resolved by any ordering. To break these circular dependencies, the notion of *maximum retardation* is introduced. At the first latch encountered in each loop, an initial value of retardation is guessed. The analysis then proceeds and stays consistent with the retardation values assumed. The clocking parameters obtained insure correct operation of the circuit but may not be optimal.

This problem can therefore be seen as an optimization problem which resembles the Linear Programming optimization problem. Indeed, the timing requirements form a set of inequalities to be satisfied which depend on the retardation values. The cost function to minimize is the sum of the interval durations, in order to obtain the shortest cycle duration.

Nonetheless, there is an essential difference between the Linear Programming optimization problem and the Cycle Duration one described below. Indeed, the function that links the stabilization time of the input and output of a level-sensitive latch is monotonic but not linear (1)(2). The new iterative procedure to determine the optimal clocking parameters in a circuit containing level-sensitive latches is described in this section.

Each iteration step in the procedure can be performed in time linearly proportional to the circuit size. At each step, a near-optimal solution is available. Indeed, the solution produced always satisfies the timing requirements and leads to the shortest cycle time, given the current retardation values set at the latches. At the end of the iterative process, the retardation values producing the optimal clocking parameters are found and the shortest possible cycle time is obtained. These retardation values represent the maximum allowable time borrowing between cycles at each latch. The retardation values bring requirements on the cycle duration to insure correct operation. Since different correct solutions are obtained depending on the retardation values, the problem is to find the values which lead to the shortest cycle time.

The analysis proceeds one interval at a time in a breath-first search manner. The latches turning *off* are processed first and determine the start time of the interval examined using equation (6). Then, the latches turning *on* are examined, once the start time of the interval is computed. The problem arises when the first latch in a feedback loop is reached. Indeed, the input signal to such a latch crosses the cycle boundary and cannot be determined until later in the cycle, since it originates in a subsequent interval.

Initially, the maximum retardation at the first latch in each feedback loop is set to 0 and the analysis proceeds to find a set of clocking parameters for this initial iteration. Then, at the end of each iteration, the retardation value is increased slightly in each feedback loop which is on the critical path. The fact that the retardation is increased delays the *out_stable* value computed at the corresponding latch from (2). Such a change on a value of *out_stable* cannot shorten the *cycle_end* computed in the first section of the procedure. Indeed, it simply delays the time at which a certain number of nodes become stable. This cannot shorten the minimum cycle time required at the latches turning off, which is given by the setup time for signals crossing the cycle boundary, (6)and (6a), and the minimum pulse width requirements at latches in the last interval of the cycle, (9). However, the increase in the retardation value at the latches on the critical path can shorten the cycle time required at these latches given by (8) (retardation requirement).

Indeed, when it is assumed that a latch has a certain retardation value, the cycle time needs to be adjusted such that the signal coming on the feedback loop satisfies this value of retardation (8). Otherwise, the *in_stable* value obtained around the feedback loop produces a larger retardation value than what was initially assumed. In such a case, the corresponding *out_stable* value (2) would have been underestimated, and correct operation is not insured with the interval durations computed for the smaller value initially assumed. If the retardation value is increased but the *in_stable* signal on the feedback loop is not delayed, the required cycle time given by (8) is shorter. Thus the following theorems hold:

theorem 1 The only portion of the cycle which can be shortened is that caused by *retardation* requirements (8). Thus, in any given iteration, a lower bound on the shortest possible cycle time is the minimum cycle time allowed by the *setup time* and *minimum pulse width* requirements, given by (6),(6a) and (9).

corollary 1 The difference between the minimum cycle time imposed by the *retardation* requirements and that imposed by the *setup* and *min pulse width* requirements is the maximum value by which the cycle time can be shortened. Thus, when this difference goes to 0, the cycle time cannot be shortened and the optimal solution is reached.

proof The iterations start with a null retardation at the first latches in each feedback loop. The minimum cycle duration required by the *setup* and *min pulse width* requirements, (6),(6a) and (9), cannot be shortened since the retardation cannot be negative. Thus, the only way to obtain a better upper bound on the minimum valid cycle duration is by increasing the retardation allowed at these latches. Indeed, changing the retardation values can affect the minimum cycle time imposed by the *retardation* requirements (8).

Increasing the retardation at a latch which is not on the critical path cannot shorten the cycle duration computed. The retardation at all the latches on the critical path must be increased in order to get any reduction on the cycle time computed. Increasing this retardation by δ can delay the time at which nodes stabilize by up to δ and correspondingly the cycle time required by the *setup* and *min pulse width* requirements by up to δ. The signals on the feedback paths can then be delayed by a value between 0 and δ, which is less than or equal to the additional retardation allowed on the critical paths. Consequently, the added retardation cannot increase the cycle time imposed by the *retardation* requirements. This brings up the following conclusions:

theorem 2 The cycle duration imposed by the *setup* and *min pulse width* requirements cannot be shortened. It increases monotonically with the retardation at the latches.

proof The equations (6),(6a) and (9) represent non-linear functions which increase monotonically with the retardation. The retardation starts at null in the first iteration and can only increase at each iteration. Thus, the lower bound on the cycle duration provided by the *setup* and *min pulse width* can only increase as the iterative solution progresses.

theorem 3 The only way to shorten the cycle duration is by increasing the retardation allowed at the latches in the critical feedback paths. The increase in retardation at these latches cannot increase the cycle duration imposed by the *retardation* requirements. This insures that the iterative solution converges monotonically towards the optimal solution.

proof The time at which the output of a latch stabilizes is at its minimum when the retardation is null at all the latches. The cycle duration imposed by the *setup time* and *min pulse width* requirements cannot be shortened by *theorem 2*. Thus, only the *retardation* requirements could possibly be relaxed to allow a shorter cycle duration. In fact, the only term in requirement (8) which can be changed is the retardation value. When the retardation is increased by δ at some latches, the time at which the latch output is stable can increase by at most δ. Consequently, the minimum cycle time required by (8) at a latch cannot increase by increasing the retardation at this latch and other latches by a given amount δ. If the *in_stable* time at the latch increases by less than δ, the minimum cycle duration required is shorter. It should be noted that the retardation at all the latches on critical feedback loops must be increased in order to get any reduction on the cycle time.

theorem 4 The optimal solution is reached when the minimum cycle duration imposed by the *retardation* requirements reaches the duration imposed by the other requirements.

proof Indeed, increasing the retardation is the only way to shorten the cycle duration. It cannot increase the duration imposed by the *retardation* requirements but does increase the duration imposed by the other requirements. Thus, until the two durations become equal, the solution can only improve. When the difference between the two durations becomes null, no more improvement is possible by **corollary 1**. Thus, the solution obtained at that point must be the optimal solution.

It is interesting to note that the "first" and "last" intervals in a cycle are an arbitrary notion because of the cyclic nature of the phenomenon. Indeed, one can in fact decide to break the cycle and start the iterative analysis in any interval. This would change the latches at which null retardation values will be initially assumed and, correspondingly, the related signals which cross the new cycle boundary. Consequently, the near-optimal solution obtained in the first iterations may change depending on how the cycle is broken. However, the shortest cycle duration reached at the end of the iterative process does not vary.

DISCUSSION

This algorithm has been implemented in TAMIA, a circuit-level timing analysis procedure developed by the authors [6]. However, in its present implementation, it performs a single iteration, and thus produces near-optimal clocking parameters. This choice was dictated by the fact that a circuit-level procedure requires a long processing time, not suitable for multiple iterations.

Circuits using different clocking strategies have been successfully analyzed using TAMIA. The accuracy obtained with the circuit-level timing model is between 5 and 10% and the processing time for the single iteration grows linearly with the circuit size and amounts to approximately 0.4s per transistor on a microVAX II computer.

The switch-level timing analysis procedure called LEADOUT developed by Szymanski [5] uses an iterative algorithm. Although LEADOUT does not determine automatically the optimal clocking parameters, it has to cope with the circular dependencies in order to verify if the circuit can operate correctly under specified clocking parameters. Interestingly, he reports that the number of iterations required is usually small (e.g. 2 or 3) in most cases but can be very large for some pathological cases [7].

CONCLUSION

A new procedure was presented that can automatically determine the optimal clocking parameters for synchronous circuits using level-sensitive latches and arbitrarily complex clocking schemes. This added functionality, not available previously, is obtained through an iterative process. To overcome the problems brought by level-sensitive latches, two major modifications had to be made. First, within each interval, the nodes are ordered to evaluate the latches turning *on* after evaluating the latches turning *off* on which they depend through the start time of the interval. Secondly, to break the circular dependencies that exist along feedback loops, initial retardation values are assumed, and impose requirements on the cycle duration to keep the analysis consistent.

This procedure is not bound to a delay model and can be applied to gate, switch and circuit-level static timing analysis procedures. An algorithm that iteratively refines the values of retardation obtained was proposed. Its convergence was proved. However, further experiments will be required to evaluate various means to obtain a rapid convergence. Even when a costly timing model is used and iterating many times is not possible, diverse heuristics may lead rapidly to a very near-optimal solution. The interval in the cycle at which the analysis starts can change the initial solutions obtained. Also, the lower bound on cycle duration provided by **corollary 1** can help to decide whether further iteration is desirable and by what amount to augment the retardation at latches on the critical path.

ACKNOWLEDGEMENTS

The authors wish to thank Tom Szymanski for stimulating discussions. This work was supported by the Natural Sciences and Engineering Research Council of Canada and the Ministère de l'enseignement supérieur et de la science de Québec (MESSQ).

REFERENCES

[1] T.M. McWilliams, "Verification of Timing Constraints on Large Digital Systems," *Proceedings of the 17th Design Automation Conference*, Minneapolis, Minnesota, 1980, pp. 139-147.

[2] L.A. Glasser, and D.W. Dobberpuhl, "The Design and Analysis of VLSI Circuits." *Addison Wesley*, Reading, Massachusetts, December 1985.

[3] J.K. Ousterhout, "Switch-level Timing Verifier for Digital MOS VLSI," *IEEE Transactions on CAD of Integrated Circuits and Systems*, vol. CAD-4, no 3, July 1985, pp. 336-349.

[4] N.P. Jouppi, "Timing Analysis for NMOS VLSI," *Proceedings of the 20th Design Automation Conference*, 1983, pp. 411-418.

[5] T.G. Szymanski, "LEADOUT: A Static Timing Analyzer for MOS Circuits," *Proceedings of the 1986 IEEE International Conference on CAD*, Santa Clara, California, November 1986, pp. 130-133.

[6] M.R. Dagenais, and N.C. Rumin, "Timing Analysis and Verification of Digital MOS Circuits," *Proceedings of CompEuro 1987*, Hamburg, Federal Republic of Germany, May 1987, pp. 242-245.

[7] T.G. Szymanski, "Private Communication," *McGill*, Montreal, Canada, April 1987.

Syntax-directed Translation
of Concurrent Programs
into Self-timed Circuits

Steven M. Burns and Alain J. Martin

Computer Science Department
California Institute of Technology
Pasadena, CA 91125

We present a method for the automatic compilation of concurrent programs into self-timed circuits. The compilation is directed by the syntax of the source language and produces linear-sized implementations of arbitrary concurrent programs. We have constructed a compiler which performs this translation.

An automatic method for synthesizing a digital circuit from an abstract specification represents an important step forward in managing the complexity of VLSI system design. By allowing the designer to think in terms of high-level programs instead of detailed circuits, systems can be designed in less time with fewer errors. In fact, if formal proofs are used to verify that the program implements the system specification, the circuits derived are correct by construction.

To take full advantage of the inherently concurrent nature of digital circuits, we have chosen a variant of Communicating Sequential Processes (CSP)[4] as the source language for our synthesis method. While self-timed circuits[10] offer several advantages over clocked circuits, we chose them as our synthesis target mainly because of the composition properties they possess. Self-timed sub-circuits may be composed together to form circuits of arbitrary size without the hazards associated with clock distribution in large synchronous designs.

In this paper, we present a mechanical method for transforming a concurrent program into a semantically equivalent self-timed circuit. This method is based on the manual circuit-compilation techniques of [6,8], but the translation mechanism follows standard syntax-directed techniques. We describe the necessary translation rules and apply these to an example. This transformation method allows any program

⟨process⟩	::=	(⟨process⟩ { ‖ ⟨process⟩ }) { ⟨channel⟩ }
		| { ⟨port⟩ } { ⟨var⟩ } ⟨sequence⟩
⟨channel⟩	::=	channel (⟨pNAME⟩ , ⟨pNAME⟩)
⟨port⟩	::=	(passive | active) ⟨pNAME⟩ (⟨rINT⟩ , ⟨rINT⟩)
⟨var⟩	::=	boolean ⟨vNAME⟩ = (true | false)
⟨sequence⟩	::=	⟨statement⟩ { ; ⟨statement⟩ }
⟨statement⟩	::=	skip | ⟨vNAME⟩ (up | down) | [⟨gcs⟩] | *[⟨gcs⟩]
		| ⟨pNAME⟩ (⟨vINT⟩) : [⟨responses⟩]
⟨gcs⟩	::=	⟨gc⟩ { | ⟨gc⟩ }
⟨gc⟩	::=	⟨expr⟩ --> ⟨sequence⟩
⟨responses⟩	::=	⟨response⟩ { | ⟨response⟩ }
⟨response⟩	::=	⟨vINT⟩ --> ⟨sequence⟩
⟨expr⟩	::=	⟨conjunct⟩ { or ⟨conjunct⟩ }
⟨conjunct⟩	::=	⟨primary⟩ { and ⟨primary⟩ }
⟨primary⟩	::=	⟨vNAME⟩ | probe ⟨pNAME⟩ | true | false
		| not ⟨primary⟩ | (⟨expr⟩)

Figure 1: BNF for Source Language

in the source language to be compiled into a self-timed circuit, and furthermore constructively proves that the size of the resulting circuit is linearly related to the size of the source program. The method has been automated and we compare the designs produced by the compiler with designs produced by manual methods.

1   Source Language

The source language is based on CSP[4] with the addition of the *probe*[9] and a new communication construct. A complete description of the language syntax is given in Figure 1. A program in this language consists of a set of sequential processes with interconnecting channels. Associated with each sequential process is a set of ports, a set of private variables, and a list of statements to be executed sequentially. Ports that do not connect to another process connect to the environment.

Only boolean variables are allowed. Variables are changed by assignment to true (x up) or to false (x down). The selection ([⟨gcs⟩]) and repetition (*[⟨gcs⟩]) constructs are based on guarded commands. We use *[⟨sequence⟩] as an abbreviation for *[true-->⟨sequence⟩].

Synchronization between two processes is accomplished by zero-slack communication actions across channels denoted by pairs of ports.

Of the two ports which make up a channel, one is declared *active* and the other is declared *passive*. The process which owns the *passive* port may determine whether the other process is waiting for a communication on this channel by evaluating a boolean condition called a probe. Probes may be used in arbitrary boolean expressions.

Though concurrently operating processes may not share variables, processes may communicate data by exchanging values from small sets during a synchronization action. The communication construct provides a syntax allowing differing sequences of commands to be executed based on the value received during a communication. When declaring a port, we specify both the send and receive sets of values—each set being represented by a single integer. For example,

```
passive L(3,2)
```

declares a *passive* port L with send set $\{0,1,2\}$ and receive set $\{0,1\}$. The communication action (on the same port, L)

```
L(1):[ 0 --> x down | 1 --> x up ]
```

sends the value 1 while simultaneously receiving either a 1 or a 0. If a 0 is received, x is set to false; if a 1 is received, x is set to true.

The output value specification of the communication action may be suppressed if the port has only one send value. Similarly, the receive value selection need not be specified if the port has only one receive value.

We illustrate the programming language and, later, the compilation procedure with the following example.

Routing Automata

By interconnecting processes of two different types, we can construct worm-hole routing systems[3] that allow processors, connected in a variety of different network structures, to communicate. The *switch* process (diamond), together with an *arbiter* process (circle), are sufficient to implement deterministic routing systems for hypercubes, tori, meshes and other networks in which the path of the message can be determined before the message is sent. Figure 2 shows a building block configuration for sending messages up and right in a two-dimensional mesh. Each processor injects messages (strings of bits) into the system through port P and extracts messages through port Q. The processor sending a message prefixes the message with a string of bits specifying the path the message will travel through the network. A

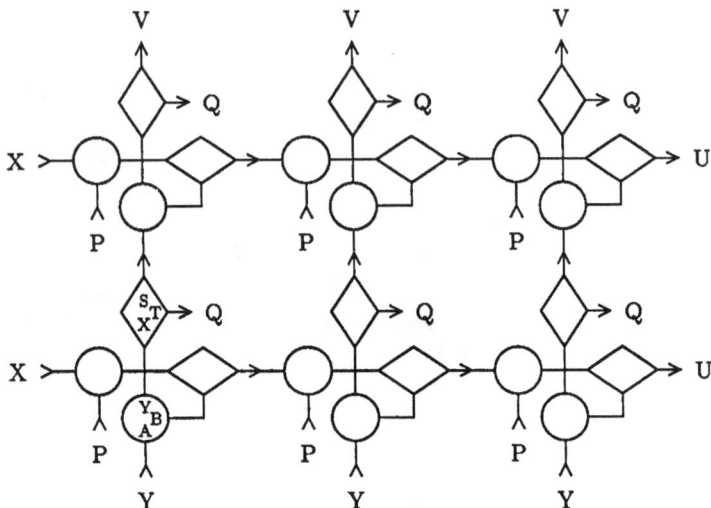

Figure 2: Building Block Interconnections for a One-way Mesh

special token is used to mark the end of the message. The packet—the combination of the header, the message, and the trailing token—cuts a path through the mesh, first to the right and then up, to its destination. The *arbiter* transmits an entire packet contiguously from an input port to the output port. Packets from different input ports are not interleaved. The *switch* consumes the first bit of the packet, and, based on whether the bit is one or zero, passes the remainder of the message out through either port S or port T.

Figure 3 shows programs for these two sequential processes. The *switch* process first communicates through port X, and raises one of two boolean flags, s or t. If s was raised, the process acts as a one-place buffer, sending the value received by port X out through port S, until the special trailing value 2 is received; in which case, flag s is lowered, and the repetition terminates. If t was raised, the process behaves in a similar manner, except T is used as the output port. The *arbiter* process performs fair arbitration[7] between the two input streams by first checking if an input is pending on port A and, if it is, then acting as a one-place buffer with output port Y until the trailing value is received. The port B will be chosen next if an input is pending, ensuring that each port is given the opportunity to transmit its packet.

```
passive X(1,3) active S(3,1) active T(3,1)
boolean s = false boolean t = false
*[ X:[ 1--> s up | 0--> t up ];
   *[ s --> X:[ 0--> S(0) | 1--> S(1) | 2--> S(2); s down ]
    | t --> X:[ 0--> T(0) | 1--> T(1) | 2--> T(2); t down ]
 ] ]
process switch(X,S,T)

passive A(1,3) passive B(1,3) active Y(3,1)
boolean a = false boolean b = false
*[ [ probe A --> a up | not probe A --> skip ];
   *[ a --> A:[ 0--> Y(0) | 1--> Y(1) | 2--> Y(2); a down ]];
   [ probe B --> b up | not probe B --> skip ];
   *[ b --> B:[ 0--> Y(0) | 1--> Y(1) | 2--> Y(2); b down ]]
 ]
process arbiter(A,B,Y)
```

Figure 3: Programs for the Switch and Arbiter Processes

2   Target Language

The target of the compilation is a self-timed circuit—a set of circuit variables (nodes) interconnected by a set of operators (gates). These circuits are designed to function correctly regardless of the internal delays of the operators. The target operator types include the combinational operators WIRE, AND, and OR; and the state-holding operators shown in Figure 4. Each operator is defined in terms of a set of production rules[6,8]. A production rule is a simple transition rule of the form $G \longmapsto S$; where G is a boolean expression and S is an assignment to true or false. All references to a circuit variable are assumed to have the same value (isochronic forks)[2,6]. A synchronizer, which cannot be represented in terms of production rules, is included to allow the implementation of programs with negated probes[7]. The synchronizer, as well as the other operators, have been implemented as CMOS standard cells.

Self-timed circuit implementations of concurrent programs are generated by implementing each sequential process as a separate sub-circuit. The sub-circuits are connected (by wire operators) only to implement communication actions. The simultaneity required in the zero-slack communications is implemented using a four-phase hand-

$$(x, y)\, C\, (z) \equiv \begin{cases} x \wedge y & \longmapsto z\uparrow \\ \neg x \wedge \neg y & \longmapsto z\downarrow \end{cases}$$

$$(\underline{x}, \underline{y})\, F\, (z) \equiv \begin{cases} \bigvee x_i & \longmapsto z\uparrow \\ \bigvee \neg y_i & \longmapsto z\downarrow \end{cases}$$

$$(x, b)\, S\, (u, v) \equiv \begin{cases} *[[x \wedge b & \longrightarrow u\uparrow; [\neg x]; u\downarrow \\ \quad |x \wedge \neg b & \longrightarrow v\uparrow; [\neg x]; v\downarrow \\ ]] \end{cases}$$

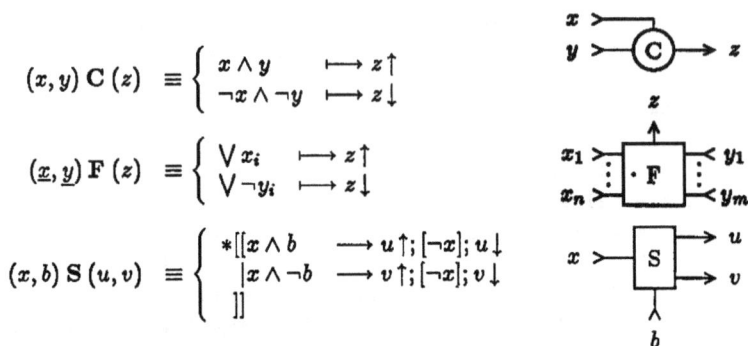

Figure 4: State-holding Operators

shaking protocol. In order to implement the exchange of data, the usual request/acknowledge pair of wires is replaced by one wire for each send value and one wire for each receive value. For example, using this unary encoding scheme, a channel, connecting the active port Y of the *arbiter* process and the passive port X of the *switch* process, is implemented using four wire operators: three pointing from Y to X and one pointing from X to Y (Figure 5).

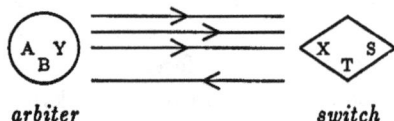

Figure 5: Implementation of a CSP Communication

The delay-insensitive nature of interprocess communication allows concurrent programs with any number of processes to be implemented. In implementating individual processes, we shall exploit this property, through a technique known as process decomposition.

3 Syntax-directed Translation

The translation from a source program into a target circuit is performed using standard syntax-directed techniques. We derive rules for generating sub-circuits that correspond to each syntactic construct of the source language, and for composing these sub-circuits to generate circuits for arbitrary programs. These translation rules are derived by applying the process decomposition transformation[6].

3.1 Process Decomposition

An arbitrary program statement β can be replaced by a single active communication C' and a new sub-process implementing β.

$$\alpha; \beta; \gamma \vartriangleright (\alpha; C'; \gamma \parallel *[[\overline{C} \longrightarrow \beta; C]])$$

The probe $\overline{C}$, of the connected passive port, is used to guard the execution of the statement β. While process decomposition introduces a new sub-process, it does not add concurrency to the implementation; the active communication C' cannot finish until its corresponding passive communication C does, and thus the strict sequencing $\alpha; \beta; \gamma$ is maintained. The original process and the new sub-process may now share variables and ports, since processes and sub-processes are never active concurrently. The variables and ports of a process are implemented by circuit variables, which must be distributed to many sub-circuits. We assume these circuit variables are distributed by isochronic forks. This assumption may be relaxed by suspending a sequential process until distribution of the variable to every reference point is detected. This refinement, described in [2], reduces all isochronic forks to size two.

Communications across channels introduced by process decomposition have simple implementations because the handshaking actions of the passive communication may be interleaved with the execution of β. Wire operators are typically sufficient to implement this special form of the passive communication action.

We apply process decomposition to generate translation rules for the BNF rules naming ⟨statement⟩. Every occurrence of ⟨statement⟩ on the right-hand side of a BNF rule is replaced by an active communication. For every BNF rule with ⟨statement⟩ on its left-hand side, a new sub-process, protected by a passive communication, is generated to implement the right-hand side.

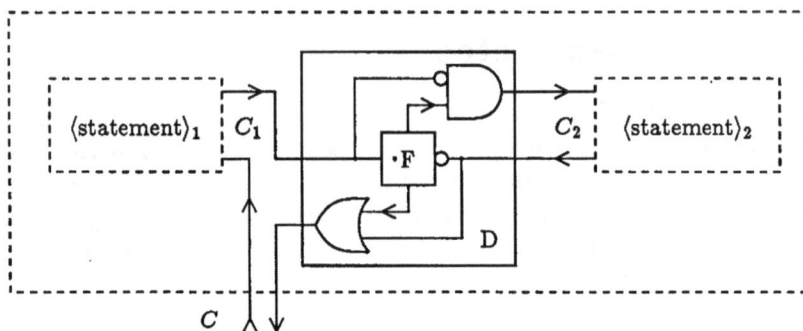

Figure 6: Translation Rule For Sequencing

We generate the translation rule for the sequencing construct by the following series of transformations:

$$*[[\overline{C} \longrightarrow \langle \text{statement} \rangle; C]]$$
$$\triangleright \qquad\qquad\qquad \{BNF \; rule\}$$
$$*[[\overline{C} \longrightarrow \langle \text{statement} \rangle_1; \langle \text{statement} \rangle_2; C]]$$
$$\triangleright \qquad\qquad\qquad \{Process \; decomposition\}$$
$$\begin{array}{ll} (& *[[\overline{C} \longrightarrow C_1'; C_2'; C]] \\ \| & *[[\overline{C_1} \longrightarrow \langle \text{statement} \rangle_1; C_1]] \\ \| & *[[\overline{C_2} \longrightarrow \langle \text{statement} \rangle_2; C_2]] \\ ) & \end{array}$$

The first sub-process is of fixed size and may be compiled into a self-timed circuit using the techniques in [6,8]. Circuits for the other two sub-processes are generated recursively by applying the translation rules. Figure 6 displays the translation rule in circuit form. The sub-circuit enclosed in box **D** is used in the implementation of several other constructs. In the following, we shall refer to instances of this circuit as instances of a **D**-element.

The other translation rules that use ⟨statement⟩ can be derived similarly. Complete derivations are described in [2]. Figure 7 shows the circuit representations of these rules. The leaf sub-processes are interconnected by several circuit variables. The named variables within the dashed boxes are connected to form the complete circuit. Each assignment sub-circuit adds a new input to the flip-flop implementing the variable. The output of the flip-flop is distributed to all sub-circuits using the variable. Each port usage adds a new input to an **OR** operator merging its output transitions. A circuit variable is

43

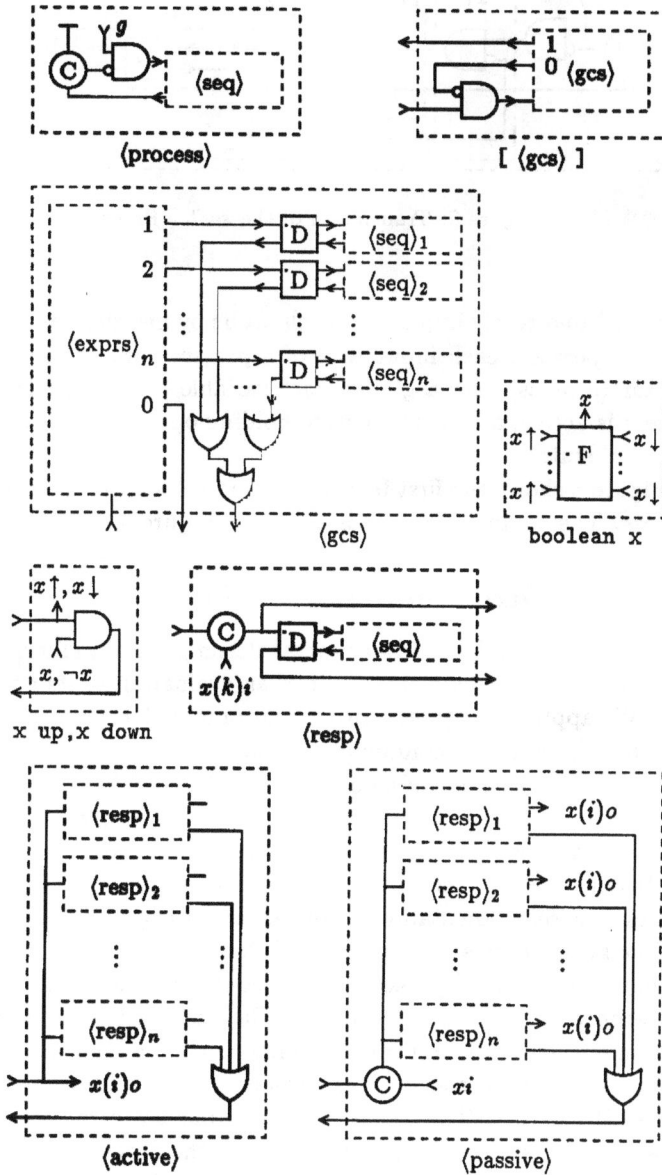

Figure 7: Circuits Implementing Statements

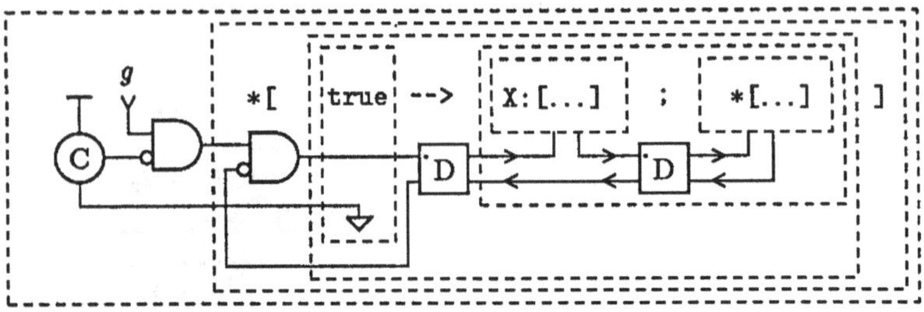

Figure 8: Translation of the *switch* Process

generated to directly implement each probe by merging together, with an **OR** operator, each input wire of a *passive* port. The translation rule for ⟨process⟩ uses a global reset variable g. When g is false, no process is active, and thus all state-holding operators can be reset to an initial state.

Figure 8 shows the first few translation rules applied to the *switch* process. The complete circuit is shown in Figure 9.

3.2 Guarded Command Evaluation

Data channels can be used in process decomposition, allowing a subprocess to communicate the result of an evaluation to its parent process. We apply this general form of decomposition to derive translation rules for guarded command evaluation.

The implementation of a guarded command set (g.c.s.) returns one of two values: 1 if a guard evaluated to true and its corresponding command was executed, and 0 if no guard evaluated to true. The repetition statement reevaluates the g.c.s. if 1 is returned. Using busy-waiting, the selection statement reevaluates the g.c.s. if 0 is returned.

The g.c.s. process is decomposed into a process for guard set evaluation; and a control process which sequences guard evaluation and the associated command execution. The guard set process selects at most one true guard, returning a value corresponding to the selected guard. This process also detects when no guard evaluates to true, and returns the value 0. The control process explicitly introduces the state variables necessary to distinguish between the program state just before guard evaluation and the program states just after a guard has been selected, thus allowing guard evaluation to complete before subsequent statements change variables and probes used in the guards.

We concentrate now on evaluating an arbitrary guard set. The se-

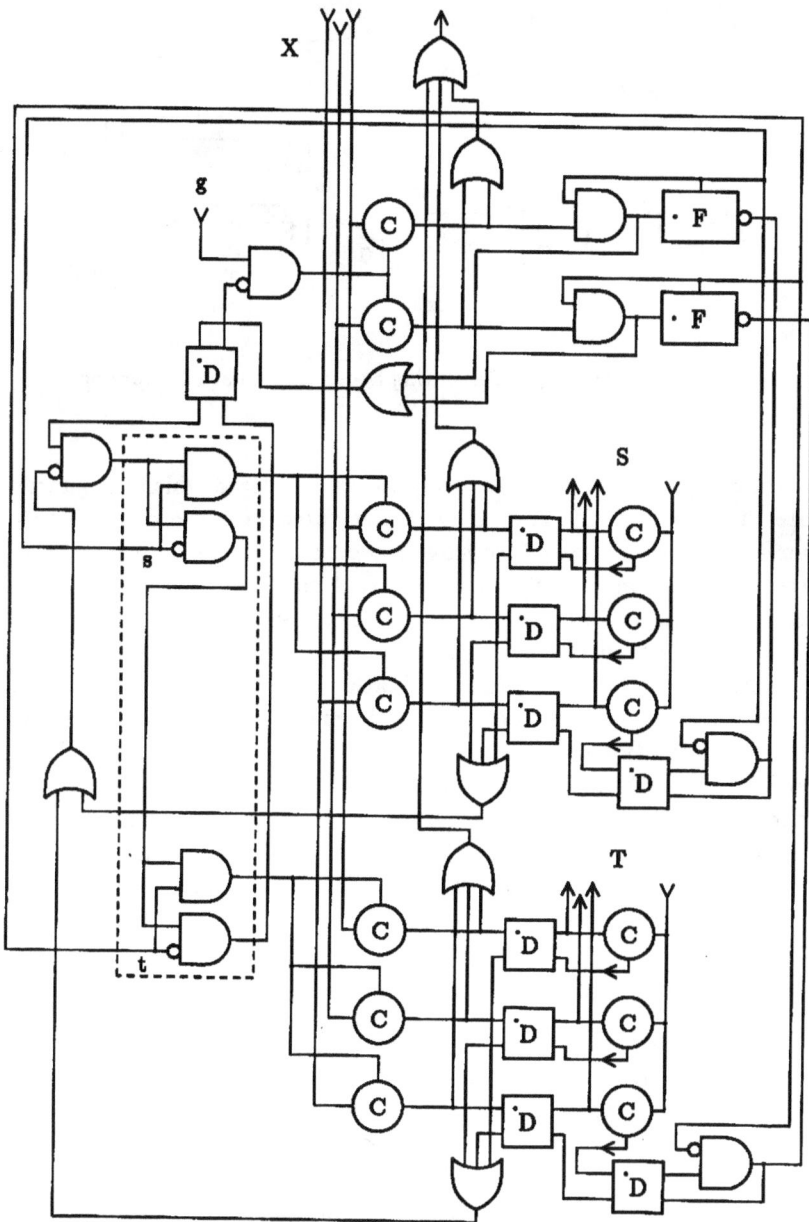

Figure 9: Optimized Circuit Implementing the *switch* Process

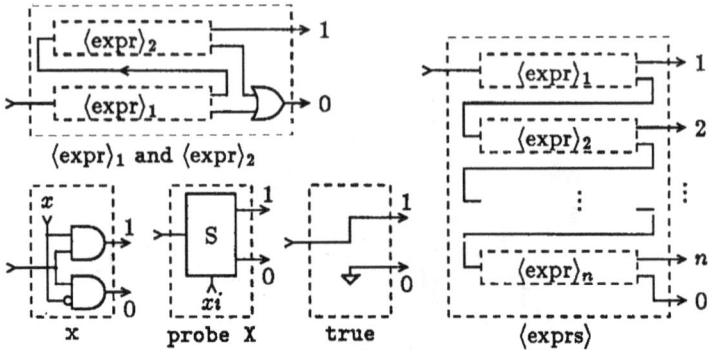

Figure 10: Circuits Implementing Sequential Guard Evaluation

mantics of the language does not specify the order in which the guards
are to be evaluated. They may be evaluated in any order or all simultaneously. We introduce three schemes for implementing guards:
sequential, *concurrent-all*, and *concurrent-one*. The *sequential* scheme
imposes a total order on the guards and generates a circuit which
conditionally evaluates each guard and each connective with a guard.
Evaluation of the guard set terminates when a guard evaluates to
true. In the *concurrent-all* scheme all guards are evaluated concurrently. Evaluation of the guard set finishes after *all* guards have been
evaluated. In the *concurrent-one* scheme all guards are evaluated concurrently, but the guards and the sub-expressions within a guard have
been strengthened so that only *one* path through the evaluation circuit will become active. For all three schemes, we must ensure mutual
exclusion among the guards and generate the all-false value.

3.2.1 Sequential Guard Evaluation

The guards are evaluated one by one until either one evaluates to
true or the last guard evaluates to false. Mutual exclusion among the
guards is accomplished implicitly by this construction.

Implementations of complex expressions are built up from implementations of simple ones. Primitive circuits exist for evaluating
variables and probes. We use conditional evaluation to implement
the logical connectives. Negation is accomplished by exchanging the
meanings of the false and true values. We show the translation rules
for sequential guard evaluation in Figure 10.

If probes are used, a further transformation is required. Because the value of a probe may change from false to true while an expression is being evaluated, all probes used more than once in a guard set are evaluated and assigned to a variable before the guard set itself is evaluated. Probes within these expressions are changed to refer to the corresponding stable variable.

The sequential scheme is applied to implement the repetition construct of the *switch* process (see outlined sub-circuit in Figure 9). The variable s is evaluated, and, if false, t is evaluated. If both are false, an all-false signal is returned.

3.2.2 Concurrent-all Guard Evaluation

All guards can be evaluated concurrently if each guard evaluation completes before the result for the entire guard set is returned. If the circuit does not wait for all the evaluations to complete, subsequent evaluations may fail because internal variables of the implementation have not been reset to a required state.

In the concurrent-all scheme, all variable and probe evaluations for the entire guard set are started simultaneously. Logical connectives are evaluated by first evaluating the two sub-expressions, and then applying the logical operation to the results. Sub-expressions may be shared among the guards, producing more efficient implementations. The number of probe references per guard set can be reduced to one, thereby eliminating the need to store each probe in a variable.

The exclusion transformation must be performed explicitly to use the concurrent-all evaluation scheme. Each guard E_i is strengthened with $\bigwedge_{j=1}^{i-1} \neg E_j$, ensuring that no two guards will evaluate to true. An all-false guard $\bigwedge_{j=1}^{n} \neg E_j$ is also created. By factoring common sub-expressions, the $O(n^2)$ increase in guard size can be reduced to adding $O(n)$ new and connectives and $O(n)$ nesting levels. The transformation also can be performed by a parallel prefix network [5], leading to $O(n \log n)$ new and connectives, but with $O(\log n)$ nesting levels, resulting in the asymptotically fastest evaluation of the three schemes.

3.2.3 Concurrent-one Guard Evaluation

As in the concurrent-all scheme, all guards are evaluated concurrently, but instead of adding circuitry to wait for all the guards to evaluate, the guards are transformed so that only one path through the evaluation circuit is active at a time. The guard set finishes evaluation when the single guard becomes true.

Each guard is transformed into AND-OR form. The AND terms are further strengthened until exclusive evaluation of each term is ensured. While exponential blow-up may occur when transforming pathological expressions, this method produces the smallest and fastest implementations of most small expressions.

This scheme cannot be used to implement all guard sets. Negated probes cannot be used, and if many variables are used in the guard set, AND-operators with several inputs are required. However, selection statements with only positive probes may be implemented without the busy-waiting iteration of the previous schemes.

4 Optimizations

Simple optimizations greatly improve the compiled circuits. *Peephole* optimization is applied to the target circuits by removing operators that can be shown redundant by a local analysis of the circuit. Other optimizations are applied at the program level. They take the form of proving that the program satisfies some invariant condition, and then using this invariant to simplify the implementation of a construct.

As an example, the explicit sequencing (**D**-element) between guard evaluation and the execution of the first statement of a corresponding command can be removed if the first statement of a command sequence does not change the value of any guard. This is always the case if the guard set consists of only constants or if the first statement is i) a skip, ii) an assignment to a variable not named in the guard set, or iii) a communication and no probes are named in the guard set.

After applying optimizations of these forms, the *switch* process compiles into the circuit shown in Figure 9. Notice the removal, from the non-optimized compilation started in Figure 8, of the initialization operators and the **D**-element protecting the constant guard true.

5 Compiler and Performance

We have implemented the translation and optimization rules described in this paper. The compiler consists of approximately 800 PROLOG clauses and is based on the program-to-machine-code compiler described in [11]. PROLOG provides an excellent environment for developing compilers and, in our case, its unification procedure provides a simple means of representing and composing circuits.

Name	Operators per process	
	Manual	Compiled
$3x + 1$ iterator	95	175
routing switch	37	44
routing arbiter	36	49
lazy stack	50	86
token ring	13	27
systolic multiplier	38	59

Figure 11: Comparison of Manual and Compiled Designs

The compiled circuits are typically no more than twice the size of those derived by hand. In Figure 11, a rough comparison between the number of operators is shown for a variety of fabricated hand designs. The extra operators needed in the compiled circuits can be explained by two interrelated factors: The current source language cannot represent the interleaving of statements within a sequential process (reshuffling of handshaking expansions), and the current program level optimizer does not detect all the cases where explicit sequencing can be removed. We expect to narrow this gap in the future.

Using 3μm SCMOS, we have fabricated several circuits derived manually from concurrent programs. We have used either a standard-cell place-and-route preprocessor for the MAGIC design tool, or the MOSIS FUSION service[1], to generate layout from an electrically optimized operator-level description of the circuit. Rudimentary electrical optimization is performed individually on each circuit variable by inserting appropriately sized drivers to minimize the transition time of each operator. Such simple techniques have met with good results; in an early implementation of a mesh router (Figure 2), the transition time of a typical operator was 2.3ns, resulting in a pad-to-pad propagation delay of 45ns for an *arbiter* and a *switch* process connected in series. We expect similar electrical performance from chips generated by the compiler.

The translation method produces correct, self-timed implementations of arbitrarily large concurrent programs, and because each translation rule is of fixed size, the size of an implementation is no worse than linearly related to the size of the source program. The translation method and the compiler provide a constructive proof that a design methodology based on programs instead of circuits is not inherently inefficient, and thus represents a practical approach to the design of VLSI systems.

Acknowledgments

We wish to thank Pieter Hazewindus for his comments on early versions of this manuscript. This research is sponsored by the Defense Advanced Research Projects Agency, ARPA Order number 3771, and monitored by the Office of Naval Research under contract number N00014-79-C-0597.

References

[1] R. Ayres, *FUSION: A New MOSIS Service*, Information Sciences Institute, ISI/TM-87-194 (1987)

[2] S.M. Burns, *Automated Compilation of Concurrent Programs into Self-timed Circuits*, M.S. Thesis, California Institute of Technology, Caltech-CS-TR-88-2 (1988)

[3] W.J. Dally and C.L. Seitz, "The Torus Routing Chip", *Distributed Computing*, 1, pp 187-196 (1986)

[4] C.A.R. Hoare, "Communicating Sequential Processes", *C. ACM* 21, 8, pp 666-677 (August 1978)

[5] R.E. Ladner and M.J. Fischer, "Parallel Prefix Computation", *J. ACM*, 27, pp. 831–838 (1980)

[6] A.J. Martin, "Compiling Communicating Processes into Delay-Insensitive VLSI Circuits", *Distributed Computing*, 1, pp 226-234 (1986)

[7] A.J. Martin, *A Delay-Insensitive Fair Arbiter*, Caltech Computer Science Technical Report, 5193:TR:85 (1985)

[8] A.J. Martin, "The Design of a Self-Timed Circuit for Distributed Mutual Exclusion," *Proc. 1985 Chapel Hill Conf. VLSI*, ed. Henry Fuchs, pp 247-260 (1985)

[9] A.J. Martin, "The Probe: an Addition to Communication Primitives," *Information Processing Letters*, 20, pp 125-130 (1985)

[10] C.L. Seitz, "System Timing," Chapter 7 in Mead and Conway, *Introduction to VLSI Systems*, Addison-Wesley, Reading MA (1980)

[11] D. Warren, "Logic Programming and Compiler Writing," *Software—Practice and Experience*, 10, 2 (1980)

Trace Theory for Automatic Hierarchical Verification of Speed-Independent Circuits

David L. Dill

Computer Systems Laboratory
Stanford University
Stanford, CA 94305

A trace theory for modelling and specifying speed-independent circuits is presented. The theory supports efficient automatic and hierarchical verification of circuits. An implementation of a verifier based on the theory is described. This verifier is applied to Martin's proposed implementation of a distributed mutual-exclusion circuit, in which it finds a bug. An amended circuit is also shown to be correct.

1   Introduction

As circuits increase in complexity, it becomes very difficult to time the signals in them. Broadcasting a global signal is particularly troublesome. Events which are supposed to be simultaneous occur at significantly different times because of transmission delays. When the global signal is a clock, this is *clock skew*, which can be a serious problem in large circuits. Additionally, the task of checking whether timing constraints are met increases in difficulty with the complexity of the circuit.

One way to address these problems is to design circuits so that they do not depend on the relative speeds of their components. Such designs are called *speed independent*. A speed-independent design can be implemented with confidence that it will function properly given any finite delays in its components. Of course, any circuit that has a clock that runs at a fixed speed cannot be speed-independent, since the clock interval places a lower-bound on the speeds of some components.

A major barrier to the use of speed-independent circuits is that they are difficult to design and analyze. In general, every execution path resulting from every possible combination of delays in the circuit must be considered. Fortunately, this barrier can be lowered by delegating to a computer the tedious task of considering all possible executions. We describe an automatic verifier based on *trace theory* that can check relatively large speed-independent control circuits. As an example, the verifier is applied to Martin's distributed mutual exclusion circuit [3].

The verifier is founded on a trace-theoretic model of speed-independent circuit operation (essentially, a formal semantics for circuits). The same formalism can also be used to *specify* circuits. This leads directly to a theory that supports hierarchical verification in a very natural and general way.

The remaining sections present these results in more detail. Section 2 defines the formal representation of a single circuit, which is called a trace structure. Section 3 shows how trace structures can be used to model certain simple building-blocks that are frequently used in speed-independent circuits. Section 4 describes operations that can be used to construct the trace structures for more complex circuits out of trace structures for simple circuits. Section 5 shows how trace structures can also be used for specifications and describes an implementation of a verifier based on this result. Section 6 describes a proposed speed-independent design of a distributed mutual exclusion circuit and describes the results of applying the verifier to the circuit at two levels of abstraction. Finally, section 7 presents conclusions and directions for future work.

The presentation of these results is relatively informal. They are described in greater detail and proved in the author's PhD thesis [1].

2   Trace Theory

The fundamental idea behind trace theory is that the behavior of a circuit can be described using sets of event sequences, called *traces*. The events of interest are *transitions*; a transition is a change of the signal on a wire from a logical 0 to a logical 1 or vice versa (rising and falling transitions are not distinguished). Since a transition is the only event that can occur on a wire, the name of the wire can be used to represent the transition.

Trace models are a good basis for simple models of some types of concurrent systems (of which speed-independent circuits are an instance). Hoare has used a trace model for a semantics of CSP [4]. Trace theory has also been used for specification of delay-insensitive circuits, by Rem, Snepscheut, Udding, and others [6,7,8,9]. The theory presented here is closely related to the theories of Rem, Snepscheut, and Udding. The primary differences are in the treatment of input and output wires and the modifications to support verification.

Circuits are modeled using *prefix-closed trace structures* (called trace structures for brevity). A trace structure T is a quadruple (I, O, S, F). Every circuit has a finite set of input wires, I, and a finite set of output wires, O. The inputs and outputs are disjoint. For convenience, A is used as an abbreviation for $I \cup O$. By convention, a trace structure T^X always has components named I^X, O^X, S^X, and F^X for any superscript X.

A novel aspect of this version of trace theory is that *several* distinct trace sets are used to describe a circuit. There is a set $S \subseteq A^*$ of *successful traces* (also called *successes*) representing the behavior of the circuit when it receives legal inputs. The set of successes is prefix-closed, meaning that if x is a member of S, so is every prefix of x, including ϵ. S is a *regular set*, which is reasonable because digital circuits have a finite number of states.

There is another set of traces that models undesirable signals: the set F of *failure traces* (also called *failures*), which is also a regular subset of A^*. It is important to note that the sets of successes and failures are not necessarily disjoint; if a trace is a member of both sets, a circuit can either succeed or fail, nondeterministically, when it exhibits that trace.

The set P of *possible traces* is defined to be the union of the S and F sets. The possible traces represent executions that *can* occur, regardless of whether they *should* occur. This distinction is important for low-level hardware, because a circuit cannot prevent an unwanted input from being sent to it. P is prefix-closed, also. There is an additional requirement on P, called *receptiveness*, that ensures that a circuit cannot control its inputs. P is receptive if appending additional inputs to any trace in P yields another trace in P. This property can be written $PI \subseteq P$ (if X and Y are sets of sequences, XY represents the set of all concatenations xy such that $x \in X$ and $y \in Y$).

3 Example Trace Structures

This section presents examples of trace structures describing some commonly-used components of speed-independent circuits.

3.1 Boolean Gate

A boolean gate is defined by a set of zero or more inputs, a single output, and a boolean function which maps a logical valuation of the inputs to an output value. The output of the gate is a sample of the value of the boolean function of the input values, delayed by an arbitrary amount.

A *hazard* occurs if the value boolean function changes and then reverts to its original value without necessarily waiting for the actual output of the gate to change. Hazards can cause severe problems in speed-independent circuits; in response to the input changes, there may be an arbitrarily brief change on the output (a spike), a change to an intermediate logical value, or no change at all. Hence, the trace structures given here for boolean gates classify all hazard-causing inputs as failures.

It is not difficult to construct the two automata for the S and F sets of a gate, given the corresponding inputs, output, and boolean function. The S and F automata are identical except for their accepting states. The states of each automaton are the possible logical valuations for the inputs and output of the gate (for a gate with n inputs, this is isomorphic to the set of all bitvectors of length $n+1$, so these are called *bitvector states*), and an additional state for accepting failures (called the *failure state*).

A *stable* state is a bitvector state in which the value of the boolean function (determined by the input values) is the same as the actual value of the output wire. An *unstable* state is a bitvector state in which they are not equal. The failure state is neither stable nor unstable. The transitions of the automata are described by a function that maps a *current state* and wire name to a *next state*. The *natural successor* of a bitvector state for a wire name a is the bitvector state which is the same as the original except that the value of a is complemented.

To define the automata for a gate, let q be a state and a be a symbol, and let q' be the natural successor of q for a. If a is an input symbol, there is a transition on a from q to q' if q is stable or if q' is unstable. Otherwise, there is a transition on a from q to the failure state. If a is an output, there is a transition from q to q' on a if q is unstable (in this case, q' is necessarily stable). Otherwise, there is not transition on a. Finally, there is a transition from the failure state to itself on every wire name.

These rules define the states and transitions of the automaton.

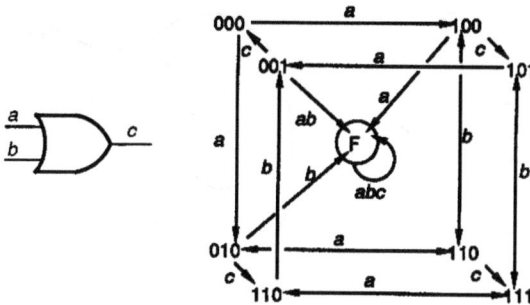

Figure 1: An OR gate and its state graph.

The start state and a set of accepting states remain to be defined. The start state depends on the initial conditions of the device; it is the bitvector corresponding to the initial values of the input and output wires. The set of accepting states for S consists of all the bitvector states. The accepting set for F consists of the failure state.

Hazards can occur only when there is an input transition from an unstable state to a stable state. In the automaton, these transitions are routed to the failure state instead of the natural successor, so any input that could cause a hazard is classified as a failure.

Figure 1 shows an OR gate and the state graph for representing its S and F sets when the gate starts out with both inputs and the output at logical 0. There is a state at each vertex of the cube, labeled with the values of a, b, and c, in order. The failure state is in the center of the cube. The states 000, 011, 101, and 111 are stable and 001, 010, 100, and 110 are unstable.

3.2   C element

A *C element* is an extremely useful primitive in speed-independent circuits. It is typically used to signal the completion of all of several concurrent computations. The output value of a C element remains constant until *all* of its inputs are equal to the complement of the output; the output then changes to its complement after an arbitrary delay. Figure 2 shows a two-input C element and the state graph for its S and F sets when the initial values of the wires are either $abc = 000$ or $abc = 111$.

In the trace structure for the C element, changes in the inputs are restricted so that when all of the inputs are equal to the complement of the output, they remain constant until the output changes. Input

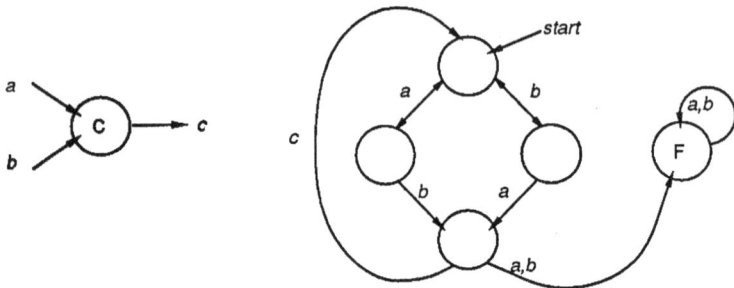

Figure 2: A C element and its state graph.

transitions that violate this restriction are failures.

3.3 Mutual-Exclusion Element

A mutual-exclusion element (ME-element) can be used to ensure that no more than one of several *users* accesses a resource at the same time. For mutual exclusion among n users, each user has a *request* wire and an *acknowledge* wire. Mutual exclusion is enforced when all parties obey a particular protocol: the user asks for the resource by signalling on its request wire; the ME element then grants the resource by signalling on the acknowledge wire; the user then releases the resource by signalling a second time on the request wire; finally, the ME element reclaims the resource by signalling on the acknowledge wire.

The ME element enforces mutual exclusion by refusing to grant the resource to a user if any other user has been granted it and has not released it. In other words, there is a *critical region* between the grant and release signals, and the ME element refuses to allow more than one user at a time to enter the critical region.

Figure 3 shows a two-input ME element and its state graph. The initial values of all wires are assumed to be 0. Each state is labelled with the values of $r1$, $r2$, $a1$, and $a2$, in that order.

4 Operations on Trace Structures

There are three operations on trace structures which can be used to describe complicated circuits in terms of simpler ones: **hide**, **compose**, and **rename**.

hide makes output wires internal, so they can no longer be connected to wires of other circuits, and so that transitions on them

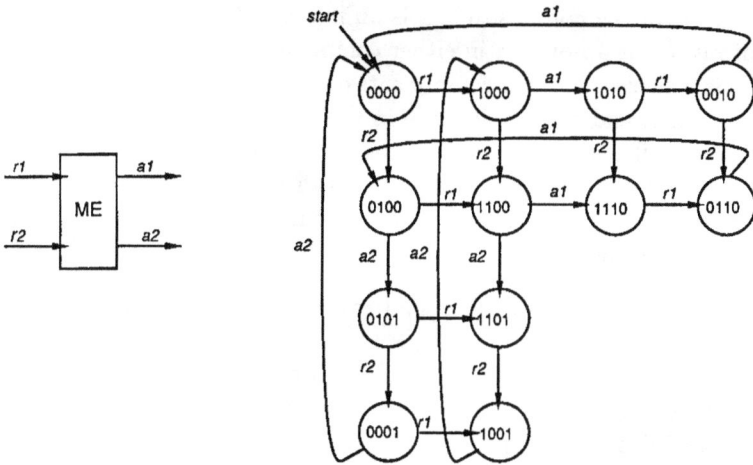

Figure 3: An ME element and its state graph.

cannot be observed. **hide** is defined using an auxilliary function on traces and trace sets: if $x \in A^*$, then $\mathbf{del}(D)(x)$ is the trace x with all D symbols deleted. If T is any trace structure, D is a subset of O, and $T' = \mathbf{hide}(D)(T)$, then $I' = I$, $O' = O - D$, $S' = \mathbf{del}(D)(S)$ and $F' = \mathbf{del}(D)(F)$. ($\mathbf{del}(D)(S)$ is the image of S under $\mathbf{del}(D)$, since S is a set of sequences. The same is true for F.)

   compose is a binary operation on trace structures which is defined only when the outputs of the two trace structures are disjoint. **compose** is defined in terms of two simpler operations. The first of these is *inverse deletion*, which essentially adds inputs to a circuit, which are then ignored. Given a set of traces $X \subseteq A^*$ and a set of wire names D that is disjoint from A, $\mathbf{del}(D)^{-1}(X)$ (the inverse image of the delete function) is the set of all traces x in $(A \cup D)^*$ such that $\mathbf{del}(D)(x) \in X$. Inverse deletion is extended to trace structures so that $\mathbf{del}(D)^{-1}(T) = (I \cup D, O, \mathbf{del}(D)^{-1}(S), \mathbf{del}(D)^{-1}(F))$. Intuitively, if a trace structure is regarded as a constraint on traces that can appear in a composition, this operation adds inputs that are not constrained at all.

   The second operation used to define composition is *intersection*, which is composition in the special case where every wire in each component is connected to a wire in the other component. If T and T' are trace structures such that $O \cap O' = \emptyset$ and $A = A'$, their intersection, $T'' = T \cap T'$, is defined so that $I'' = I \cap I'$, $O'' = O \cup O'$, $S'' = S \cap S'$, and $F'' = (F \cap P') \cup (P \cap F')$. In other words, a wire is

an input in the composite if it is an input in both components or an
output if it is an output in either component; and a trace is a success
if it is a success in both components or a failure if it is a failure in one
component and a possible trace in the other.

The full composition operation can be defined by using inverse
deletion to add the minimum necessary set of inputs to each component to make the sets of wires of the components the same, and then
intersecting them:

$$T \,|\, T' = \mathbf{del}(A' - A)^{-1}(T) \cap \mathbf{del}(A - A')^{-1}(T').$$

rename does a one-for-one substitution of a set of new wire names
for old wire names. The renaming is specified by a *renaming function*
r which bijectively maps old wire names A to a set of new wire names
A'. If T is a trace structure and **r** is a renaming function, then $T' =$
rename(r)(T) has $I' = \mathbf{r}(I)$, $O' = \mathbf{r}(O)$, $S' = \mathbf{r}(S)$, and $F' = \mathbf{r}(F)$.

Using obvious notations for sets of wire names, renaming functions, and so on, *expressions* can be written using these three operations. These expressions can be used to describe a circuit structure in
a textual form that is convenient for processing by computer. The literals in the expressions are trace structures for primitive components.

5 Verification with Trace Theory

Trace structures can be used not only to model the actual behavior of
circuits, but also to specify desired behaviors. Moreover, it is possible
to define formally the proper relation between a specification and
an implementation and to check automatically whether the relation
holds.

We regard a specification as a description of an idealized component. Larger circuits can be designed which will be correct if the
idealized component is used in them. Any trace structure which preserves the correctness of the larger circuit when substituted for the
specification is an implementation of the specification. A substitution
of one component for another that preserves the correctness of the
larger circuit is called a *safe substitution*.

A notion of verification based on safe substitution has an two important advantages. First, it is simple because a single formalism can
be used both for descriptions and for specifications. Second, it facili-
tates *hierarchical verification*: specifications at one level of abstraction
can be treated as descriptions of implementations at higher levels of

abstraction. In this way, irrelevant implementation details can be suppressed in moving from one level of abstraction to the next. Hierarchical processing is essential in order to contain the computational complexity of verification.

To formalize safe substitution, we define a *context* to be a trace theory expression with a free variable. Such an expression is written $\mathcal{E}[\,]$; $\mathcal{E}[T]$ represents the same expression with the trace structure T substituted for the free variable. A trace structure is said to be *failure-free* when its set of failures is empty. A trace structure T *conforms to* another, T', if for every context $\mathcal{E}[\,]$, if $\mathcal{E}[T']$ is failure-free, then so is $\mathcal{E}[T]$. This relation is also written $T \preceq T'$.

This definition of conformation is not implementable because there are infinitely many possible contexts. However, it is possible to find a single worst-case context that will be a counterexample to $T \preceq T'$ if the relation does not hold. This context is a composition of the free variable with a trace structure T'^M, called the *mirror* of T'. The inputs and outputs of T'^M are exactly the opposite of the inputs and outputs of T'. T'^M is defined to be the maximum trace structure under the $\preceq$ ordering such that $T' \cap T'^M$ is failure-free.

In general, it is not easy to compute the mirror of a trace structure. However, it happens that every trace structure is equivalent to a *canonical trace structure*, and that it is very easy to compute the mirror of a canonical trace structure. All of the example trace structures above are canonical.

In a canonical trace structure, the F set is a function of the S set and can be made implicit. If T is a canonical trace structure and $S \neq \emptyset$, then F is always the set $(SI - S)A^*$. We define a *choke* trace to be a member of $SI - S$; in essence, a choke occurs when a circuit receives an unwanted input. The failures in a canonical trace structure are the traces that have choke traces as prefixes.

Consequently, a canonical trace structure can be considered to be a triple (I, O, S). For a canonical trace structure, the mirror can be computed simply by exchanging I and O. The following theorem is the basis for automatic verification using trace structures:

Theorem 1 *If T is a trace structure and T' is a canonical trace structure, $T \preceq T'$ if and only if $I = I'$, $O = O'$, and either $S' = \emptyset$ or $T \mid T'^M$ is failure-free.*

The trick of composing with the mirror is due to Ebergen [5]. However, the relation to safe substitution and the underlying semantic justification are the author's.

A verifier based on these ideas has been implemented in Common Lisp. The verifier takes as input a specification, which must be a canonical trace structure, and an expression describing the implementation. If the implementation conforms to the specification, the program reports this fact; otherwise, it prints a failure trace to help diagnose the problem.

The implementation is straightforward. Regular sets are represented as finite automata. The verifier composes the implementation with the mirror of the specification, reduces the resulting expression to an n-way composition of simple prefix-closed trace structures, and then searches in the product state graph for failures. Product states are created only as needed, which is crucial in practice because it is often easy to find a problem by examining a few states even when the entire state graph is huge. (This was a deficiency of a previous verifier based on Ed Clarke's model checker [2].)

The program is actually more general than what is described here. In addition to checking conformation, it can act as an interactive trace theory interpreter. The operations **compose**, **rename**, and **hide** can be applied to trace structures so that the results can be inspected. Also, trace structures and expressions can be tested for equivalence.

6 An Example: Distributed Mutual Exclusion

As an example, we apply the verifier to a circuit proposed by Alain Martin for achieving mutual exclusion among n users (for fixed n) [3]. Martin gave a general token-passing solution to the problem, in which a circuit to handle n users could be constructed using n standard cells connected in a ring. He also gave an implementation of a cell in terms of lower-level components, which was derived from a specification in a CSP-like language.

To illustrate hierarchical verification using trace structures, the circuit is checked at two levels of abstraction. At the higher level of abstraction, it is checked whether collections of cells properly implement mutual-exclusion elements for one, two, and three users. At the lower level, it is checked whether Martin's proposed implementation conforms to the specification of the cell.

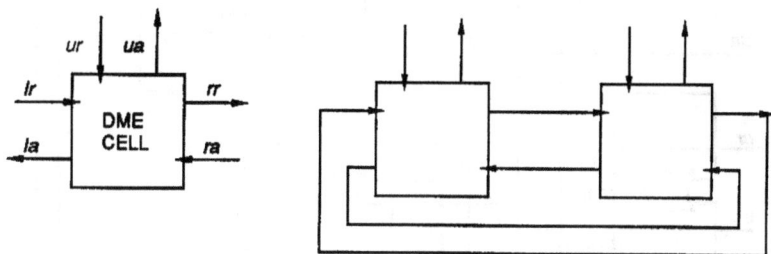

Figure 4: Distributed mutual exclusion.

6.1 High-level verification

A mutual-exclusion circuit for n users consists of n cells, connected in a ring. There is exactly one cell for each user. There is a single token, representing the resource, which inhabits at most one of the cells at any one time. Generally, requests for the token propagate to the right and the token propagates to the left. Figure 4 depicts a single cell and a ring of two cells.

In more detail, each cell has three request/acknowledge interfaces: one for its user (ur and ua), one for the cell to the left (lr and la), and one for the cell to the right (rr and ra). The protocol between the user and its cell is the same as the protocol for the mutual exclusion element described in the example above. The protocol for communicating between cells is somewhat similar: if a cell wants the token, it signals on the rr wire. The cell to the right gives it the token by signalling on ra (la of the right cell). Then the cell signals a second transition on rr and the right cell acknowledges on ra (these transitions do nothing except return both wires to their original logical values, which simplifies the implementation of the circuit).

The specification of a cell was written using an ad hoc Lisp macro that constructs a trace structure using state variables and production rules. One-, two-, and three-way mutual exclusion elements (as described above) were used as specifications, and were compared with implementations consisting of rings of one, two, and three cells. In each case, the program reported that the implementation conformed to the specification. The program examined four states for the one-cell circuit, 208 states for the two-cell circuit, and 2496 states for the three-cell circuit. The results of the high-level verification provide some confidence that the scheme works properly and that the specifications of the cells were what was desired.

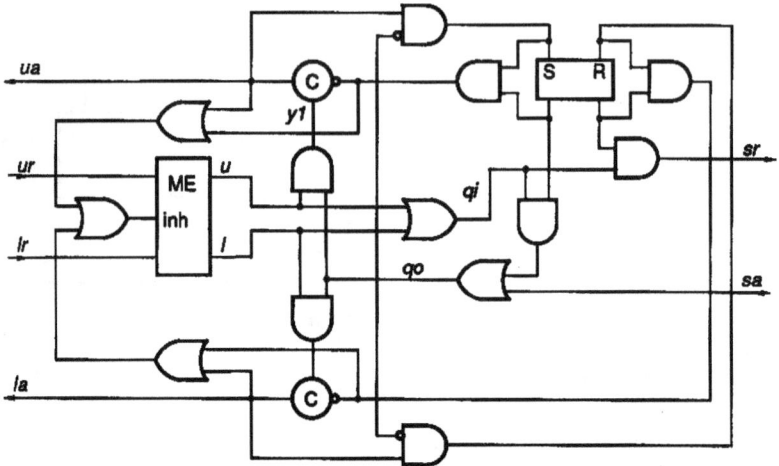

Figure 5: Original cell implementation.

6.2 Low-level verification

In the high-level verification, the trace structure for the cell was treated as a description of a primitive component. In the low-level verification, it is regarded as a specification to be implemented.

Martin's proposed implementation is depicted in Figure 5. In this implementation the ME element has an additional input *inh*; when this input is high, it inhibits the granting of the resource to a user. This input is used to prevent processing of a request from the user or the left cell until processing of the previous request has completely finished. The trace structure for this variant of the ME element is similar to the original ME element.

When this implementation is compared with the specification, the verifier reports that the implementation does not conform to the specification (upon examining the twelfth state in the product graph). The verifier provides the example failure

$$lr; ur; u; qi; rr; ra; qo; y1; ua; ur; u; l.$$

The failure is a hazard in the OR gate that has u and l as inputs. In brief, the fundamental problem is that there are three OR gates that compute the *inh* signal of the ME element. If the delays in these gates are relatively long, the inhibit signal can arrive too late, allowing the processing of the other interface to begin prematurely.

The presence of a bug in this design is surprising, since it was

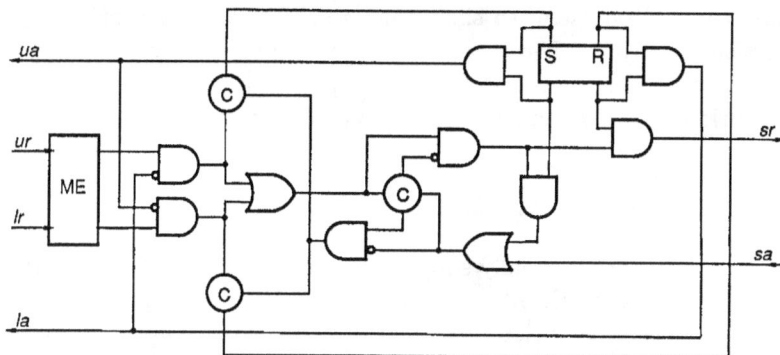

Figure 6: Amended cell implementation.

derived from a correct program using correctness-preserving transformations. However, a careful reading of Martin's original paper reveals that the offending OR gates were introduced without using the transformations.

Martin also discovered the problem after publishing the design, and has derived a new design, which is shown in Figure 6. The amended circuit uses an ordinary ME element; additional components are used to delay processing of requests on one interface until requests on the other have been completely processed. The verifier reports that this implementation *does* conform to the specification of the cell. It examines 154 states to do this.

7   Conclusions and Future Work

The automatic verifier described here promises to be a practical tool for fully speed-independent circuits. It is especially circuits of the complexity of the previous example. There is no clear relation between the size of a circuit and the number of states that must be examined to verify it; the program can handle some circuits that have many more components, and it runs out memory on examples with relatively few components.

There are many issues that remain to be explored. On the practical front, the most pressing need is for better languages for describing regular sets (e.g. program notations or bounded petri nets). The program can probably be made an order of magnitude more efficient in both time and space by known improvements in the implementation.

A deeper issue is the verification of *liveness properties*; for example, it would be useful to show that a circuit inevitably produces a particular output. The theory described here has been extended to handle general liveness and fairness properties [1]; unfortunately, this extension cannot be practically implemented in its current form.

Acknowledgements

This research was partially supported by NSF Grant MCS-82-16706.

The author would like to thank Alain Martin for providing the amended implementation of the DME cell and for clarifying the assumptions under which it was to be verified.

References

[1] David L. Dill. *Trace Theory for Automatic Hierarchical Verification of Speed-Independent Circuits*. PhD Thesis, Computer Science Department, Carnegie-Mellon University, August 1987.

[2] D.L. Dill and E.M. Clarke. Automatic Verification of Asynchronous Circuits Using Temporal Logic. *IEE Proceedings, Pt. E* 133(5):276–282, September, 1986.

[3] Jo C. Ebergen. Private Communication. 1984.

[4] C.A.R. Hoare. *Communicating Sequential Processes*. Prentice-Hall, Inc., 1985.

[5] Alain J. Martin. The Design of a Self-timed Circuit for Distributed Mutual Exclusion. In Henry Fuchs (ed), *1985 Chapel Hill Conference on Very Large Scale Integration*. pp. 245–260, Computer Science Press, Inc., 1985.

[6] Martin Rem, Jan L.A. van de Snepscheut, and Jan Tijmen Udding. Trace Theory and the Definition of Hierarchical Components. In Randal Bryant (ed), *Third CalTech Conference on Very Large Scale Integration*. pp. 225–239, Computer Science Press, Inc., 1983.

[7] Jan L.A. van de Snepscheut. Deriving Circuits from Programs. In Randal Bryant (ed), *Third CalTech Conference on Very Large Scale Integration*. pp. 241–256, Computer Science Press, Inc., 1983.

[8] Jan L.A. van de Snepscheut. *Trace Theory and VLSI Design*. PhD Thesis, Department of Computing Science, Eindhoven University of Technology, October, 1983.

[9] Jan Tijmen Udding. A formal model for defining and classifying delay-insensitive circuits and systems. *Distributed Computing* 1(4):197–204, 1986.

Invited Talk

The Memory Refresh Problem

Nicholas Pippenger
IBM Almaden Research Center
San Jose, California 95120-6099

The storage of large amounts of information for long periods of time, with small amounts of space and energy per bit, leads to the need for a "refresh" or "restoration" mechanism to counteract the accumulation of errors. Such a mechanism must rely on redundancy in the storage of information. Replication, the simplest form of redundancy, requires voting for restoration; more efficient error-correcting codes require more sophisticated computations. The devices performing these computations also introduce errors; thus, the problem of information storage in the presence of noise leads to problems of computation in the presence of noise.

The problem of devising voting schemes for the restoration of replicated information was formulated by von Neumann in the 1950's. It was not until the 1970's, however, that a rigorous solution was provided Dobrushin and Ortyukov, and not until the 1980's that explicit constructions were devised by Gács and Pippenger. These explicit constructions draw upon a long line of work concerning "expanding graphs" or, equivalently, graphs with certain spectral properties.

The use of error-correcting codes to reduce redundancy to its minimum leads to still more formidable problems. In the 1970's, Kuznetsov showed how the problem of storage and restoration can in principle be solved; his solution uses expanding graphs of a strength for which no explicit construction is yet available. Furthermore, since the information is stored in coded form; there are problems of initial encoding and final decoding that need to be addressed. Our main goal in this talk is to describe what a satisfactory solution to this problem should look like, and to survey the progress that has been made toward achieving it.

The Fluent Abstract Machine

Abhiram G. Ranade
Sandeep N. Bhatt
S. Lennart Johnsson

Department of Computer Science
Yale University
New Haven CT 06520.

Underlying every general programming model is a shared address space. Every process can potentially access any object in this space in one step. While this allows tremendous expressive power, it poses an enormous challenge to the communications hardware. This conflict between ideal programming models and real architectures has traditionally been resolved by supporting a less general model which restricts the possible patterns of access.

The Fluent abstract machine supports a very powerful programming model. In addition to arbitrary access patterns, the instruction repertoire of the Fluent machine also includes the multiprefix operation and high-level set operations.

The Fluent machine consists of over one hundred thousand processors interconnected by a butterfly network. The efficiency of the Fluent machine derives from a very simple router, which effectively eliminates the possibility of congestion. The routing hardware is extremely simple, inexpensive, and provably efficient.

1 Introduction

We envisage building a Fluent machine with over one hundred thousand processors. Except for highly structured computations, such a large computer must necessarily spend a good deal of time communicating messages between its processors. As long as the total communication time does not swamp the total computation time, high performance is guaranteed.

Large parallel computers are also difficult to program. The situation becomes intolerable if the programmer must explicitly manage communication between processors. For this reason it is necessary to have a powerful programming model (abstract machine) which abstracts away concerns not directly relevant to the problem being solved. For overall performance, the abstract machine must be efficiently supported on the underlying machine.

Of the programming models proposed thus far, shared memory models have been the most attractive. The most general shared memory models in the literature, the concurrent-read concurrent-write parallel random-access machines (CRCW PRAMS) allow an arbitrary number of processors to read or write a common memory location in one time step. Complex communications operations, broadcast and multicast for example, can be implemented in one step. Abstracting complex communications patterns into unit steps greatly simplifies the tasks of designing algorithms and writing programs. For this reason, CRCW PRAM models are favored over weaker abstract machine models for which most, if not all, of the programming effort is spent synchronizing the movement of data.

How do we implement a shared memory model on a machine with processors and memories distributed throughout an interconnection network? The solution is to devise an efficient router which emulates shared memory operations and hides details of the communications network from the user. This is precisely what recent machines such as the Thinking Machines Corporation's Connection Machine [8,9], the BBN Butterfly [2] and Monarch, the IBM RP3 [13], and the NYU Ultracomputer [6] aim to achieve.

These machines emulate abstract machines of varying generality and power. The Connection Machine CM2 has hardware support for concurrent read as well as concurrent write operations with combination. The Connection Machine and the NYU Ultracomputer/RP3 efficiently support the scan operation [4]. The Ultracomputer and RP3 also support the fetch-and-add operation, but the switching hardware is expensive and experiments reveal poor performance because of "hot spots" [11,14]. It thus becomes difficult to argue that the abstract machine operations are performed in unit time.

The Fluent abstract machine subsumes each of the abstract machines mentioned above. In fact, the *multiprefix* primitive of Fluent requires arbitrarily many primitive operations on the other abstract machines. The Fluent instruction set also includes basic set operations. With its rich instruction set, the Fluent abstract machine is readily suited as an intermediate language for compiling very high level languages.

The Fluent abstract machine can be supported efficiently and inexpensively in hardware. The heart of the Fluent machine is the router which is based on the recent work of Ranade [16]. In contrast with the Ultracomputer and RP3, the hardware requirements are minimal. More importantly, we can prove that each Fluent instruction is implemented quickly. This justifies our thesis that large Fluent machines will be less expensive, faster and easier to program than existing parallel machines.

The remainder of this extended abstract is organized as follows. Section 2 describes the Fluent abstract machine and contrasts it with other models. Section 3 outlines the implementation of the abstract machine on the butterfly network. Section 4 outlines a design for the routing switch. Section 5 describes the Fluent machine, presents results of timing simulations, and cost and performance estimates. Section 6 concludes with some of the important research issues that need further study, and outlines our ongoing work.

2 The Fluent Abstract Machine

This section describes the primitive instructions of the Fluent abstract machine, and contrasts the Fluent programming model with other models. In later sections we show how every instruction is supported efficiently in hardware. As a consequence, the time–complexity of a Fluent program can be easily estimated as the maximum number of primitive instructions executed by one processor.

The Fluent abstract machine has N (virtual) processors indexed $1, 2, \ldots, N$ which are connected to a shared memory holding M variables indexed $1, 2, \ldots, M$. The processors of the abstract machine operate synchronously in discrete time cycles. Every primitive instruction is executed in one time cycle; executing an instruction at time T (in the Tth time cycle) has the effect of changing the state that existed at the start of time cycle T.

The Fluent abstract machine is characterized by two types of primitives — *multiprefix* and *set operations*. The multiprefix operation is a fully general prefix operation and subsumes the fetch-and-op primitive on the NYU Ultracomputer [7], as well as the scan operation on the Connection machine [4]. Set operations are not supported as primitives on these machines. With its primitive set operations, the Fluent machine can be programmed at a very high-level of abstraction.

2.1 The Multiprefix Operation

The multiprefix operation has the form $MP(A, v, \otimes)$ where A is a shared variable, v is a value, and $\otimes$ is a binary associative operator. At any time step a processor can execute a multiprefix operation, with the constraint that if P_i and P_j execute $MP(A, v_i, \otimes_i)$ and $MP(A, v_i, \otimes_j)$, then $\otimes_i = \otimes_j$. The semantics of the multiprefix operator is as follows:

> At time T let $P_A = \{p_1 \dots p_k\}$ be the set of processors refering to variable A, such that $p_1 < p_2 < \dots < p_k$. Suppose that $p_i \in P_A$ executes instruction $MP(A, v_i, \otimes)$. Let a_0 be the value of A at the start of time T. Then, at the end of time cycle T, processor p_i will receive the value $a_0 \otimes v_1 \otimes \cdots \otimes v_{i-1}$ and the value of variable A will be $a_0 \otimes v_1 \otimes \cdots \otimes v_k$.

Thus, when a set of processors perform a multiprefix operation on a common variable, the result is the same as if a single prefix operation were performed with the processors ordered by their index. For example, suppose that processors numbered 25, 32 and 65 execute the instructions $MP(A, 4, +), MP(A, 7, +)$ and $MP(A, 11, +)$ respectively at time T, and suppose that variable A initially contains the value 5. Then, at the end of the Tth cycle, processor 25 will receive 5, processor 32 will receive 9, processor 65 will receive 16, and the variable A will equal 27.

The fetch-and-$\otimes$ operation [7] also calculates a set of prefixes, but the order of inputs is undetermined before execution. Multiprefix is a determinate implementation of the fetch-and-$\otimes$, and is more powerful. The *scan* operation [4] is a special case of the multiprefix, one in which the set S includes all N processors. Scan does not allow multiple prefixes over all collections of disjoint subsets, whereas the multiprefix does.

For convenience we include two more primitives — READ and WRITE. READ(A) returns the value of A to the requesting processor. WRITE($A, v, \otimes$) is equivalent to $MP(A, v, \otimes)$ except that no value is returned to the processor executing the instruction. Both operations are special cases of multiprefix, as has been observed earlier [7].

2.2 A Fast Radix Sort using Multiprefix

In this section we present a radix sort based on the multiprefix instruction. The program is considerably simpler than Batcher's bitonic sort [1] and comparable in performance when the number of keys is very large.

When each key to be sorted is less than $\log N$ bits in size, fetch-and-add can be used to sort N keys in a constant number of steps. Unfortunately, this idea cannot be used iteratively to sort longer keys because the fetch-and-add, being non-deterministic, is not stable [4].

With the multiprefix we can implement a stable iterative radix sort. As we show below, N keys, each $k \log N$ bits long, can be sorted in $O(k)$ Fluent instructions. When k itself is small, the number of Fluent instructions executed is constant. In contrast, no other programming model supports such a concise sort even for short keys.

Theorem 1 *N keys, each of size $k \log N$, can be sorted in $O(k)$ steps on the Fluent abstract machine.*

Proof. We first describe a stable scheme for N keys of length $\log N$, one key per processor. The total number of distinct key values is N. Below we give the program for each processor. The keys to be sorted are stored in an array $KEY[*]$. The idea is to first count the number of occurrences[1] of $KEY(i)$ that lie in processors indexed less than i, then add to that the cumulative sum of the counts for keys less than $KEY[i]$.

```
SHORTSORT:
    COUNT[*]       := 0
    CUMULATIVE[*] := 0
    TEMP           := 0
    MP(COUNT[KEY[*]], 1, +)
    CUMULATIVE[*] := MP(TEMP, COUNT[*], +)
    return MP(CUMULATIVE[KEY[*]], 1, +)
```

Because the multiprefix operation is ordered by processor indices, the simple sort above is stable. We can iterate shortsort to sort larger keys in blocks. The primitive operation $LSBLOCK(w, j)$ below returns the least significant jth block of $\log N$ bits of location w, that is, bits $(j-1) \log N + 1$ through $j \log N$.

```
SORT:
    RANK[*]     := 0
    KEYPTR[*]   := *      (initialize pointer to self)
    FOR j=1 to k DO
        KEY[*]   := LSBLOCK(KEYPTR[*], j)
        RANK[*]  := SHORTSORT[KEY[*]]
        KEYPTR[RANK[*]] := KEYPTR[*]
    ENDDO
```

---

[1]This simple histogram computation cannot be done in a constant number of steps on the scan-model [4].

2.3 Set Operations

Sets are a fundamental data abstraction. Traditionally, sets have not been supported as primitive objects, but instead have been built on top of lower level structures such as lists, arrays, trees and tables. The Fluent abstract machine includes set operations as primitives:

- INSERT (x, S) Insert element x into set S.

- DELETE (x, S) Delete element x from set S.

- MEMBER? (x, S) Is x an element of the set S?

- APPLY (S, f) Apply the function f to the elements of set S. Note that f may change the values of the elements in S.

- REDUCE (S, f) Evaluate f with arguments that are elements of S. Note that f must be a binary associative operator.

In addition, set union, intersection, difference, prefix, and enumerate are also supported.

Every Fluent processor can execute a set instruction, so that many sets can be manipulated simultaneously. For example, several processors may simultaneously insert elements, possibly into the same set. The result of concurrent set operations is as if the individual instructions were executed atomically in some arbitrary unspecified serial order. The implementation however is completely parallel, and provably efficient. The ability to simultaneously update multiple sets is costly on existing machines.

3   Implementing Fluent Instructions

This section describes how the Fluent abstract machine is implemented on the butterfly network. The routing algorithm used is extremely simple and provably efficient, and forms the basis of the Fluent machine proposed in Section 5.

3.1   The Fluent Network

The nodes of the Fluent machine are interconnected in the butterfly (FFT) pattern. There are 2^n nodes in each of $n + 1$ levels, for a total of $N = (n + 1)2^n$ nodes. Each node is labelled with a string $\langle c, r \rangle$ $(0 \leq c \leq n, 0 \leq r < 2^n)$ formed by concatenating the binary representations of the level number c and the index r of the node within the level. Each node $\langle c, r \rangle$ $(c < n)$ is connected by forward links to the nodes $\langle c+1, r \rangle$ and $\langle c+1, r \oplus 2^c \rangle$, where $\oplus$ denotes bitwise exclusive or. Each node (except for levels 0 and n) thus has four connections: two connections to the next higher level and two to the previous level.

Each node in the butterfly contains a processor, a memory module and 6 routing switches. Each switch has 2 inputs and 2 outputs. Every input into a switch enters a first-in first-out queue, which has the capacity to buffer a small number (2 or 3) of messages in transit.

3.2 The Address Map

The shared variables of a Fluent program are distributed among the local memories of the nodes using an appropriately chosen address map. If the Fluent program does not involve run-time address computation then the physical address of each shared variable can be embedded within the program of each processor. Otherwise, we must compute addresses quickly at run time.

We propose to distribute the M shared variables randomly among the processors, each processor being assigned M/N variables. With a random hash function, memory bottlenecks are unlikely because the accessed variables will be distributed throughout the network. Suppose that we have chosen such a hash function $\mathcal{H}$ [2]. This function maps a $\log M$ bit address to a $\log N$ bit node address. A second function $\mathcal{M}$ computes the address ($\log(M/N)$ bits) within the memory of node $\mathcal{H}(x)$. The physical address of shared variable x is given by the concatenation $\langle \mathcal{H}(x), \mathcal{M}(x) \rangle$.

3.3 Message Structure and Path

Suppose that processor $\langle c, r \rangle$ wishes to access variable x. It generates a REQUEST, a message of the form $\langle \text{dest}, \text{type}, \text{data} \rangle$. The destination dest is $\langle \mathcal{H}(x), \mathcal{M}(x) \rangle$, the physical address of varible x. The type field denotes the kind of access requested, e.g. READ, WRITE, or MP. Other possible values include EOS or GHOST, which are used internally by the communication algorithm as we will see shortly. The REQUEST is injected into the network. It will reach node $\mathcal{H}(x)$ and return with the required data.

The path from node $\langle c, r \rangle$ to node $\mathcal{H}(x) = \langle c', r' \rangle$ and back involves 6 phases through the butterfly. Every other phase is a forward phase,

---

[2]Our simulations show that simple first degree polynomials perform well in practice. A random $O(\log N)$ degree polynomial provably works well [10,16].

Processor $\langle c, r \rangle$　　Module $\langle c', r' \rangle$

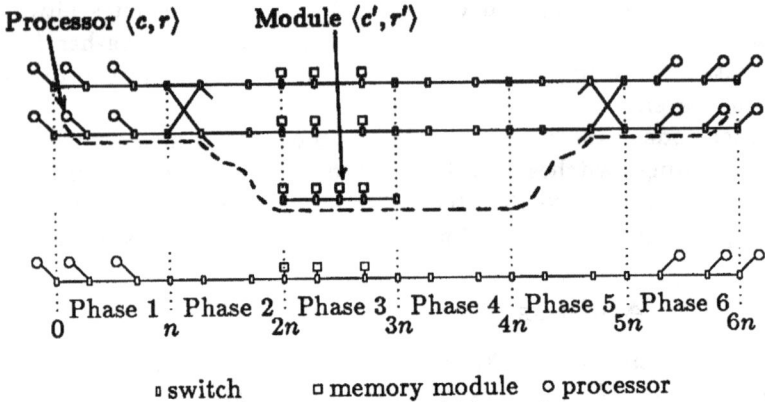

□ switch　　□ memory module　○ processor

Figure 1: Logical Network

and these are interleaved with backward phases. Figure 1 shows the 6 phases.

In the first phase, the message issued at node $\langle c, r \rangle$ is directed to node $\langle n, r \rangle$. In Phase 2, the message follows the unique (backward) path in the butterfly from node $\langle n, r \rangle$ to node $\langle 0, r' \rangle$. This path is determined at each switch by looking at the appropriate bit of dest. In Phase 3, the message reaches the node $\langle c', r' \rangle$, where it acquires the required data. The next 3 phases simply retrace the path traced thus far, back to the source processor $\langle c, r \rangle$. The access is now complete.

For convenience, we describe the routing mechanism in terms of the logical network of Figure 1 instead of the butterfly. The correspondence between the two is clear and each butterfly node does the work of 6 switches in the logical network.

3.4　How to Combine Messages

At the heart of the Fluent machine lies the routing strategy [16]. The key idea is a simple way of combining instructions that reference a common variable. Consider the case when several processors READ a common variable. The paths of these messages form a tree, as in Figure 2. Each message moves along the directed path from its source to the destination.

There is, however, no need to send more than one request along any branch of this tree. Each tree node forwards a request only when it "knows" that no future incoming request will have the same destination. The key idea here is that each node forwards requests in ascending order of destination addresses. Each node receives messages along two incoming edges and places them into the corresponding FIFO

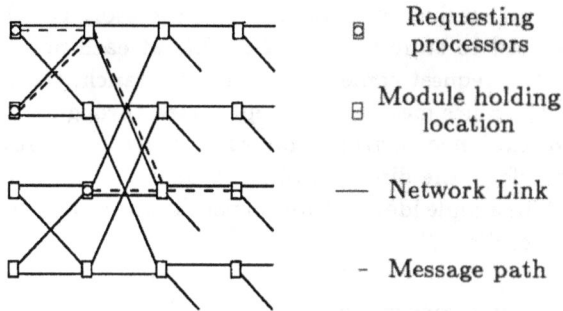

Figure 2: Message paths to a common location form a tree

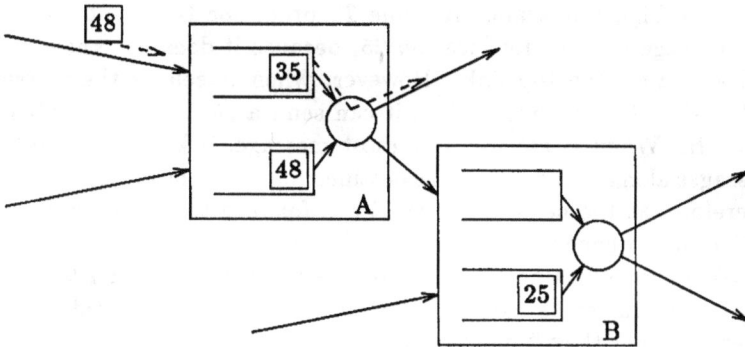

Figure 3: Combining Messages by Merging Streams

queues. At each step the node compares the destination addresses of the messages at the heads of the two queues. The message with the smaller destination address is transmitted forward. If both messages are destined for the same location, they are combined and only one request is sent out. Finally, if only one queue has a message waiting and the other queue is empty, no message is sent out. (If the message were sent, the next message along the other edge could conceivably have a smaller destination, thus violating the sorting requirement).

In our snapshot at time T, node A in Figure 3 selects the message destined for location 35. Then it waits until the message to location 48 arrives, at which point it discovers that the messages at the heads of both the queues are to location 48, and can be combined.

3.4.1 Reply routing

How do we return the data to all requesting processors? The reply message, upon reading the data, returns backwards along each edge of the tree and reaches every requesting processor. For the backrouting we only need to store two *direction bits* at each node. The bits say whether the request came along the top branch, the bottom one, or along both. Since messages are kept sorted throughout the six phases, *replies at each node arrive in the same order as the requsts were sent out*. Therefore, the direction bits can be stored in a 2-bit wide FIFO queue. This simple idea is more efficient than the associative memories proposed earlier [7].

3.4.2 Ghost messages

The simple idea of keeping message streams sorted has one deficiency. Consider Figure 3 again. At time T, processor B cannot transmit the message it holds for location 25, because it does not know what will arrive on the top link. However, when A selects the message to location 35 for transmission, it can send a *ghost* message labelled 35 to B. When B receives the ghost message, it knows that future messages along that edge will be destined for locations greater than 35. Therefore, at the next time step B can forward the message waiting in the lower queue.

Ghost messages notify nodes of the minimum location to which subsequent messages can be destined. Ghosts are not used for any other purpose, they "keep the system fluent."

3.4.3 Flow control

It is possible that a switch S is ready to transmit a message forward but the input queue for next switch is full. When this happens, S simply retains the message and tries in the next clock cycle. Of course, if the message S tried to transmit was a *ghost*, it can be dropped without any loss of information.

Many routing algorithms which adopt such a holding policy give poor performance because congested buffers back up buffers upstream. For our algorithm the probability of such degradation is provably miniscule, and the algorithm is always deadlock-free.

3.4.4 Termination

Immediately following a request, each processor also issues an end-of-stream EOS message. The dest field of every end-of-stream message is ∞. An EOS notifies a switch that no more requests will follow. The switch can now safely forward the requests on the other edge, and eventually forward the EOS messages themselves. EOS messages form a wavefront which guarantees that every instruction will terminate.

3.4.5 Performance

Following Ranade [16], we can show that this routing algorithm is close to optimal.

Theorem 2 *Assuming a perfect random address map, the probability that any memory reference takes more than* $15 \log N$ *steps is less than* N^{-20}.

Every routing algorithm must take at least $4 \log N$ steps. Observe that the provable performance is only slightly far from this lower bound, and considerably faster than previous algorithms for routing on butterflies of reasonable size.

Figure 7 gives timing results from simulations of the routing algorithm. We experimented with a number of different memory access patterns, e.g. matrix access, trees of different types, shuffles, random permutations etc. In no case was the time taken more than $11 \log N$, even with queues of size 2. Increasing queue size did not appreciablly affect performance. We found that simple hash functions (shared variable x mapped to physical address $ax + b \bmod M$) were satisfactory. Section 5.1 describes more simulation experiments.

3.5   Multiprefix instructions

We first describe the implementation for fetch-and-add proposed in [7]. Let s be an arbitrary switch in phase 1 (or 2). Suppose that the messages at the heads of the queues are $m_1 = \langle l, \text{fetch-add}, v_1 \rangle$ and $m_2 = \langle l, \text{fetch-add}, v_2 \rangle$ respectively. As shown in [7] the switch must forward a message $m = \langle l, \text{fetch-add}, v_1 + v_2 \rangle$ in place of m_1 and m_2. If the reply to m is a value v, then the corresponding switch in phase 6 (or 5) returns v as a reply to m_1, and $v + v_1$ as a reply to m_2. Thus the switch must remember the value v_1 received on its top queue for each pair of fetch-and-add messages that it combines.

Notice that this is equivalent to a serial execution of the message received on the top input (m_1) before the message received on the bottom input (m_2). Thus if we ensure that messages received on the top input always originate in a processor with a smaller number than those received at the bottom input, we effectively have an implementation for the multiprefix operation, with addition replaced by the prefix operator. We show how to do this by numbering the processors appropriately.

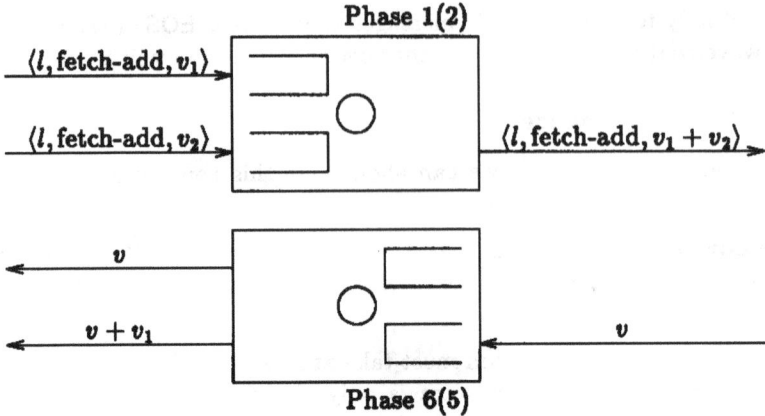

Phase 1(2)

$\langle l, \text{fetch-add}, v_1 \rangle$

$\langle l, \text{fetch-add}, v_2 \rangle$

$\langle l, \text{fetch-add}, v_1 + v_2 \rangle$

v

$v + v_1$

v

Phase 6(5)

Figure 4: Fetch-and-add

Theorem 3 *The multiprefix operation will, with overwhelming probability, terminate in $O(\log N)$ steps.*

Proof: We present the required numbering for the processors and switch inputs. Processor $\langle c, r \rangle$ is numbered $nr + c$. A switch $\langle c, r \rangle$ in phases 1 or 2 receives its inputs i_0, i_1 from switches $\langle c - 1, r_0 \rangle$ and $\langle c - 1, r_1 \rangle$ respectively. If $r_0 < r_1$, then we shall label i_0 as the top else we label i_1 as the top. ∎

As noted earlier, the only extra requirement over a read instruction is that, in addition to the two direction bits, each switch must remember a value (*partial sum*) for every combination that occurs at that switch. Figure 5 shows a pair of switches with the required queues.

3.6   Processor synchronization

> "It is always 4 o' clock here," said the March Hare to Alice.
> —*Lewis Carroll*, Alice in Wonderland

We use EOS messages to implement a distributed global clock. Recall that one EOS message per instruction passes through each switch. By maintaining a count of the number of EOS messages that have passed through, each switch keeps its version of the global time.

Different switches may indeed have different counts or versions of the global time, but that is perfectly alright. If two instructions access a common location in the same time step, then the one that arrives first will have to wait for the slower one to reach an intermediate switch for combination. Because we keep messages sorted by tag, and

Phase 1(2)

Partial Sum Queue ⟶ ⟵ Direction Bits

Phase 6(5)

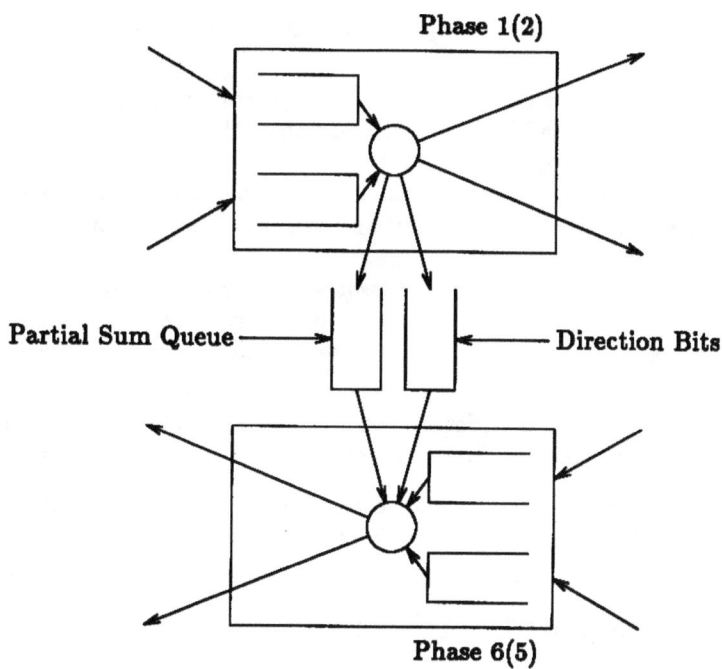

Figure 5: Internals of a pair of switches

TIME	PROCESSOR 1	PROCESSOR 2
1	A=20	B=10
2	Read B	Read A
3	A=30	B=40

Figure 6: Synchronization Guarantee

we guarantee that only one request for access will be passed into the memory module which holds the variable, the effect is the same as if all the processors were operating synchronously. For example, our implementation guarantees that for the code of figure 6 processor 1 and 2 will respectively read 10 and 20, provided no other processor writes a and b in the meantime. This is guaranteed in spite of the fact that both processors might issue all 3 instructions without waiting for any to complete. This is a very strong synchronization condition requiring special primitives on most other programming models.

This implementation also allows each processor to stop the global clock if necessary, if it detects an error for example. This is done by withholding the end-of-stream message.

For lack of space we will not describe how set operations are implemented. The interested reader is referred to [15].

4   The Routing Switch

In this section we outline a bit-serial design for the routing switch and estimate its layout requirements. The design extends to wider data paths in a straightforward manner.

Although Section 3 presented the routing algorithm with the implicit assumption that messages were transmitted in atomic packets, this is not necessary. In particular, each message can be transmitted bit-serially in a pipelined manner. This is analogous to the wormhole router of Dally and Seitz [5]. Message transmission can be pipelined because:

1. Address comparison can be done bit-serially, provided the addresses are received most significant bit first.

2. Message combination can be done bit-serially; for operators like +, the data must be transmitted least significant bit first. Also see on-line arithmetic [17].

3. When a message leaves a switch, the corresponding GHOST message (whose dest is identical to the real message) can be generated bit-serially.

Each message is transmitted with the dest field first (most significant bit leading), followed by the type field, and finally the data field (least siginficant bit leading). A switch begins operating when: (1) each input queue contains at least one message, and (2) the input queues of the receiving switches are not full.

We now describe the operation of a switch in phase 2. Switches in other phases can be specified similarly.

1. **Transmit dest**: The minimum of the destinations of the two messages in the input queues is transmitted along both outputs. The minimum is discovered only after the transmission, so till then both destinations must be retained in the input queues.

2. **Transmit type**: While transmitting the destination, the switch detects which output link the request must be routed on. This requires checking one fixed bit in the destfield. The type of the message with the minimum destination is transmitted on the that output, while on the other, type GHOST is transmitted.

3. **Transmit data**: This is relevant for messages like MP-$\otimes$ or WRITE. In either case, the message type indicates how messages must be combined when necessary. Again, the data fields can be combined and transmitted as they arrive.

The ability to pipeline messages speeds up message delivery considerably when there are no queueing delays. The message delivery time reduces from (network latency) $\times$ (message length) to (network latency) + (message length). We expect the latency of each switch to be about 4 (message enters an input queue, passes through the ALU, is sent to the output queue, and then transmitted), giving a total latency of $4 \times 6n$ for the logical network. Assuming 100 bit long messages and 4-bit wide data paths, the time for a 13 dimensional butterfly is $(4 \times 6 \times 13) + 100/4 = 337$ steps.

We now estimate the area requirements for the routing switches per node. Each switch consists of message queues, an ALU (for address comparison, message combination, etc.), counters to maintain the message FIFO queues, memory for storing partial sums, and direction bits for reply routing. In the following we assume that messages are 100 bits wide, and that partial sums are 64 bits wide.

Switches in phases 2 and 5 have two input queues, while others only have one input queue. The total number of message queues per node

Feature size for VLSI	1μ ($\lambda = 0.5\mu$)
Chip size	$100mm^2 = 400$ Mλ^2
Pins per chip	≈ 150
Printed circuit boardsize	.5 m × .5 m
Off board connections	512

Table 1: Technology for the Fluent-I

is therefore 8. Simulations (section 5.1) indicate that for 100,000 node machine each message queue need hold only 3 messages. The total memory requirement for message queues thus equals $8 \times 3 \times 100 = 2400$ bits, or roughly 1.2 Mλ^2(at $500\lambda^2$per bit[3]).

Simulations also strongly indicate that no switch will ever transmit more than 40 messages along its outputs. For reply routing we need 2 bit wide direction queues, and 64 bit wide partial sums. Long partial sum queues are maintained only in phase 2 so that the total memory requirement adds up to $40 \times 64 + 6 \times 2 \times 40 = 3040$ bits, or 1.52 Mλ^2.

Each queue requires 3 counters, except for the message queues which require 4. Assuming 8 bit wide counters, the total memory is 424 bits. With 3000 λ^2per counter bit, total area requirement is 1.28 Mλ^2.

Assuming 8 bit wide data paths, each ALU requires around 1.2 Mλ^2, for a total of 7.2 Mλ^2per node.

The total area requirement is thus approximately 11.2 Mλ^2. Including miscellaneous overhead, 15 M$\lambda^2$is a conservative estimate for 6 switches per node.

5 The Fluent Machine

This section presents an outline for a Fluent machine which can be constructed within the next few years with conservative technology. Table 1 summarizes our assumptions about the technology available. Needless to say, breakthroughs in packaging technologies will have the largest impact.

The Fluent-I is organized as a 13–dimensional butterfly, with 2^{13} nodes in each of 14 ranks for a total of 114,688 nodes. These nodes are divided into 256 boards, each housing a 6–dimensional butterfly. The network is partitioned into 2 planes of boards, arranged in the manner

---

[3]The estimates for the different components are from [12].

switches	30 Mλ^2
2 32-bit RISC Processors	40 Mλ^2
Floating point unit	100 Mλ^2
128 Kbytes memory per processor	200 Mλ^2
Total area requirement per chip	370 Mλ^2

Table 2: Chip Specification

Processors	114,688
Floating Point Units	57,344
Memory	16 Gbytes
Cycle time	50 ns
Peak Floating Point Rate	2.3 Tflops

Table 3: Fluent-I Highlights

suggested by Wise [18]. Each board has 448 nodes, divided among 224 chips, with 2 nodes per chip. In addition to the 2 processors, each chip also has routing switches for the two nodes, one floating point unit (multiplier and adder), and memory. Table 2 summarizes the breakup of chip area, using estimates as in the previous section.

Data paths between nodes vary in width depending on whether the path is on-board or across boards. Each board has 128 4-bit wide data paths out (64 nodes in the last rank of a 6-dimensional butterfly, each with 2 forward links). On-board paths are 8 bits wide. The butterfly can be partitioned so that each chip requires 16 data paths so that 128 pin connections suffice.

This variation in data path widths was not considered in the previous sections. The performance of the routing algorithm changes somewhat with narrow channels. The off-board channels also have to be multiplexed over the 6 phases of the logical network, while on-board channels are replicated. At worst one would estimate that the 4-bit wide off-board channels would slow the system by a factor of 12 (the other channels are 8 bits wide), but our simulations show that this is wildly pessimistic.

5.1 Simulation Results

We performed timing simulations of the communication network on the Connection Machine. The objectives were to observe the sensitivity of the routing scheme to variations in queue size, address maps, and memory reference patterns. A final objective was to study the effect of using multiplexed, narrower offboard channels.

Our conclusions in brief:

1. Simple hash functions perform well. We tried various linear congruential maps: variable x placed in location $ax + b \bmod M$, where M, the size of the address space, is a prime, and a and b are constants.

2. Routing time varies little with access pattern. We tried several patterns: matrix access, binary trees, shuffle permutations, random accesses etc. Random patterns took slightly longer in all cases.

3. Concurrent access is faster than exclusive access. The extreme case is when all processors read the same variable. The number of steps reduces from 154 (see Figure 7) to 85 because there is no buffering delay. This assumes that messages are no wider than channels (see below for further discussion).

4. Queue–size 3 is adequate. While queue size 1 degrades performance drastically, queue sizes 2 or more give similar performance.

5. Figure 7 plots the routing time when off-board channels are multiplexed, with *narrowness* being the ratio of the width of the offboard channels to the onboard channels. Switches in lower phases are given higher priority in accessing channels. Each channel first allows phase 0 messages to pass, followed by phase 1 messages, and so on. From the plot we can conclude that the performance degrades by a factor of 1.7 over the ideal case (no narrow channels and no multiplexing). The time goes up from 154 steps as in Figure 7 to about 260 steps (extrapolated for 114,688 processors from Figure 7).

5.2 Router performance

Suppose that messages are 100 bits long (64 data bits, 32 address bits, and 4 type bits). If every channel was 8 bits wide, sending a message across one link would require $100/8 \approx 13$ steps. From the

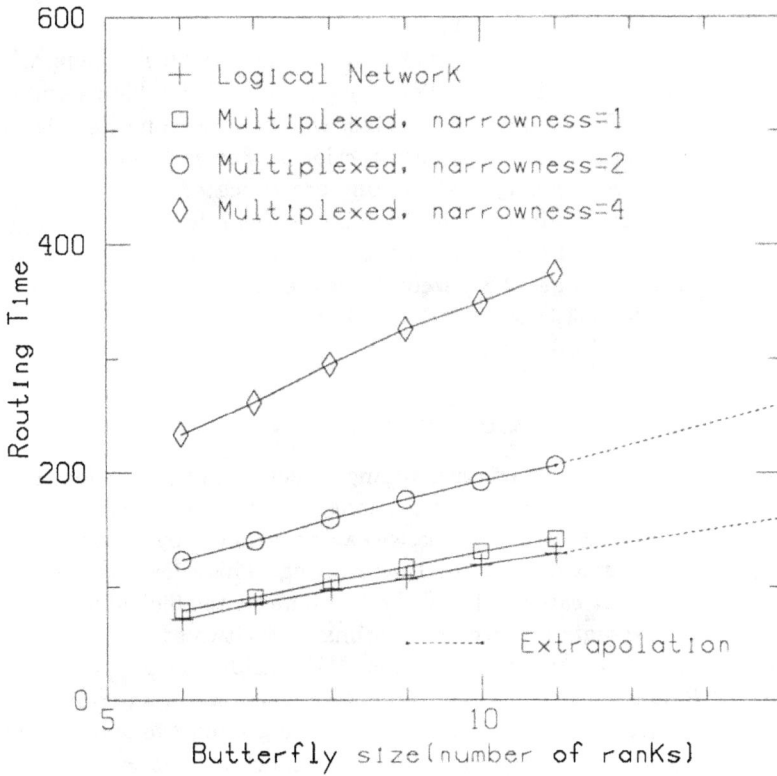

Figure 7: Average routing time (50 randomly chosen access patterns)

results of the previous section we can therefore estimate that, with narrow channels and multiplexing, an arbitrary permutation can be routed in $260 \times 13 = 3380$ steps. With a 50 nanosecond clock rate, the time is about 169 μsec.

If all processors access a single variable, then the time is just 337 cycles (section 4), or about 17 μsec.

As an example, suppose that we wish to sort 16-bit numbers, with 32K numbers in each node. There are roughly 3.5 billion numbers being sorted. On the Fluent machine, we only need one iteration of the procedure SHORTSORT from Section 2. For each number being sorted, 3 shared memory instructions are executed (the others are local). However, the instructions can be packed into 50 bit messages. The total number of steps required is $3 \times 32K \times 3380/2$, or about 169 million for a total time of 8.5 seconds. If the numbers are 32 bits long, the time is about 17 seconds. Note that this is the time to sort the entire contents of memory.

5.3 Structured Computations

Much work has been done on mapping structured computations onto butterfly networks. These computations do not need the generality of shared memory. Better performance can be achieved by direct nearest neighbor communication rather than routing. This allows us to utilize the floating point capabilities of the machine more efficiently.

Table 3 presents performance estimates for two structured problems: FFT and Matrix multiplication. We considered a 2^{30} point complex FFT, and used the standard mapping. We obtain a performance of between 1.2 Tflops and 2 Tflops depending upon the assumptions made about local memory bandwidth. Batcher's bitonic sort (N numbers) on the butterfly takes $2\log^2 N$ steps. With 32K 16 bit numbers per node, and each communication step requiring 4 cycles, the total time is $4 \times 2 \times 289 \times 32K \approx 75M$ cycles. At 50 ns clock this gives a time of 3.7 seconds. While this estimate is lower than that of the shared-memory radix sort, extracting the extra peformance requires non-trivial and tedious low level fine tuning.

Besides nearest neighbor communication, performance gains can also be achieved by partitioning structured problems into blocks, and doing block computations locally within each node. This reduces the number of shared memory instructions. For matrix multiplication there are no good mappings into the butterfly [3]. Instead, we partition a large matrix into block submatrices, each of which is stored in one node. Instead of mapping blocks randomly to nodes, we use a

Estimated multiprefix time	169μsec
Radix Sort $3.5 \cdot 10^9$ 16 bit numbers	8.5sec
Bitonic Sort $3.5 \cdot 10^9$ 16 bit numbers	3.7sec
Matrix multiplication	0.8 Tflops
FFT	1.2 Tflops

Table 4: Fluent-I Performance

simple hierarchical approach: decompose the matrix into large blocks, and map these into random boards. Next, decompose the large blocks into smaller blocks and map them randomly into nodes. This allows us to exploit locality at the processor and board levels, and reduces the communication load on the off-board channels.

6   Conclusions and Extensions

Powerful models of parallel computation need neither be expensive nor slow – this is what we wish to demonstrate by building a Fluent parallel computer. In this extended abstract we have presented the Fluent abstract machine which is more powerful than any other abstract shared memory model, and shown that it can be implemented inexpensively on the Fluent machine.

We are continuing simulation experiments. By programming different applications we hope to get more insight into the expressive power of the Fluent programming model. We also expect to identify various tradeoffs, and adjust design parameters accordingly. For example, by providing even wider data paths on board, at the expense of reducing the number of switches per node (by multiplexing them) we expect that overall performance can be improved.

In this abstract we have not considered many issues in processor/chip design, and have mostly presented very conservative estimates for area requirements. We expect to begin detailed design of the router and communications harware following our experiences with the simulator. Future work will throw more light on issues such as SIMD vs. MIMD organization, processor complexity/wordlength, and operating system issues.

Acknowledgements

We thank Nick Carriero, David Greenberg, Tom Leighton, Charles Leiserson, and the students of the Fall '87 CS445 (Parallel Algorithms and Architectures) class at Yale for stimulating discussions. We are especially grateful to Carlton Geckler for helping us access the Connection Machine at NPAC, Syracuse University. Lennart Johnsson and Abhiram Ranade were supported by ONR grant N00014-86-K-0564. Sandeep Bhatt was supported by NSF grant MIP 8601885.

References

[1] K. Batcher. Sorting networks and their applications. In *AFIPS Spring Joint Comp. Conf.*, pages 307–314, 1968.

[2] *Butterfly Parallel Processor Overview*. BBN Laboratories Inc., 1985.

[3] S.N. Bhatt, F.R.K. Chung, J-W. Hong, F.T. Leighton, and A.L. Rosenberg. Optimal simulations by butterfly networks. In *Proceedings of the ACM Symposium on Theory of Computing*, 1988. to appear.

[4] Guy Blelloch. Scans as primitive parallel operations. In *Proceedings of the International Conference on Parallel Processing*, pages 355–362, 1987.

[5] William Dally and Charles Seitz. *Deadlock Free Message Routing in Multiprocessor Interconnection Networks*. Technical Report 5206:TR:86, California Institute of Technology, 1986.

[6] A. Gottlieb, R. Grishman, C. Kruskal, K. McAuliffe, L. Rudolph, and M. Snir. The NYU Ultracomputer - Designing an MIMD Shared Memory Parallel Computer. *IEEE Transactions on Computers*, C-32:175–189, February 1983.

[7] A. Gottlieb, B. D. Lubachevsky, and L. Rudolph. Coordinating large numbers of processors. In *1981 International conference on Parallel Processing*, 1981.

[8] W. Daniel Hillis. *The Connection Machine*. The MIT Press, 1985.

[9] W. Daniel Hillis and Jr. Guy L. Steele. Data parallel algorithms. *CACM*, 29(12):1170–1183, December 1986.

[10] Anna Karlin and Eli Upfal. Parallel hashing - an efficient implementation of shared memory. In *Proceedings of the Symposium on Theory of Computing*, 1986.

[11] Gyungho Lee, Clyde P. Kruskal, and David J. Kuck. The effectiveness of combining in shared memory parallel computers in the presence of 'hot spots'. In *Proceedings of the International Conference on Parallel Processing*, pages 35–41, 1986.

[12] John Newkirk and Robert Mathews. *The VLSI Designer's Library*. Addison-Wesley Publishing Co., 1983.

[13] G. F. Pfister, W. C. Brantley, D. A. George, S. L. Harvey, W. J. Kleinfelder, K. P. McAuliffe E. A. Melton, V. A. Norton, and J. Weiss. The IBM Research Parallel Processor Prototype (RP3): Introduction and Architecture. In *Proceedings of International Conference on Parallel Processing*, pages 764–771, 1985.

[14] G. F. Pfister and V. A. Norton. Hot-spot contention and combining in multistage interconnection networks. In *Proceedings of International Conference on Parallel Processing*, pages 790–797, 1985.

[15] Abhiram G. Ranade. Fluent Data Structures for Sets. in preparation.

[16] Abhiram G. Ranade. How to emulate shared memory. In *Proceedings of the IEEE Symposium on Foundations of Computer Science*, 1987. Also available as Yale Univ. Comp. Sc. TR-578.

[17] K. S. Trivedi and M. D. Ercegovac. On-line algorithm for division and multiplication. *IEEE Transactions on Computers*, C-26:681–687, July 1977.

[18] David S. Wise. Compact layouts of Banyan/FFT networks. In H. T. Kung, Bob Sproull, and Guy Steele, editors, *Proceedings of CMU conference on VLSI systems and computations*, pages 186–195, 1981.

The Design and Testing of MIPS-X

Paul Chow[1] and Mark Horowitz

Computer Systems Laboratory
Stanford University
Stanford, CA 94305

1   Abstract

MIPS-X is a high performance RISC microprocessor that can run with
a peak throughput of 16 MIPS. The design began in the spring of 1984,
and in the summer of 1987 we were able to execute Pascal programs
on a prototype system. This paper describes the design of MIPS-X examining the strengths and weaknesses of our tools, methodology and
design environment. We found that consistency within a representation and across representations of the design was an important issue
and that having an executable description of the design in the form
of a functional simulator was key to the success of our project. We
describe a technique we call comparison-mode simulation that made
final verification and system integration much easier.

2   Introduction

Designing any large system inevitably becomes an experiment in design methodology and CAD tools. The Stanford MIPS-X processor
project was no exception. The methodology and tools that we used
in its design were successful because we were able to design a high-
performance 32-bit processor and which worked on first silicon. However, they were not perfect because the path to a working system was
neither straightforward nor smooth.

MIPS-X[7,6,4,1] is a second generation RISC microprocessor built
at Stanford University that follows the original MIPS machine[11]. It
has a simple instruction set (four instruction classes and 37 instructions), a 5-stage pipeline and uses a software object code reorganizer

---

[1]Paul Chow is now with the Department of Electrical Engineering at the University of Toronto, Toronto, Ontario, Canada, M5S 1A4.

to handle pipeline interlocks. MIPS-X is implemented in a conservative 2μm, 2-level metal, n-well CMOS technology. The design target was to use a nonoverlapping, 2-phase, 20 MHz clock and issue one instruction every cycle. First silicon worked at a clock rate of 16 MHz. A 2K-byte on-chip instruction cache satisfies 90% of all instruction fetches and reduces the instruction bandwidth requirements by a factor of six[6]. MIPS-X has a peak operating rate of 16 MIPS and has an effective throughput of 10 MIPS when pipeline stalls and the internal and external cache misses are included. The processor contains 150K transistors on an 8mm x 8.5mm die.

The project began with a small set of requirements. The primary goal was to build a machine that would be an order of magnitude faster than the MIPS processor. It took over a year to transform this goal into a description that we could begin to implement. It took us an additional year and a half to get from this point to working silicon. Figure 1 shows the basic flow of the design. During the latter half of the design as many as seven hardware designers were working concurrently on the project. It was important that we had a flexible environment and a methodology that allowed them to work independently most of the time. This flexibility meant consistency between the various representations of the machine became a very important issue.

We begin by describing how we defined what we wanted to build and then how we partitioned and managed the large implementation effort. For the detailed implementation we chose to partition the system into seven large sections and assigned each section to a designer. Maintaining consistent interfaces between the sections was an important problem. We explain how we managed this problem by writing a function model of the entire chip and forcing all the section interfaces to be the same as the ones in the functional model. During final verification and testing we found a few bugs that managed to slip through most of the tools and we examine how or if better tools could have prevented these problems. Next, we look at the issues of maintaining consistency among the various representations of the chip. Finally, we give a chronology of the project to quantify the effort in each of the phases.

Architectural
Specification

↓

Microarchitecture and
Circuit Design

Writing Functional
Simulator Models

Layout

Testing Functional
Simulator Models

Comparing Simulator
Representations

↓

Final Integration

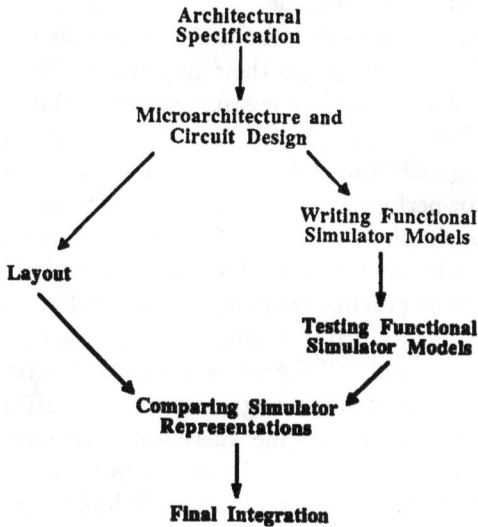

Figure 1: Design flow for MIPS-X

3 Project Specification

MIPS-X took about two years from the beginning of the project until the design was complete and the layout sent for fabrication. Over half of this time was spent deciding what to implement. These times are very similar to those for the original MIPS microprocessor[5,10]. Although the implementation methods were different, both projects spent over a year, and more than half their design time, before work on the details of the circuit design and layout began. We believe that this phase of design is essential to a successful research project in computer architecture, since it is during this time that different computer organizations are proposed and discussed. It is interesting to note that almost no CAD tools were used in this initial phase of the project except to support a small amount of exploratory layout to determine key cell sizes. Most of the architectural decisions for MIPS-X were made using data gathered from the original MIPS software system and instruction-level simulator, and good old-fashioned back-of-the-envelope estimates.

Our project started with the goal of designing a high performance RISC microprocessor. We felt that the key to achieving the best performance was to issue one instruction every cycle and to squeeze the clock as short as possible. Since we knew that the machine would be pipelined, we spent the first few months evaluating various pipelines and trying to judge their implementation complexity and performance. Very subjective measures were used during this period, and we relied heavily on the judgment of the original MIPS designers. In the end, simplicity won out. Since no one pipeline organization had clearly superior performance, the original MIPS team convinced us to start with the simplest machine. They assured us that the machine would turn out to be more complex than it looked.

After the basic pipeline structure was settled, we moved on to the instruction set. Trace-driven evaluation using the compilers and simulators for the original MIPS processor was used to refine the instruction set until it was stable enough to write the MIPS-X instruction-level simulator. During this time, there was much interaction between the software group and the hardware group to discuss the tradeoffs of implementing various instructions. We looked at many different aspects of a processor, such as internal cache organizations, branch schemes, coprocessors, multiplication and division support, and went through many possible instruction sets[4].

Quick paper designs were used to estimate the amount of hardware needed and keep a crude check on the changes to the critical path that new features caused. Tools that could help automate this part of the design are needed but are non-existent. A tool that simply kept track of assumptions made in the design would have been helpful, and one that made deductions from the assumptions would have been invaluable. For example, sometime during the initial design we added up all the estimates for the sizes of the datapath elements and found that the datapath was going to be much too long. It was over 7mm long, more than 1mm longer than the floorplan. We actually had all the information needed to do this for sometime, but no one thought to look at the data. Once the problem was uncovered, we began looking at instructions and hardware that we could delete to bring the machine closer to the desired size. These design iterations improved the quality of the machine, but they took time. Innovative designs will always need these iterations to find a system that can best satisfy the given specifications. Thus design time should be reduced not by eliminating the loops, but rather by providing methods that reduce the time needed to evaluate new design ideas.

4 Implementation

Once the basic machine architecture and micro-architecture were fixed, work on the detailed design of the processor began in earnest. We partitioned the design according to functional units: the instruction cache, the instruction register, the external interface, the register file, the execute unit, the PC unit and the tag unit. The last four units comprise the datapath of the machine and are roughly equal in size. Each section was assigned to a designer, and that designer was responsible for everything from building a functional description to doing the layout and verification of the section. This meant that each person became the expert for his part of the chip and did not need to know details about the other sections except for the timing of the signals at the boundaries. Even with the use of separate independent controllers for each section, there still were a number global signals that needed to be managed. These signals, along with other global design issues, like power distribution and chip floorplan, were the responsibility of one of the designers.

This tall thin design style worked very well for us. The designers had a chance to work at many different levels and see the effect of some of their design decisions. By forcing each designer to implement all the features that he added, there was a great incentive to keep the machine simple. The key ingredient to making this design style successful was stable interfaces. We were extremely careful to make sure everyone understood what signals they would get at the boundaries and the timing of those signals. For a while, we maintained a file in each designer's area that defined all the signals that crossed in and out of his section. This signal list was eventually replaced by the netlist used in the functional simulator.

4.1 Funsim: The Functional Simulator

During the initial design period questions often arose about whether the processor behaved correctly for different combinations of exceptional conditions, for example, how interrupts affected the handling of internal cache misses. Most of these issues were easily resolved, but on occasion it took a few days until the designers agreed on what the processor would do in each situation. These discussions make it clear that we needed a more precise specification of the machine. To fill in this hole we wrote a functional simulator for MIPS-X.

The functional simulator, Funsim, was used primarily as a means to keep track of the definition of the machine. Having an executable version of the design also gave the designers confidence that nothing had been missed in the logical and functional design of the machine. We no longer needed to argue about what the machine would do, we could simulate the situation and look at the results. Funsim was written specifically for MIPS-X because most of the available functional simulators would not run in our environment (4.2 BSD Unix), and because we wanted the flexibility to add features to the functional simulator as needed capabilities became clearer. We used this flexibility to great advantage and Funsim became the hub for all of the testing tools.

Funsim is a zero-delay, event-driven simulator written in Modula-2[13]. We chose zero-delay models because it seemed to be the easiest way to build a quick implementation. In our timing philosophy all global signals had to be valid by the end of one of the clock phases so we did not need to worry about more detailed timing or scheduling in the simulator. Funsim consists of basically two parts. The first part is the kernel that handles the basic simulator functions such as scheduling of model evaluation, managing the net lists for all of the connections, and providing the user interface. The second part is the functional descriptions of the pieces of the design, written as a set of models and their interconnection.

We used Modula-2 mainly because of the strong type and array bounds checking and because we thought that the availability of coroutines might be advantageous. Each device or functional model was a separate coroutine. In retrospect, the coroutines could have easily been replaced by procedure calls but the type checking did make it easier to build a robust program. The kernel, consisting of over 6000 lines of code, was running after nine weeks and only minor bugs have been found in that code. None of them delayed or prevented the use of the simulator. About 12000 lines of code were written for the functional descriptions. Bugs in the functional descriptions are a separate story.

Many of the problems that we encountered in writing the functional descriptions had similar roots. One class of errors involved signal timings. We used a strict timing and labeling discipline in MIPS-X. Signals were labeled as being valid, stable, or qualified during a clock phase. Unfortunately, some of the designers did not initially understand how these signals needed to be generated in the functional simulator. For example, a stable Phi 1 signal really meant that the value for that signal had to be computed during the previous

Phi 2 so it would remain unchanging during Phi 1. Initially many of these signals were updated on Phi 1, which could, and often did, cause another section of the design to malfunction.

Another class of errors was caused by the need to evaluate a functional model many times during a clock phase. Since Funsim is an event-driven simulator, a functional model is reevaluated each time one of its inputs changes. Many of the models initially contained an implicit assumption that they would be evaluated only once. For example one model contained a counter, and the counter would increment when Phi 1 was high. If the model was evaluated twice the counter would increment by two. The real problem was the programmers did not understand that the internal variables were essentially the outputs of flow latches, and not D flip-flops. Finding these errors was also difficult because the order of functional model evaluation affected when the model broke. Both of these problems would have been much less frequent if we had distributed a set of example functional descriptions before the models for the processor were written. Our problems were caused in part because we were learning how to write functional models while we were writing the simulator[2].

4.2 Circuit Design and Layout

Once Funsim was running we concentrated our efforts on completing the circuit design and layout of the processor. Since we were without a good schematic capture system, most of the circuits were not simulated before layout. Our standard procedure was to design a circuit, check it out manually, and then go to layout. Once the layout was done, we could back extract the circuit description and simulate it. On some of the more critical circuits, like the ALU and the standard latch, schematics were entered by hand and then simulated at both the circuit-level and the switch-level, but this was the exception rather than the rule.

Even during this stage of the design we used only a small number of tools. For layout, we started with Caesar[8] to design the instruction cache and then switched to the Magic[9] layout system from Berkeley

---

[2]In the process of writing the functional models, we found that Modula-2 was not a good language for specifying logic. It was too easy to make mistakes describing simple logical functions. Moreover, once this logic was specified it had to be transcribed into another form for PLA or standard cell generation. Other researchers at Stanford have taken the good ideas in Funsim and written a general purpose functional simulator called Thor[2]. The Thor system is being used for new chips designed at Stanford.

when it became available. Magic proved to be an excellent layout tool; the incremental DRC and integrated extractor were particularly helpful. We did need to port the program to the three IRIS 1400 color workstations that we owned, and later to the X windows system. In general, Magic had very few bugs, and the developers were very good at sending fixes for problems that we found.

For switch-level simulation we initially tried both MOSSIM II[3] and RSIM[12]. Neither worked particularly well on some of the more advanced circuits that we used. After deciding to use RSIM, we modified both the timing models it used and the way it calculated the logical value at a node to make it more robust. Even with these changes we needed to make a few small modifications to the netlist to enable RSIM to simulate the entire chip. These changes allowed RSIM to simulate the 6-transistor exclusive-or gate[3]. The main advantage of RSIM over MOSSIM was its ability to calculate timing information, and handle ratioed gates. The timing was particularly important since MIPS-X contains some self-timed circuits.

We wrote a number of little programs and makefiles[4] to help get around small problems that we encountered, like the problem with the XOR gates. Other problems arose simply because of the size of the design. For example, when we tried to generate a flat netlist from the hierarchical netlist that the Magic system generates, the flattener ran out of virtual memory. The string space needed to store the long node names generated by Magic was too large. The solution was a simple program and a makefile that would read the Magic files and shortened cell instance names. Even with the shortened names we still needed 32 MBytes of virtual memory to flatten the chip without the instruction cache RAM cells.

4.3   Comparison-Mode Simulation

Initially each designer manually created a set of input vectors to test the functionality of the circuits that she/he designed. As these small circuits were combined into larger units it became more difficult to generate good input test cases. The functional simulator provided a solution to this problem. Since Funsim was the definition of the machine, we could use it to generate inputs and check outputs of a switch-level simulator. Funsim was modified so it could run RSIM

---

[3]We have continued our work on RSIM and these circuit changes are no longer needed to simulate exclusive-or gates.

[4]Makefiles are used by the Unix utility 'Make'.

in parallel, and do the checking automatically. The user specified
the corresponding names in the two descriptions for input and outputs. We call this comparison-mode simulation, as opposed to mixed-
mode simulation. In mixed-mode simulation a piece of the functional
model is replaced by a switch-level representation. In comparison-
mode simulation, the two models are run in parallel and checked for
consistency. Comparing the two descriptions makes bugs very easy
to find, since the simulation stops as soon as a difference is detected.
We found many bugs in Funsim only after we started the comparison-
mode simulations. These bugs were usually subtle timing errors in
Funsim, that did not cause it to fail, but did cause problems in the
switch-level circuit it was driving.

Comparison-mode simulation imposes some restrictions on the
functional models used for simulation. Every node that needs to be
driven or checked at the switch-level must be explicitly visible in the
functional simulator. This requirement causes the interfaces of the
functional models to be roughly the same as the switch-level models.
However, the partitioning is less restrictive than a mixed-mode simulator where they have to be exactly the same. Several times we found
during the circuit design that it was easier for a signal to be generated in a different place than the one that the functional simulator
used. As long as the needed signals existed in Funsim we could do
our comparisons. There really were no "boundaries" in the functional
model that needed to be respected.

5   Verification

It took us roughly four months to take the almost complete layout of
the chip and produce a layout that was extensively tested and ready
for fabrication. Most of this time was not spent simulating the chip
and fixing bugs. During this four month period RSIM's timing and
final value models were changed and Funsim was modified to drive
the comparison-mode simulation. We also needed to finalize some
dangling chip design issues. Mostly these issues related to testing
the part and special test hardware that had been discussed but never
completely thought out. Finally we needed to get all three representations of the chip to agree. It was only at this point that we forced
the instruction set simulator, Funsim, and the real chip to execute the
same instruction set and brought the compiler up to date.

For most of the design we could not use the current compiler system to generate code that would run on the functional simulator or the physical description because the software people did not want to constantly track the small changes we would make to the machine. Initially the changes were to the actual instruction set, while later in the design they were mostly changes to the bit encoding of the instructions. Although simply changing the assembler would not have been hard, changing the whole software system (simulator, analysis programs, etc.) was a fair amount of work. Since these changes were not important to their work, they were not done until the very end of the design when the machine was stable. This is one example of a consistency problem between representations that should be addressed in future designs. Early in the design the hardware group would have been better off if they had a version of Funsim that ran the same instructions as the instruction set simulator. Then the hardware group could have been verifying the basic machine while doing the detailed implementation. Instead Funsim tracked the hardware a little too closely, making it hard to generate code to run on it. Initially we only could use several hand-coded machine language programs and could not exercise all of the instructions. It was only during the last few months before tape-out that we could compile Pascal programs that would run on the functional simulator.

Once comparison-mode simulation started it became crucial that the functional simulator and the layout remain consistent. Changes had to be made quickly or else simulations would stop. The pieces of the chip were simulated in comparison mode for about three weeks until we felt that they were behaving correctly. Within a day after initial integration of the major sections of MIPS-X, we could execute a few instructions on the switch-level model of the full design and a few days later we were running Pascal programs.

The tests consisted of some Pascal programs and a few assembly language programs to exercise instructions and features that we knew would not be tested by the Pascal programs. In addition to simply running programs, we also loaded a simple interrupt handler and ran programs with random patterns of interrupts. During this period we found many bugs in both the simulator and the layout. Almost every test that exercised a new feature would point up some problem, usually minor, but still something that needed to be fixed. Debugging of the full design was done by two people but when problems were localized to particular sections, the designer of that section was called in to make the necessary changes.

A major problem with the tool set was the amount of time it took to find a bug, fix it and then rerun the simulation. During the end of the design, it would take over three CPU hours on a DEC Microvax II to generate and initialize a new simulation run after the layout was changed. We usually let this process run overnight. Even the simulation speed was something of a problem. The full switch-level simulation required about one minute per clock cycle on a Microvax II. This meant that finding a bug could take a few hours of simulation time. We became very careful about having RSIM leave "statefiles" during the simulation to allow us to quickly go back to a time just before an error occurred and find what caused the problem. These long compute times really slowed down debugging. A set of incremental tools that would allow the designer to change the circuit or layout and quickly discover the effect of the change would be a great advance over the current tool suite.

5.1 Layout Checks

Near the end of the design, some team members became concerned that the Magic DRC and extractor would not find all the possible errors in the layout. We began an informal survey of experienced chip designers to find out what final checks they used before shipping a chip, and what problems they have had on their recent chips. The information that came back was quite helpful. We were able to use Magic (probably in ways it was never intended to be used) to find a number of previously hidden problems. The major checks were for connectivity, well taps, and zener diodes.

The Magic extractor connects all nets with the same global name together, whether or not the nets were electrically connected[5]. This made global names (names with a '!' at the end) very dangerous to use. To ensure MIPS-X did not have any problems we wrote a simple script that found all the global labels in the design, sorted them, and then looked for duplicates.

We were also careful about power and ground routing. One designer warned us to check that transistors were connected to power or ground by metal and not through a high resistance well or substrate connection. Since the extractor does not use resistance values it would not expose this problem. We implemented this check by creating a technology file for Magic that made these high resistance layers non-conducting and then used this technology file to generate the netlist

---

[5]Version 5 of Magic now warns users of unconnected global nodes with the same name.

that we simulated. In this way only true connections would make it to the simulator.

We were also able to create a technology file for Magic that let us check for well diffusion without a connection to a power supply. The key to this test is understanding how Magic generates node names for unlabeled nodes. These nodes are given a three digit name, such as #1_#2_#3, where the first digit is the 'plane' that the object is on. Since the well layer exists on a separate plane, we could check for floating wells by simply making wells electrically connect to well contacts, giving well a high area capacitance, and then extracting the entire layout. The capacitance causes the extractor to output a capacitor between the well and ground. Tied wells would be aliased to the name of the power supplies, leaving only the floating wells with their original node names. Floating wells could be found by searching through the netlist for nodes that started with the number of the well plane. We found only a few floating wells, but one would have been enough to cause the chip to latch up.

We also used Magic to find accidental zener diodes. This structure occurs when a well-contact diffusion is butted into normal diffusion. The design-rule checker does not catch this error since it is legal if the two diffusions are shorted by metal, as shown in Figure 2. Again we used the Magic extractor to find these errors. The circuit needs to be extracted twice, once with a technology file that connects well-contact diffusion to normal diffusion when they abut and once with them not electrically connected. If zener diodes do not exist then the two extracted versions of the cells will be the same. If the two extracted versions are different then a zener diode exists. When we ran this test on MIPS-X we were surprised to find a number of problems that needed to be fixed.

Once the correct technology files were created it was relatively easy to run these checks, but it did take a large amount of CPU time. As a result these checks were run infrequently to ensure no new errors had crept into the design.

5.2 The Last Bugs

During final simulation of the whole chip, we found two classes of errors that pointed up major holes in our tools. The first class of errors was caused by timing problems. MIPS-X does not follow our two-phase clocking rule in a few places. For performance, the machine occasionally starts driving a signal a little early, for example on Phi 1

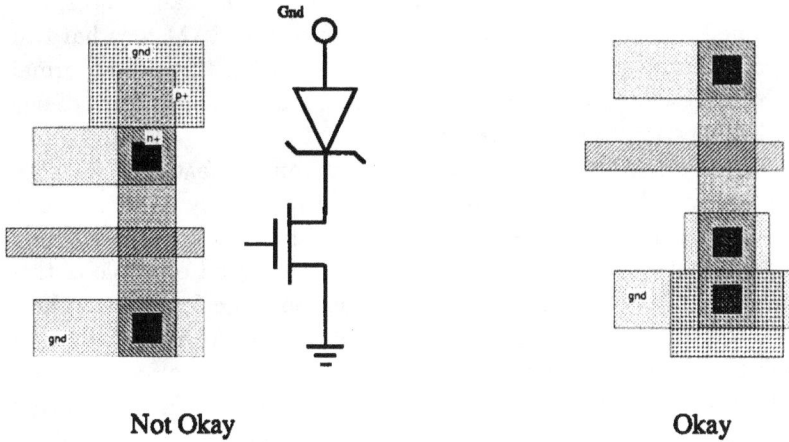

Not Okay   Okay

Figure 2: Accidental zener diode.

falling rather than Phi 2 rising. These small cheats caused a number of problems. In the register file, the bit lines are latched on Phi 2 falling, but this edge also causes them to begin to precharge. If the precharge is too fast, it will change the bit line value before it is latched, causing the machine to fail. This race was only detected by accident, when the machine was extracted using the wrong technology file. In the correct technology the race was won by a nanosecond, so normal simulations did not show the error. Other timing related errors that appeared during final simulations were caused by one functional unit glitching a line when another unit was still using the wire. For example, we tried to speed up stores by driving the store value on the Memory Bus a little early, on Phi 1 falling. Unfortunately, the Instruction Register was latching the bus on Phi 1 falling and sometimes the writes drove the bus too soon. Because of these problems almost all the cheats were removed. They were not worth the risks.

The other class of error that we found during final verification was with features added during the last few months of the design. Many of these errors were related to the testability features. The testing hardware was designed on a tight schedule and sometimes was not completely thought out when the implementation began. Although we knew very early in the design we needed a way to test the instruction cache, the details were never worked out. The simple method we came up with was to force the processor to increment the pc, generating sequential addresses to the cache, while the data pads were connected to the I/O port of the cache RAM. This method basically works, but it turned out that the read/write control of the RAM was hard to directly access, and so the cache had to be written through a normal cache miss sequence. This realization forced us to change some of our implementation to make the feature work.

Bug fixes were another source of last minute features. Once a bug was found the solution only would be verified by hand before it was implemented. As a result it was not unusual to find another bug caused by the last bug fix. The PC PLA was a good example of this situation. The PLA in the PC Unit was too large for the standard domino structure used in the rest of the design. After realizing the problem late in the design we changed the PLA to be self-timed, only to find out that this would not work either. More hacking converted the PLA to be a pseudo-nMOS design which finally worked. In the second revision of the chip, this entire section has been replaced by a small standard-cell layout. In hindsight, it would have been better to think through each of the changes more carefully, and even implement them in the functional simulator first before committing them to hardware. We ultimately would have saved time that way.

6 Testing

Funsim was also used to check the silicon when the chips returned from fabrication. We modified it once again, this time to produce vector files for the low speed functional tester that we had. The initial parts that were fabricated were diced but not packaged, so we used a probecard to probe individual die. After discovering numerous problems in the test setup, including incorrect wiring and poor probe contact, we were able to find functional parts. The test setup included the ability to change from clocking the chip with the tester to using external function generators to run at higher speeds. Using this feature we could load a small program into the instruction cache and then change to the external clock generators to speed test the part. However, this could not test two of the possible critical paths: the external memory access and instruction cache miss detection. It did give us a feel for how well the rest of the chip was working and we were able to achieve over 16 MHz operation at room temperature. This was very close to the expected result of around 14 MHz that we predicted using RSIM. We have also used microprobes to look at the internal signals in the part, and they are close the values that RSIM predicted. In general the parts are 20-30 percent faster than RSIM predicted.

After performing more complete testing of the chips, we found two errors in the design; both were in parts of the the die that were never exercised in our test programs. One was located in the instruction cache, and the other involved the coincidence of multiple exceptional conditions. The latter was simply a testing oversight; we forgot to test that combination of events. A little board logic was able to get around the problem. The instruction cache problem is more interesting. In the memory array, one of the decoder outputs was shorted to a power supply. The array of 16K memory cells was not simulated, (since they are a regular array of 100K transistors) but the decoders were simulated, since they are not regular. Unfortunately, Funsim modeled the instruction cache at too high a level, and did not check the decoder outputs. In fact, if we had tried to look at the decoder outputs we would have found the problem, since one of the outputs would have been missing. It was aliased to ground.

7 Managing Consistency

Managing the consistency of a design in a multiple representation, multiple person, multiple machine environment is tricky but extremely important. There are two aspects to the consistency issue: One aspect deals with multiple representations of the design. We had software development using an instruction-level simulator representing the machine at a very high level. At an intermediate level was the functional simulator and at the lowest level was the transistor description. We dealt with this aspect when we discussed the activity during final verification in Section 5. The other consistency problem occurs because several people are working within one level of a representation. This problem is compounded when the database becomes distributed and is not solved even when there is only one database and the Network File System (NFS) from Sun Microsystems is used.

When we were writing the functional simulator everything was fine when we were each working on our own individual sections. We used RCS, a source code control system available on many Unix systems, to manage the files. After the full simulator was put together we began to have problems when changes and fixes had to be made. Updating one section without changing the rest of the simulator usually caused it to stop working. Who would be allowed to have access to make changes? At first this was handled informally by looking to see if anyone was editing or talking to people, but occasionally this would fail and it sometimes took over a day to realize what had happened. We finally instituted a physical lock on the simulator by creating the *conch*[6]. This was a pink index card emblazoned with the word conch and only the person in possession of the conch was allowed to make changes. Release of the conch also meant that all the regression tests had been run successfully.

Before we had NFS running, maintaining consistency of the physical design database was done by agreeing that the master source be on one machine and people were responsible for copying the updated files back to that machine when they were finished. This worked sufficiently well when we were working on the individual sections, but once the design was put together, many consistency problems arose. Some problems remain even after we started using NFS. Although people worked on completely independent sections, they were still working in the MIPS-X hierarchy, and cells that they did not directly edit would have to be updated because of the way Magic used timestamps to figure out when to run its incremental DRC. If a child cell changed, the parent was rewritten since its timestamp had to change. Thus it was possible that someone's real changes could be wiped out by another person writing a cell for timestamps. We solved this problem by not editing the whole machine at once. For example, when changes were made to the PC Unit only the PC Unit was edited instead of editing it as a subcell of the entire design. The editing of subsections eliminated the problem of losing data, but it did cause Magic's DRC program to do unnecessary work. During the final verification, an informal means was used to lock down the database so that only one person was making changes at any time. All these problems point to the need for an intelligent distributed database management system that can keep track of different versions of the design and multiple user accesses.

NFS did work well for the simulation of the chip since we could run multiple simulations on different machines without having to copy the database around and we knew that each machine was simulating the same version. In general, each machine was assigned to run one particular test program, so we did not have problems with multiple machines trying to write the same file.

---

[6]This access mechanism was first observed at Xerox PARC locking access to a plotter.

8 Where the Time Went

The MIPS-X project began in the early spring of 1984. During the first eight months the initial architectural definition of MIPS-X was developed. At this point the machine was considered stable enough for us to write an architecture manual. The next phase of the project was refinement of the architecture and work on the microarchitecture. By the end of 13 months we had begun to write the functional simulator and look at circuit issues. At this time we felt that the design was solid enough to hold a design review with some people from industry and other research laboratories to get feedback on our design. The next three months were used to get the functional simulator up and running, and then the designers began work on circuit design and layout. Near the end of 1985 we had enough layout done to start simulation of some pieces of the chip to exercise the verification tools.

During the first three months of 1986 we spent most of the time preparing the verification tools and finishing the last details of the layout. About three weeks were spent verifying the sections of the chip and the last two weeks were spent simulating the full chip before a version was frozen and prepared for tape-out.

We waited for five months until the first die came back and then began testing. By this time most of the designers had gone on to other work and very little was done beyond the testing of the die. During the Spring of 1987 a design class at Stanford designed and built a test board for the processor. This board came up during the summer, ten months after the die came back and almost 3.5 years after the project started.

Determining the total man-months spent on the project is a very difficult task. This is mainly because most of the people had other commitments during this time such as teaching or doing course work so it is hard to judge what fraction of their time was actually spent on the project. Table 1 gives an indication of how much time was spent on the project. The time period is from March 1984 to the end of April 1986, when the layout was completed. There are two columns giving a weighted and a non-weighted time in man-months and a column showing the number of people involved. The non-weighted figure can be interpreted as an upper bound and more closely measures the time that people were participants. The weighted figure takes the time for students and assumes that they worked half time on MIPS-X and half time on course work. This is probably closer to the truth but it is difficult to say by how much. The first two rows give the breakdown of

	Non-weighted Man-months	Weighted Man-months	No. People	Months Spent
Hardware	176	136		
Software	86	43		
Arch. Def.	73	52	10	8
Transition	82	54	10	8
Simul. & Layout	69	54	7	10
Final Software	38	19	4	10

Table 1: Man-months spent on MIPS-X

the time in terms of hardware and software effort. The former includes the time for architectural definition and circuit design and the latter is the time spent on compiler development. The last rows break the project time into three phases: architectural definition, a transition phase, and finally, simulation, layout and software development. The transition phase is the time when architectural definition is still taking place but simulation and circuit design has started. One column shows the months spent in each phase. If half of the transition time is spent on architectural definition, then about half of the total hardware time is spent defining what to build and the remainder of the time is spent doing implementation.

9 Summary

The MIPS-X project, like many processors before it, took about two years to design. Much of this time was spent trying to decide what exactly the processor should be. During this period we experimented with cache organizations, pipeline structures, instruction sets, and microarchitectures. The time spent in this phase of the design allowed us to produce a simple, yet high performance, processor. Reducing design time by limiting the number of alternatives considered during this period is probably a bad idea. Instead tools need to be developed to make these evaluations less painful by reducing the time for each iteration.

For the detailed implementation, our choice of partitioning the design and creating experts for each section worked very well. Each designer ended up with a manageable piece to design. Maintaining strict rules about how signals at the interface worked and were changed was key in making this partitioning work. These global signals were kept consistent to some degree by the functional simulator. Using the functional simulator as the specification of the machine and the basis of verification and testing made it a very powerful tool. Even with the functional simulator support, we needed to have one designer who was well versed in all areas of the design. He helped maintain the global signals and settled issues that involved multiple blocks.

Finally, the MIPS-X design showed us the importance of simulation, and the need for consistent representations for leveraging the work of other people. Our version of Murphy's Law was, "If you did not test it, it does not work," and "If you think you tested it, you probably forgot something."

10 Acknowledgements

The MIPS-X research effort was supported by the Defense Advanced Project Research Agency under Contract No. MDA903-83-C-0335. Paul Chow was partially supported by the Natural Sciences and Engineering Research Council of Canada. The authors gratefully acknowledge J. Gasbarro and E. McCreight of Xerox for providing fabrication support.

The rest of the MIPS-X team deserve special thanks for their contributions: John Acken, Anant Agarwal, C.Y Chu, Glenn Gulak, John Hennessy, Scott McFarling, Steve Przybylski, Steve Richardson, Arturo Salz, Rich Simoni, Don Stark, Peter Steenkiste, Steve Tjiang and Malcolm Wing. Tom Blank, Martin Freeman, Giovanni DeMicheli and their EE392C class are to be commended for the design and construction of the board.

References

[1] Anant Agarwal, Paul Chow, Mark Horowitz, John Acken, Arturo Salz, and John Hennessy. On-Chip Instruction Caches for High-Performance Processors. In *1987 Stanford Conference on Advanced Research in VLSI*, pages 1–24, Stanford, California, March 1987.

[2] Tom Blank et al. *The Thor User's Manual.* Technical Report In preparation, Computer Systems Laboratory, Stanford University, 1988.

[3] Randy Bryant, Mike Schuster, and Doug Whiting. *MOSSIM II: A Switch-Level Simulator for MOS LSI User's Manual.* Technical Report 5033:TR:82, California Institute of Technology, Computer Science, January 1983.

[4] Paul Chow and Mark Horowitz. Architectural Tradeoffs in the Design of MIPS-X. In *The 14th Annual International Symposium on Computer Architecture*, pages 300–308, IEEE, Pittsburg, Pennsylvania, June 1987.

[5] John Hennessy, Norman Jouppi, Steven Przybylski, Christopher Rowen, and Thomas Gross. Design of a High Performance VLSI Processor. In *Proceedings of the Third Caltech Conference on VLSI*, pages 33–54, March 1983.

114

[6] Mark Horowitz, Paul Chow, Don Stark, Richard Simoni, Arturo Salz, Steven Przybylski, John Hennessy, Glenn Gulak, Anant Agarwal, and John Acken. MIPS-X: A 20 MIPS Peak, 32-Bit Microprocessor with On-Chip Cache. *IEEE Journal of Solid-State Circuits*, SC-22(5):790–799, October 1987.

[7] Mark Horowitz, John Hennessy, Paul Chow, Glenn Gulak, John Acken, Anant Agarwal, Chorng-Yeung Chu, Scott McFarling, Steven Przybylski, Steve Richardson, Arturo Salz, Richard Simoni, Don Stark, Peter Steenkiste, Steve Tjiang, and Malcolm Wing. A 32b Microprocessor with On-Chip 2Kbyte Instruction Cache. In *ISSCC Digest of Technical Papers*, pages 30–31,328, February 1987.

[8] John K. Ousterhout. The User Interface and Implementation of an IC Layout Editor. *IEEE Transactions on Computer-Aided Design*, CAD-3(3):242–249, July 1984.

[9] John K. Ousterhout, Gordon T. Hamachi, Robert N. Mayo, Walter S. Scott, and George S. Taylor. The Magic VLSI Layout System. *IEEE Design & Test of Computers*, 2(1):19–30, February 1985.

[10] Steven Przybylski. The Design Verification and Testing of MIPS. In *1984 Conference on Advanced Research in VLSI, M.I.T.*, pages 100–109, Boston, MA, January 1984.

[11] Chris Rowen, Steven A. Przybylski, Norman P. Jouppi, Thomas R. Gross, John D. Shott, and John L. Hennessy. A Pipelined 32b NMOS Microprocessor. In *ISSCC Digest of Technical Papers*, pages 180–181, February 1984.

[12] Christopher Terman. *Simulation Tools for Digital LSI Design*. Technical Report MIT/LCS/TR-304, MIT, September 1983.

[13] Niklaus Wirth. *Programming in Modula-2*. Texts and Monographs in Computer Science, Springer-Verlag, New York, 1983.

Invited Talk

From First Silicon to First Sales:
The Other Half of the Design Cycle

Patrick Bosshart

VLSI Design Laboratory
Texas Instruments
Dallas, Texas

This talk will describe the tools used and lessons learned during the productization phase of a 553K-transistor LISP processor chip. During this phase, problems with functionality and speed must be resolved, together with electrical, packaging and reliability problems, while simultaneously breaking new ground in complexity and clock rate during testing. In every aspect of the design, it is more productive to build in quality from the start than to retro-fit it later; some steps toward this goal will be presented.

Invited Talk

The Application of Wafer-Scale Technology to Neuromorphic Systems[*]

Jack I. Raffel
Lincoln Laboratory
Massachusetts Institute of Technology
Lexington, Massachusetts, 02173-0073

Abstract

A design approach is proposed for building wafer-scale neuromorphic systems in which the application specific architecture is laser-programmed using additive linking structures to interconnect synaptic weights realized with multiplying analog-to-digital converters. Chip level arrays have been fabricated using two different circuit configurations and small networks have been built and studied. Major unresolved issues include determination of resolution for synaptic weights during learning, development of circuits for on-chip learning , and optimization of tradeoffs among component speed, parallelism and power dissipation.

Introduction

Restructurable VLSI is an approach to silicon system design which is aimed at providing mechanisms for deployment of redundant resources to circumvent defects and for rapid, post-fabrication customization.[1] The basis for this approach is a set of techniques for forming or removing low resistance connections on processed chips or wafers using a laser under positioning control of a specialized CAD system. This technology makes use of three alternative silicon device structures built into circuits at predetermined sites which are specially assigned for linking. At Lincoln Laboratory some half dozen wafer-scale systems have been built using these techniques and are operational, and a like number are in various stages of design. More recently, work has been undertaken to apply this approach to the electronic implementation of neuromorphic systems. The massive parallelism and unprecedented complexity of interconnect of such systems make them ideal candidates for wafer-scale implementation.[2] Equally important, however, is the attractiveness of providing a programmable interconnect technology suitable for rapid prototyping of neural architectures in which the application specific connectivity (analogous to the evolutionary structuring in living systems) can be programmed using laser linking.

---

[*]This work was sponsored by the Department of the Air Force. The views expressed are those of the authors and do not reflect the official policy or position of the U. S. Government.

Restructurable VLSI Technology

The additive linking technology developed at Lincoln Laboratory uses a vertical microweld between two levels of metal separated by an insulating dielectric. This structure is shown in Figure 1. Two insulators have been used successfully to build working circuits. The first is a sandwich of amorphous silicon enclosed top and bottom by 100 Å oxide barriers to prevent interdiffusion during a high temperature sinter. Recently, in an attempt to substitute more conventional processing, a silicon-rich silicon nitride has been used successfully,[3] thereby eliminating the need for oxide barriers and obviating the special requirements for amorphous silicon. This device has impedances an order of magnitude higher than the typical 1 ohm level achieved with amorphous silicon but still quite acceptable. A third structure, aimed at foundry fabrication using completely standard MOS fabrication, is a lateral link formed between two substrate diodes as shown in Figure 2.[4] Here the structure, which uses a separation between the diffusions comparable to the laser beam diameter of 2 to 4 microns, depends upon the ability to melt the silicon thereby causing diffusion of the dopant across the gap. The process requires the successful recrystallization of the silicon so as to preserve the diode isolation to the substrate of the resulting single merged diffusion. The resulting connection, for a device the size of a typical transistor, has a resistance of about 100 ohms for standard source-drain doping. Lower resistances could, of course, be obtained with higher doping densities.

Wafer-scale Design Methodology

All of the wafer-scale systems designed thus far have been built around a few cell types, the basic replaceable modules, embedded in a two dimensional grid of two level metal interconnect with programmable cuts and links sited so as to permit ready replacement of pretested defective cells. Cell redundancy of between 50% and 100%, which directly determines required cell yield, is thought to produce the best compromise between cell size and restructuring overhead. In addition to the pretesting of logic modules, the RVLSI approach allows access to all interconnect metal for pretesting as well, since all wafer length tracks on first and second level metal are brought out to test pads at the wafer periphery. All cell i-o stubs, which can be linked to the wafer length interconnect, are themselves connected to the cell pads used for cell functional pretest so they can also be tested for shorts and opens. Finally, it is possible to verify during the restructuring process that all links and cuts have been properly made by measuring laser induced photocurrents to determine correct flow in restructured nets.

Design of Neuromorphic Wafer-Scale Systems

It is important to recognize that there is a wide range of architectures that are loosely grouped under the heading of neural networks. These can differ enormously in connectivity and node processing functionality. At one extreme are locally connected arrays such as those for low-level image processing; at the other, fully connected arrays such as those studied by Hopfield.[5] In between these extremes are various architectures which combine assemblages of nodes having very low internal connectivity with very high external connectivity between these assemblages. For instance, multilevel perceptrons have no connections between nodes at the same level but high fanout between levels.

In addition to architectural variety, which is largely reflected in connectivity, the specific electronic implementation approach will have far-reaching effects on wafer-scale design. These technologies can be categorized by the devices which form the sum of weighted inputs, the techniques used to store and program weights, the amplification and thresholding devices at each node and the electronic quantity used to represent activation. The most critical technology decision of all is determining which parts of the system should be built in analog and which in digital form. Here the economies of analog computation and the uncertainty as to the precision requirements of highly redundant, adaptive systems argue strongly for the careful study of the linearity and uniformity requirements for these devices using realistic net simulations. A wide variety of device technologies and circuits have been proposed for implementing programmable weights including MNOS-controlled CCD's,[6] voltage-switched, once-programmable amorphous silicon links,[7] gate charge modulation of transistor conductance[8] and multiplying digital-to-analog converters (mdac's).[9] This last provides the most straightforward approach to achieving early implementation of nets of significant size, combined with the ability to diagnose experimental systems using the flexibility and programmability of existing powerful digital testers. Furthermore, it allows for a ready interface with off-chip learning computation and the gradual incorporation of these capabilities on-chip.

Figures 3 and 4 show two versions of mdac's which have been used to build chip level circuits in the MOSIS silicon foundry.[2] The circuit of figure 3 converts a signed four bit digital weight, stored on-chip in programmable static RAM, to an analog conductance between 0 and 15 and produces a current equal to the product of the input analog voltage and the conductance. The sign bit determines which of two summing wires, Ie (excitatory) or Ii (inhibitory) is connected to the mdac output. Many such programmable weights, corresponding to all the synapses feeding a common neuron, are connected together and the two summing currents are then subtracted by first inverting the inhibitory current in the right hand operational amplifier and connecting its output to the exciting current. The net current, Ie - Ii, is then fed to the transimpedance amplifier which converts the current to an output voltage thereby providing a compatible input to other similar synapses. This device has been used to build a small network which implements a Gaussian classifier for the recognition of monosyllabic spoken digits. One of the disadvantages of this circuit is that it requires two operational amplifiers per neuron and relies on control of the linear portion of the transistor curve. In addition, the output voltage from one neuron has to supply the full fanout current to the large number of inputs it may feed.

The alternative circuit in figure 4 uses a current mirror in which transistors are operated in saturation. The mirror allows direct subtraction of the exciting and inhibiting currents, only requiring a single, simple amplifier to adjust the gate voltage on the output transistor so as to sink the net current, Ie - Ii. The cascode circuit is effective in eliminating the effects of Early voltage by holding all the mdac's at the same Vds. Figure 5 shows a plot of input voltage vs. output voltage for a neuron in which the full transformation from gate voltage to transistor current and back to a gate voltage is shown for a rising and falling voltage ramp. Tracking is nearly perfect within a range limited by the threshold voltage and the minimum voltage drop across the transistor for satisfactory operation as the gate voltage approaches the upper power supply rail.

Figure 6 shows in schematic form a possible generalized wafer-scale neural net architecture whereby groups of mdac outputs are connected to a vertical line representing a single neuron with its associated summing amplifier. Horizontal lines represent input distribution wires carrying a signal either generated externally or emanating from an output from another neuron and fanned out to all commonly connected synaptic inputs. Both horizontal and vertical lines traverse the entire wafer and are, therefore, available for testing at the periphery. Similarly, both mdac's and amplifiers are also testable before restructuring. Large circles represent link points which may be programmed to connect horizontal and vertical lines thus allowing flexible interconnection of mdac's to form arrangements of multilevel nets of variable depth and width. The extension of this technique to wafer-scale dimensions containing perhaps 100,000 weights at 1 micron design rules involves scaling currents to the 10 microamp range, consistent with wafer power dissipation of the order of tens of watts. Speed of operation in these systems is controlled completely by charging time of interconnect wiring capacitance.

Further Work and Outstanding Issues

In the world of digital systems functional testing is both unambiguous and unforgiving. In the analog domain linearity and uniformity requirements become exceedingly difficult to determine for complex, multilevel, thresholded systems. For example, extensive computer simulation of the back propagation model for modifying weights indicates significantly greater resolution is required during the learning phase than during normal operation.[10] Furthermore, the number of bits required is affected by such factors as initialization and round-off strategy. Learning algorithms for multilevel systems must deal with the so-called credit assignment problem, i.e. how to distribute weight changes to synapses at different levels in response to measured output error. Algorithms such as back propagation have been developed to meet convergence criteria as determined by digital computer simulation. Electronic implementation has only recently become a consideration and presents substantial difficulties. Eventually, realizability in hardware will play a significant, if not commanding, role in determining learning strategies.

One approach to adaptation, which is realizable and is presently being built into experimental circuits, is competitive learning. This general category of system has been studied in a number of variations[11,12] which all depend in part on mutual inhibition and require the use of a winner-take-all circuit of the type which has been built previously using the mdac technology described above. More recently, a system has been fabricated which implements Kohonen's algorithm for a feature classifier[13] which organizes a network so as to cluster neurons in vector space (i.e assign similar weight patterns) to physically proximate neurons. This algorithm is now being applied to problems in speech recognition and represents our first attempt to build in on-chip learning.[14] Here, in order to simplify the computation for updating weights, unary encoding was adopted allowing a simple increment-decrement scheme to be used. This circuit affords an excellent example of the kind of compromise which will probably be necessary to expedite the implementation of on-chip learning.

Finally, the inherent speed disparity between electronic and living system response implies that a direct copy of neural architectures is unlikely to yield the most efficient utilization of computational resources. What we do not know at the present time is how to incorporate time-sharing of resources to take advantage of this speed mismatch and at the same time mimic what is demonstrably effective in nature in order to preserve the best features of evolutionary development.

References

1. J.I. Raffel, A.H. Anderson, G.H. Chapman, K.H. Konkle, B. Mathur, A.M. Soares, and P.W. Wyatt, A Wafer-Scale Digital Integrator Using Restructurable VLSI, IEEE Journal of Solid-State Circuits, Vol. SC-20, No. 1, February, 1985.

2. J.I. Raffel, J.R. Mann, R. Berger, A.M. Soares, and S. Gilbert, A Generic Architecture for Wafer-Scale Neuromorphic Systems. IEEE First International Conference on Neural Networks, San Diego, CA, 21-24 June 1987.

3. J.A. Burns, G.H. Chapman, B.L. Emerson, Low Resistance Programmable Connections Through Plasma Deposited Silicon Nitride, ECS Abstract #321, San Diego, CA, Vol. 86-2 October 1986.

4. J.M. Canter, G.H. Chapman, B. Mathur, M.L. Naiman, and J.I. Raffel, A Laser Link for Wafer-Scale Integration Using Standard CMOS Processing. Device Research Conference, University of Massachusetts at Amherst, 23-25 June 1986.

5. J.J. Hopfield, (1982) Neural Networks and Physical Systems with Emergent Collective Computational Abilities. Proceedings of the National Academy of Sciences, USA, 81, 3088-3092.

6. J.P. Sage, K. Thompson, and R.S. Withers, An Artificial Neural Network Integrated Circuit Based on MNOS/CCD Principles, American Institute of Physics, 381-385, 1986.

7. T. Daud, A. Moopenn, J.L. Lamb, R. Ramesham, and A.P. Thakoor, Neural Network Based Feed-Forward High Density Associative Memory, IEDM 87, 107-110, IEEE 1987.

8. Y. Tsividis, S. Satyanarayana, Analogue-Circuits for Variable-Synapse Electronic Neural Networks, Electronics Letters 1313-1314, 19 November 1987.

9. J. Alspector, R.B. Allen, A Neuromorphic VLSI Learning System, Advanced Research in VLSI, Proceedings of the 1987 Stanford Conference.

10. S. Gilbert, M.S. thesis, MIT, in preparation.

11. S. Grossberg, Adaptive Pattern Classification and Universal Recoding: Part 1., Parallel Development and Coding of Neural Feature Detectors. Biological Cybernetics, 23, 121-134.

12. D.E. Rumelhart, D. Zipser, Feature Discovery by Competitive Learning, Parallel Distributed Processing, Explorations in the Microstructure of Cognition, Volume 1: Foundations, 151-193, MIT Press, Cambridge, MA, 1986.

13. T. Kohonen, Self-Organization and Associate Memory, Springer Verlag, 1984.

14. J.R. Mann, R. Lippman, R. Berger, J.I. Raffel, A Self-Organizing Neural Net Chip, Custom Integrated Circuits Conference 1988, to be published.

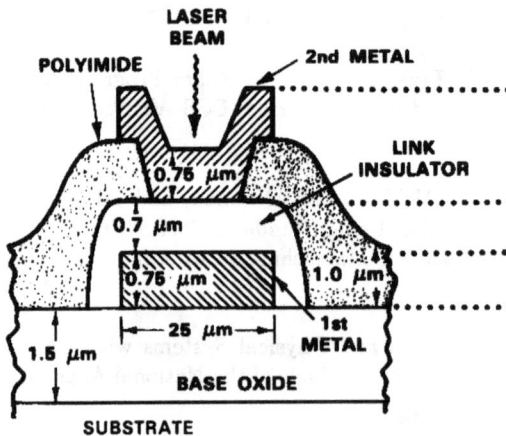

Figure 1: Vertical Link

Figure 2: Lateral Diffused Link

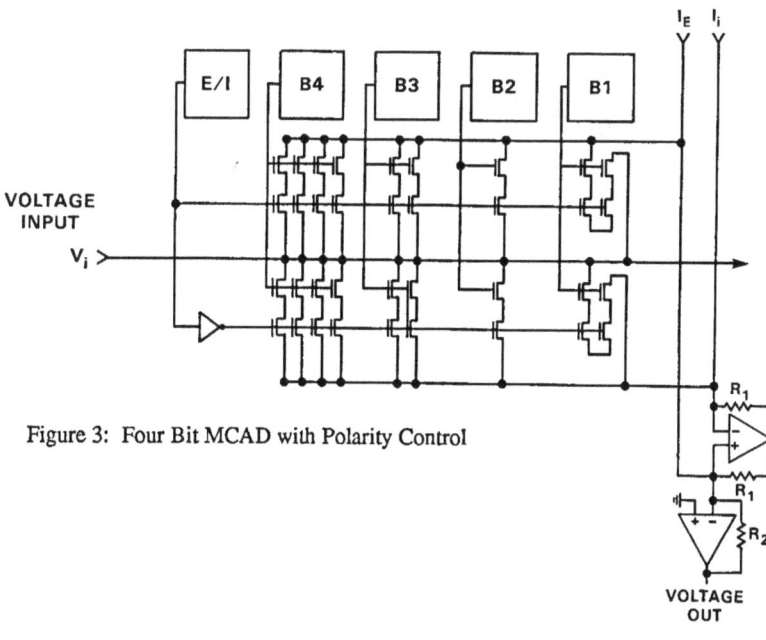

Figure 3: Four Bit MCAD with Polarity Control

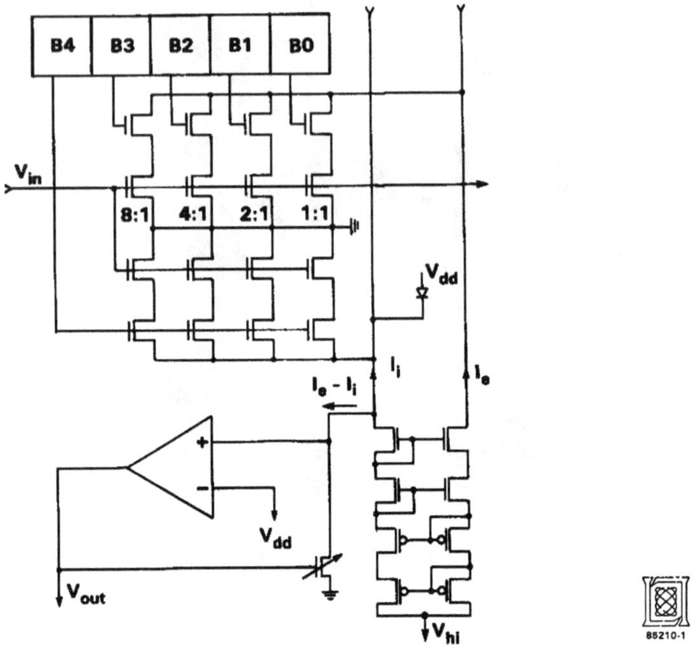

Figure 4: Current Mirror MDAC

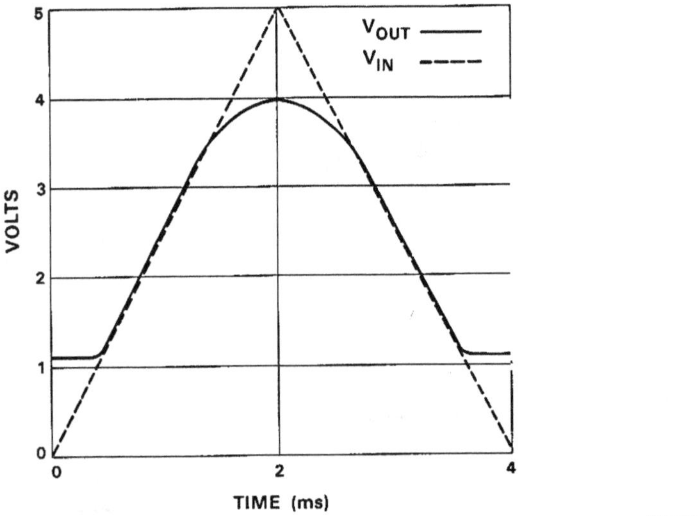

TIME (ms)

Time (MSECS)
Figure 5: VOUT vs. Vin for Current Mirror MDAC

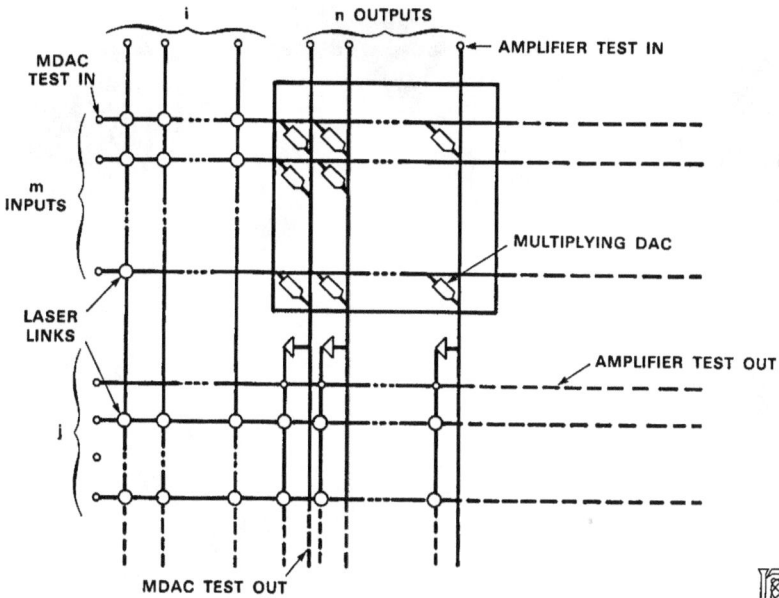

Figure 6: General Wafer-Scale Network Architecture with Programmable Interconnect

An Evaluation of Reconfiguration Algorithms in Fault Tolerant Processor Arrays

Graham E. Farr and Heiko Schröder

Computer Sciences Laboratory
Australian National University
GPO Box 4, Canberra, ACT 2601
Australia

1 Introduction

In this paper a new performance measure for fault tolerance of highly parallel systems is introduced. This classification is as relevant to fault tolerant systems as time and area complexity are in the realm of algorithm analysis.

The particular parallel systems we consider are mesh-connected networks of processors (in case these processors are small it is more appropriate to call them processing elements or cells). Such architectures are of relevant interest in the area of VLSI and WSI, where *systolic arrays* [4] and *instruction systolic arrays* [3], [5], [9] have been designed to implement a wide range of different algorithms.

We model failure of processors by assuming that all processors have the same failure probability p, and that processor failures are independent. This probability p will be expressed as a suitable function $p(N, \alpha)$ of the number N of processors in the array and a constant α. (In the situations of interest to us, $p(N, \alpha) = 1/N^\alpha$.)

In this paper we analyse algorithms which achieve fault-tolerance through reconfiguration. The reconfiguration algorithms analysed aim at guaranteeing cooperation of the surviving elements even after a number of processors fail. Two different approaches are considered here. We either allow degradation of the network, trying to keep the network as large as possible or in order to prevent degradation of the array size we make use of spare cells which are provided as an additional row and an additional column of the array. The specific algorithms we consider were introduced in [8].

The performance of a reconfiguration algorithm can be measured by its probability of getting stuck in a fatal situation, that is to say, the probability that the used algorithm does not find a suitable reconfiguration.

To this end we find it useful to consider the largest number α_c such that if $\alpha < \alpha_c$ then (under our model, with $p = 1/N^\alpha$) the reconfiguration algorithm will almost certainly get stuck in a fatal situation. Thus, the better a reconfiguration algorithm is, the smaller will be its corresponding "critical constant" α_c. We apply this new theory to compare some fault tolerant systems suggested in the literature, and show how to derive the critical constant in a number of cases.

The connection networks the reconfiguration algorithms are based on consist of local connections only. Thus even for the optimal algorithms there are small clusters of faults that cannot be overcome. For a fixed probability of cell failure the probabilty that such fault clusters occur approaches one as the network size goes to infinity. Thus there will be always a critical size of the network that cannot be overcome unless either better fault tolerance techniques become available or the probability of single cell failure decreases. The performance measure given in this paper classifies reconfiguration algorithms according to the speed (expressed in terms of the degree of the dominating term of a function) at which the critical size of the array is approached.

To get more insight into different fault tolerance techniques the reader is referred to [8] and [6].

2   Basic Notation

We adopt the notation and terminology used in [8]. We are given a square array of identical cells c_{ij}, $1 \leq i,j \leq n$. We assume that each cell is self-testing and that a signal e_{ij} states its working/faulty status.

Reconfiguration is performed by global renaming procedures that map the array functions onto the working cells. Let there be given a regular $n \times n$ array of processing elements c_{ij} to which a pattern of spare cells has been added. An $n \times n$ array of logical indices indicated as (i',j') is mapped into the set of working cells such that cells with neighbouring logical indices (i.e. cells which differ by one in exactly one of their indices, being equal in the other) are physically connectible.

The fault-tolerant array consists of three parts:

i Array cells and spare cells.

ii An interconnection pattern consisting of data paths and of devices (multiplexers or switches) controlling path selection on the basis of reconfiguration signals.

iii A control network, computing reconfiguration signals on the basis of error signals. It is directly related to the particular algorithm adopted.

In order to keep the number of added interconnections reasonably low, reconfiguration alternatives are kept limited and there are fault distributions, which are fatal for any algorithm based on the given interconnection pattern. There are also fatal fault distributions that are fatal only for the given reconfiguration algorithm, though they could be overcome by another (better) algorithm without using a different interconntection pattern.

For our theoretical treatment we embed our final processor array in the infinite square lattice graph $\mathcal{Z} \times \mathcal{Z}$, which we denote simply by L. A *pattern* is a finite set of vertices of L. Two patterns A_1 and A_2 are *isomorphic* if there is a translation of L which maps A_1 to A_2. A *pattern type* is an isomorphism class of patterns, that is, a pattern and all its translates in L. A *pattern family* is a union of pattern types, that is, a set of patterns closed under translations of L.

We are dealing with patterns on a finite square grid where the vertices are processing elements and patterns refer to subsets of these nodes whose members are faulty. Such a grid can be regarded as a subgrid of L and so one can speak of patterns occurring in finite grids in the natural way. Referring to a given reconfiguration algorithm we will use the term *atomic fail patterns*, which is a fail pattern that cannot be overcome by the given algorithm, while removing any node from this pattern makes it solvable.

Let $\mathcal{A}$ be a pattern family. (Think of $\mathcal{A}$ as the set of atomic fail patterns of some reconfiguration algorithm.) We assume that we have an $n \times n$ grid G of $N = n^2$ vertices (which we think of as processors). It is sometimes convenient to think of the grid as coordinatized in the obvious way (see Figure 1).

Assume each vertex fails with probability p, and that vertex failures are independent. We are interested in the probability that some

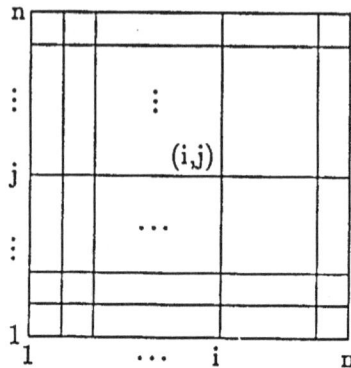

Figure 1: The grid G

member of $\mathcal{A}$ is contained in the set of failed vertices of G. To obtain results on how this probability depends on p, we will use some standard techniques from the theory of random graphs [1][7]. We do not assume prior knowledge of this area.

We are going to introduce several reconfiguration algorithms and look at their sets of atomic fail patterns. Then we do some probability theoretical analysis which serves as a basis for the comparison of their asymptotic behaviour.

The first algorithms analysed here are based on direct reconfiguration, i.e. a faulty cell c_{ij} is replaced along column i or row j (see [8]).

The next algorithms we look at, are more flexible in the sense that they allow to propagate configuration through the array changing the direction of replacement (see [8]).

3 Reconfiguration Algorithms

3.1 Reconfiguration by direct horizontal replacement

Direct reconfiguration algorithms as defined in [8] mark each fault either as a horizontal or as a vertical fault. They fail if in the course of their execution in any row or column more than one cell is marked as horizontal or vertical.

The simplest algorithms one can think of is:

Algorithm 1: There is only one column of spares available. Each fault is declared horizontal.

This algorithm runs into a fatal situation if and only if there is a row with more than one fail. Thus there are $\Theta(n^3)$ different fatal atomic fail patterns of size 2 in an $n \times n$ grid (see Figure 2 ($\mathcal{A}_1$)).

3.2 Optimal reconfiguration algorithms in H2

We allow degradation of the network and use horizontal replacement only. Furthermore we restrict ourselves to a connection pattern with maximal wire length 2. Then even the best algorithm would not be able to cope with couples of fails in horizontal neighbourhood. We refer to such an optimal algorithm as the *H2-algorithm*. For the H2-algorithm (it is rather complex, but its details do not concern us and do not fit into the scope of this paper), there is a fail pattern type of size 2 which is fatal. Furthermore it can be shown that such an algorithm also has fatal fail patterns which do not contain any of these atomic patterns of size 2.

3.3 Reconfiguration by direct horizontal and vertical replacement

The following algorithm makes use of horizontal and vertical replacement.

Algorithm 2:

   i Scan each column upwards (for $1 \leq i \leq n$); as soon as a faulty cell is found, it is marked as a vertical fault.

   ii Classify all other faults as horizontal faults.

   iii If no row has more than one horizontal fault, reconfiguration is possible. Horizontal reconfiguration and related renaming are performed; logical index $j' = 0$ is associated with each horizontal fault.

   iv Vertical reconfiguration and final renaming are performed; logical index $i' = 0$ is associated with each vertical fault.

It is easy to see that in *iii* the "if" can be replaced by an "if and only if". Thus each fail pattern that leads to a fatal situation for this algorithm must contain a subpattern that can be generated in the following manner: Pick an arbitrary row (not the lowest one though), place two faults in arbitrary positions in this row, place a fault in arbitrary position in both columns below the two faults. Thus there are $\Theta(n^5)$ different fail patterns of size 4 in an $n \times n$ grid (see Figure 2 ($\mathcal{A}_2$)).

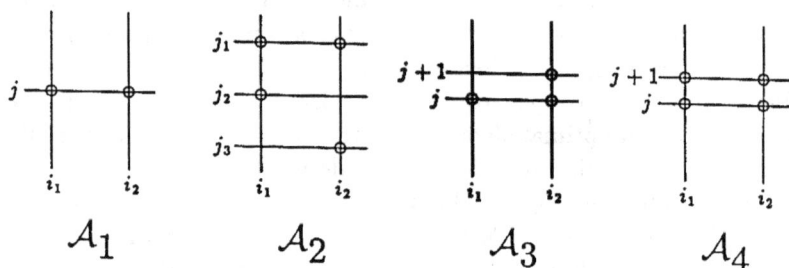

Figure 2: Typical members of the pattern families $\mathcal{A}_1$, $\mathcal{A}_2$, $\mathcal{A}_3$ and $\mathcal{A}_4$

3.4 Fixed stealing

The idea of "stealing algorithms" is that reconfiguration is done row by row working from the bottom row up. Faulty cells can "steal" a neighbouring cell in the row above to do their job. Thus this stolen cell is from then on handled similar to a faulty (not available) cell, to establish the next row.

Algorithm 3:

   i Scan the array South-North.

   ii Scan the row under construction West-East.

   iii The leftmost fault is replaced along the row to the East.

   iv All other faults "steal" the cell to their North, which from then on are treated like faulty cells.

Fatal fault patterns for this algorithm can be generated in the following manner: Pick an arbitrary position in the array (not in the western column nor in the northern row). Let this cell and its northern neighbour and any neighbour to its left be faulty. Then this cell would have to steal a faulty cell, which is fatal. Thus here we have fatal fault patterns of size 3. There are $\Theta(n^3)$ different such fail patterns in an $n \times n$ grid (see Figure 2 ($\mathcal{A}_3$)). In this case there are also other fail patterns of larger size which are fatal but do not contain the above atomic patterns of size 3. We do not consider them here.

3.5   Variable stealing

Variable stealing differs from fixed stealing only in that it does not always declare the leftmost fault in a row to be a horizontal fault. In case there is a fault in the row with another fault to its immediate north, then this fault becomes the horizontal fault (while the leftmost fault does stealing from its north). We refer to variable fault stealing as *algorithm 4*.

Clearly we run into a fatal situation here, if there is a row with two faults each having a faulty neighbour to its immediate north. There are $\Theta(n^3)$ different fail patterns of size 4 which are fatal for this algorithm (see Figure 2 ($\mathcal{A}_4$)). Again not all fatal fail patterns contain the above atomic patterns.

3.6   Optimal reconfiguration algorithms in S2

Now we look for optimal reconfiguration algorithms. We again allow to create a degraded network. This way we never get stuck because we run out of spare cells, but only because reconfiguration is impossible.

The reconfiguration algorithms introduced in [8] make use of all connections between cells of length≤ 2 and in addition to them some connections of length 3. Each connection in the connection pattern used by [8] is fixed in its direction. We restrict ourselves to connections of length two or less. We call this connection pattern $S2$ and assume that all wires can be used in arbitrary direction. An algorithm which creates a reconfiguration for all fail patterns that can be overcome in $S2$ - it will be referred to as the *S2-Algorithm* - will still run into a fatal situation whenever one of the atomic fail pattern shown in Figure 3 occurs.

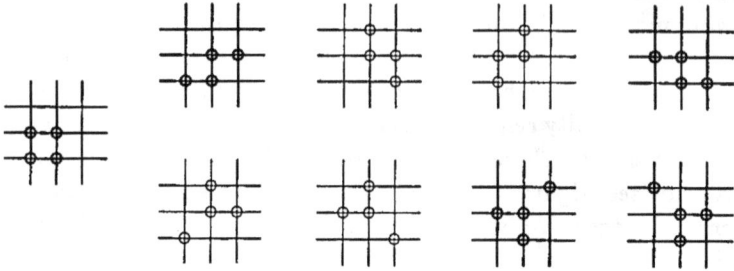

Figure 3: fatal atomic fault patterns for $S2$

4 Performance measures for reconfiguration algorithms

Recall that the vertex failure probability p is assumed to be given by $p = p(N, \alpha) = 1/N^\alpha$. We will show that, for certain sets of atomic fail patterns, there is a *critical constant* α_c such that if $\alpha > \alpha_c$ then almost certainly there is no atomic fail pattern while if $\alpha < \alpha_c$ then almost certainly there will be an atomic fail pattern. In the latter case, the relevant reconfiguration algorithm will almost surely fail. Note that in our context we say that an event ξ occurs *almost certainly* if $Pr(\xi) \to 1$ as $N \to \infty$.

An $\mathcal{A}-pattern$ of G is a set of vertices $A \subseteq V(G)$ such that $A \in \mathcal{A}$. A *failed $\mathcal{A}$-pattern* is an $\mathcal{A}$-pattern in which all vertices fail. Let the random variable $F_\mathcal{A}$ be equal to the number of failed $\mathcal{A}$-patterns in G. For certain pattern families $\mathcal{A}$ we show that (referring to the preceding paragraph) if $\alpha > \alpha_c$ then $E(F_\mathcal{A}) \to 0$ as $N \to \infty$ so that as $N \to \infty$, $Pr(F_\mathcal{A} = 0) \to 1$. We also show that if $\alpha < \alpha_c$ then $E(F_\mathcal{A}) \to \infty$ as $N \to \infty$; however more is needed to establish that $Pr(F_\mathcal{A} = 0) \to 0$ as $N \to \infty$. We use the "second moment method" [1][7], based on Chebyshev's inequality ([2], p.233); this inequality implies that if $E(F_\mathcal{A}) \neq 0$ then

$$Pr(F_\mathcal{A} = 0) \leq \frac{E(F_\mathcal{A}^2)}{E(F_\mathcal{A})^2} - 1.$$

Thus we find ourselves considering $E(F_\mathcal{A}^2)$ in detail. Finally we consider what happens *at* the critical constant α_c, again using this inequality.

For each N, let $\gamma_{\mathcal{A},N}(k)$ be the number of $\mathcal{A}$-patterns of k vertices in the grid of N vertices.

For each $\mathcal{A}$-pattern A in G, define a random variable

$$X_A = \begin{cases} 1, & \text{all vertices in } A \text{ fail}; \\ 0, & \text{otherwise.} \end{cases}$$

Then $Pr(X_A = 1) = p^{|A|}$, so that $E(X_A) = p^{|A|}$. Now obviously

$$F_{\mathcal{A}} = \sum_{\substack{A \subseteq V(G) \\ A \in \mathcal{A}}} X_A.$$

Consider the expectation of $F_{\mathcal{A}}$.

$$E(F_{\mathcal{A}}) = E\left(\sum_{\substack{A \subseteq V(G) \\ A \in \mathcal{A}}} X_A \right) = \sum_{\substack{A \subseteq V(G) \\ A \in \mathcal{A}}} E(X_A)$$

$$= \sum_{\substack{A \subseteq V(G) \\ A \in \mathcal{A}}} p^{|A|} = \sum_k \gamma_{\mathcal{A},N}(k) \cdot p^k. \tag{1}$$

We will also be concerned with the second moment $E(F_{\mathcal{A}}^2)$. Now,

$$F_{\mathcal{A}}^2 = \left(\sum_{\substack{A \subseteq V(G) \\ A \in \mathcal{A}}} X_A \right)^2 = \sum_{\substack{(A,B) \\ A,B \subseteq V(G) \\ A,B \in \mathcal{A}}} X_A X_B = \sum_{\substack{A \subseteq V(G) \\ A \in \mathcal{A}}} X_A^2 + \sum_{\substack{(A,B) \\ A,B \subseteq V(G) \\ A,B \in \mathcal{A} \\ A \neq B}} X_A X_B.$$

Certainly $X_A^2 = X_A$ always, so the first sum is just $F_{\mathcal{A}}$. Also it is convenient to split the second sum up according to the size of $|A \cap B|$. For convenience and brevity we write

$$\sum_{(t)} \text{ for } \sum_{\substack{(A,B) \\ A,B \subseteq V(G) \\ A,B \in \mathcal{A} \\ A \neq B \\ |A \cap B| = t}}.$$

Thus for example $\sum_{(0)}$ indicates summation over ordered pairs of disjoint $\mathcal{A}$-patterns. Considering expectations, then, we have

$$E(F_{\mathcal{A}}^2) = E(F_{\mathcal{A}}) + \sum_{(0)} E(X_A X_B) + \sum_{t \geq 1} \sum_{(t)} E(X_A X_B). \tag{2}$$

Now $E(X_A X_B)$ is just the probability that every vertex in $A \cup B$ fails: $E(X_A X_B) = p^{|A \cup B|}$. Hence (2) yields

$$E(F_{\mathcal{A}}^2) = E(F_{\mathcal{A}}) + \sum_{(0)} p^{|A|+|B|} + \sum_{t \geq 1} \sum_{(t)} p^{|A \cup B|}. \tag{3}$$

At this point we introduce some notation which will be useful later on. If A is an $\mathcal{A}$-pattern in G, we write $S_t(A)$ for the number of $\mathcal{A}$-patterns (excluding A itself) which meet A in exactly t vertices:

$$S_t(A) = |\{\, B \ \mid \ B \in \mathcal{A}, \ B \subseteq V(G), \ |A \cap B| = t, \ B \neq A \,\}|.$$

Thus the number of pairs (A, B) over which the sum $\sum_{(t)}$ is carried out equals

$$\sum_{\substack{A \subseteq V(G) \\ A \in \mathcal{A}}} S_t(A).$$

Having obtained general expressions for $E(F_A)$ and $E(F_A^2)$, it is time to consider the cases which are of interest to us.

5 $\mathcal{A}$ has finitely many pattern types

Let k_1 (respectively k_2) be the size of the smallest (largest) member of $\mathcal{A}$. If $k_1 \leq k \leq k_2$, let a_k be the number of k-vertex patterns in $\mathcal{A}$.

Our result here is the following.

Theorem 5.1 *If $\mathcal{A}$ has finitely many types and the vertex failure probability is $p = 1/N^\alpha$, then:*

   i If $\alpha > 1/k_1$ then almost certainly G has no failed $\mathcal{A}$-patterns.

   ii If $\alpha < 1/k_1$ then almost certainly G has a failed $\mathcal{A}$-pattern.

   iii If $\alpha = 1/k_1$ then for all $\varepsilon > 0$ the probability that G has no failed $\mathcal{A}$ patterns is bounded above by $1/a_{k_1} + \varepsilon$ for all sufficiently large N.

The proof can be found in the appendix.

A few points are worth mentioning. Firstly the "shapes", in G, of the patterns in $\mathcal{A}$ are irrelevant; only their size is really important. Furthermore the critical constant α_c depends on the smallest member of $\mathcal{A}$ only and is independent of how many types of smallest members $\mathcal{A}$ has (though of course this latter number determines our upper bound on the probability of no failed $\mathcal{A}$-patterns if $\alpha = \alpha_c$). The proofs of our results are virtually independent of the exact structure of G; the results would also hold if G was, for example, a circuit on N vertices.

Theorem 5.1 can be applied to the H2-algorithm and the S2-algorithm. For the H2-algorithm we get that its critical value for α is $\alpha_{H2} \geq 1/2$ and for the S2-algorithm we get $\alpha_{S2} \geq 1/4$. Thus we have got lower bounds for α_c for the two connection patterns $H2$ and $S2$. In both cases the authors try to prove that these bounds are tight, i.e. to show that the probability of fatal fail patterns which do not contain the atomic fail patterns is negligibly small.

6   Some cases where $\mathcal{A}$ has infinitely many pattern types

If $\mathcal{A}$ has infinitely many pattern types, possibly of many different sizes, the analysis can become very complicated. However we do not need too much generality to handle the cases that interest us. Motivated by actual atomic fail patterns for some reconfiguration algorithms, we consider four specific pattern families $\mathcal{A}_1$, $\mathcal{A}_2$, $\mathcal{A}_3$ and $\mathcal{A}_4$ (described below) and show that for each family a result similar to Theorem 5.1 holds.

To this end, we say that a pattern family $\mathcal{A}$ has the *transition property with critical constant* α_c if the following all hold:

i If $p = 1/N^\alpha$ and $\alpha > \alpha_c$ then almost certainly G has no failed $\mathcal{A}$-patterns.

ii If $p = 1/N^\alpha$ and $\alpha < \alpha_c$ then almost certainly G has a failed $\mathcal{A}$-pattern.

iii There exists $K > 0$ such that if $p = K/N^{\alpha_c}$ then the probability that there is no failed $\mathcal{A}$-pattern is bounded above by some constant < 1.

Thus, Theorem 5.1 (and the remark immediately following its proof) asserts that a finite pattern family has the transition property with critical constant $1/k_1$, where k_1 is the size of a smallest member of the family.

Our four pattern families are the following (see Figure 2).

i $\mathcal{A}_1$ consists of all patterns of the form
$\{(i_1, j), (i_2, j)\}$, $i_1 \neq i_2$.

ii A_2 consists of all patterns of the form
$\{(i_1, j_1), (i_1, j_2), (i_2, j_1), (i_2, j_3)\}$, $i_1 \neq i_2$, $j_1 > j_2$, $j_1 > j_3$.

iii A_3 consists of all patterns of the form
$\{(i_1, j), (i_2, j), (i_2, j+1)\}$, $i_1 < i_2$.

iv A_4 consists of all patterns of the form
$\{(i_1, j), (i_1, j+1), (i_2, j), (i_2, j+1)\}$, $i_1 \neq i_2$.

It will be shown that A_1, A_2, A_3 and A_4 each have the transition property with a suitable critical constant. We begin with some remarks which apply to all four families, and many other families as well.

Suppose that A is a pattern family all of whose members have size k, and that the number of A-patterns in a grid G of N vertices is given by $\gamma_{A,N} = aN^\beta$ where a and β are positive constants. From (1) we have

$$E(F_A) = \gamma_{A,N} \cdot p^k.$$

Putting $p = 1/N^\alpha$ and $\gamma_{A,N} = aN^\beta$, we have

$$E(F_A) = a \cdot N^{\beta - k\alpha}. \tag{4}$$

It is evident that the critical value of α is $\alpha_c = \beta/k$. If $\alpha > \alpha_c$ then $E(F_A) \to 0$ as $N \to \infty$, so $Pr(F_A = 0) \to 1$ as $N \to \infty$. Hence in this case there are, almost certainly, no failed A-patterns.

We did exactly the same thing earlier, for the special case where A has finitely many pattern types and hence $\beta = 1$. This is the "easy" part of showing that a transition property holds. For the rest, the second moment method is used.

Theorem 6.1 *The pattern families A_1, A_2, A_3 and A_4 have the transition property with corresponding critical constants $\alpha_{c1} = 3/4$, $\alpha_{c2} = 5/8$, $\alpha_{c3} = 1/2$, $\alpha_{c4} = 3/8$.*

The proof can be found in the appendix.

A consequence of Theorem 6.1 is that algorithm 1 can be characterized by $\alpha_{c1} = 3/4$ and algorithm 2 by $\alpha_{c2} = 5/8$. For algorithm 3 we get $\alpha_{c3} \geq 1/2$ and for algorithm 4 the result is $\alpha_{c4} \geq 3/8$. We have not proved the equality in the last 2 cases, since we did not consider all possible fatal fail patterns for these 2 algorithms.

7 Conclusions

We have developed a new performance measure for reconfiguration algorithms. It classifies reconfiguration algorithms according to their effectiveness.

In Figure 4 we have plotted the probability of single fails against the size of the array that could be achieved. This is done for the six sets of atomic fail patterns that are analysed in this paper.

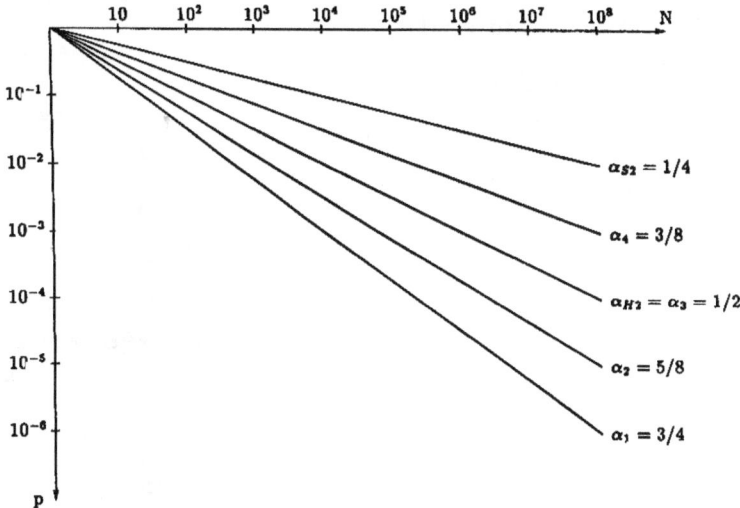

Figure 4: The effectiveness of reconfiguration algorithms

As mentioned earlier the curves for $\alpha = 3/4$ and $\alpha = 5/8$ characterize algorithm 1 and 2. The curves for $\alpha = 1/2$ and $\alpha = 3/8$ only say that the algorithms 3 and 4 cannot be better than these curves indicate, as in both cases there are atomic fail patterns that are fatal for these algorithm but have not been taken into account. The curves for $\alpha = 1/2$ and $\alpha = 1/4$ give bounds for optimal reconfiguration algorithms for the connection networks $H2$ and $S2$.

This analysis, like standard techniques of complexity analysis of algorithms, does not take care of any constant factors involved.

8   Appendix: Proofs

8.1   *Proof of Theorem 5.1*

Firstly we claim that

$$\gamma_{\mathcal{A},N}(k) = a_k \cdot N - o(N) \tag{5}$$

where $k_1 \le k \le k_2$. To see this, let b_0 be the minimum side length of a square planar grid which contains (at least) one $\mathcal{A}$-pattern of each type. Since $\mathcal{A}$ has finitely many types, b_0 is a constant independent of N. It is not difficult to see that

$$a_k(n - b_0)^2 \le \gamma_{\mathcal{A},N}(k) \le a_k n^2,$$

and the claim follows.

It then follows from equation (1) that

$$E(F_{\mathcal{A}}) = \sum_{k=k_1}^{k_2} \left(a_k N^{1-k\alpha} - o(N^{1-k\alpha}) \right). \tag{6}$$

Put $\alpha_c = 1/k_1$; this is our critical constant. If $\alpha > \alpha_c$ then $1 - k\alpha < 0$ for all $k \ge k_1$, so by (5) $E(F_{\mathcal{A}}) \to 0$ as $N \to \infty$. Therefore $Pr(F_{\mathcal{A}} = 0) \to 1$ as $N \to \infty$. This establishes (i).

The second moment method will be used to analyse the cases $\alpha < \alpha_c$ and $\alpha = \alpha_c$. Let us therefore consider $E(F_{\mathcal{A}}^2)$ (see equation 3).

Since $\mathcal{A}$ has only finitely many pattern types and each member of $\mathcal{A}$ is finite, it is clear that if A is an $\mathcal{A}$-pattern in G and $t \ge 1$ then $S_t(A)$ is bounded above by a constant independent of N.

Consider $\sum_{(0)} p^{|A|+|B|}$. The number of $\mathcal{A}$-patterns B disjoint from a given $\mathcal{A}$-pattern A is, by the previous paragraph,

$$\sum_{k=k_1}^{k_2} \left(\gamma_{\mathcal{A},N}(k) - \Theta(1) \right).$$

Hence

$$\sum_{(0)} p^{|A|+|B|} = \sum_{\substack{A \in \mathcal{A} \\ A \subseteq V(G)}} \left(\sum_{k=k_1}^{k_2} \left(\gamma_{\mathcal{A},N}(k) - \Theta(1) \right) \right) p^{|A|+k}$$

$$= \left(\sum_{\substack{A \in \mathcal{A} \\ A \subseteq V(G)}} p^{|A|} \right) \cdot \left(\sum_{k=k_1}^{k_2} \left(\gamma_{A,N}(k) p^k - \Theta(1) p^k \right) \right)$$

$$= E(F_A) \cdot \left(E(F_A) - \Theta(1) \sum_{k=k_1}^{k_2} p^k \right).$$

However if $p = 1/N^\alpha$ (for any fixed α) then $p^k \to 0$ if $N \to \infty$ for each k, $k_1 \le k \le k_2$. Hence $\Theta(1) \sum_{k=k_1}^{k_2} p^k \to 0$ as $N \to \infty$. Therefore

$$\sum_{(0)} p^{|A|+|B|} \to E(F_A)^2 \quad as \ N \to \infty. \tag{7}$$

Now consider $\sum_{(t)} p^{|A \cup B|}$. Choose a constant b independent of N such that $S_t(A) \le b$ for all $t \ge 1$ and for all $\mathcal{A}$-patterns A; this is possible since $S_t(A) = O(1)$ and $\mathcal{A}$ has just finitely many types. Note that if $A \neq B$ then $|A \cup B| \ge k_1 + 1$ so $p^{|A \cup B|} \le p^{k_1+1}$.

By these observations,

$$\sum_{(t)} p^{|A \cup B|} \le \sum_{\substack{A \subseteq V(G) \\ A \in \mathcal{A}}} b \cdot p^{k_1+1} \le \hat{a} \cdot N \cdot b \cdot p^{k_1+1}$$

by (5), where $\hat{a} = \sum_{k=k_1}^{k_2} a_k$. Therefore

$$\sum_{t \ge 1} \sum_{(t)} p^{|A \cup B|} \le (k_2 - 1) \cdot \hat{a} \cdot b \cdot N \cdot p^{k_1+1}.$$

Putting $p = 1/N^\alpha$ and dividing by $E(F_A)^2$ (see 6), we have

$$\frac{1}{E(F_A)^2} \sum_{t \ge 1} \sum_{(t)} p^{|A \cup B|} \le \frac{(k_2 - 1) \cdot \hat{a} \cdot b}{a_{k_1}^2 \cdot N^{2 - 2k_1\alpha}}.$$

$$\frac{N^{1-(k_1+1)\alpha}}{\left\{ 1 - o(1) + \sum_{k=k_1+1}^{k_2} \left((a_k/a_{k_1}) N^{-(k-k_1)\alpha} - o\left(N^{-(k-k_1)\alpha}\right) \right) \right\}^2}$$

$$= \frac{(k_2 - 1) \cdot \hat{a} \cdot b \cdot N^{(k_1-1)\alpha - 1}}{a_{k_1}^2 \cdot \left\{ 1 - o(1) + \sum_{k=k_1+1}^{k_2} \left((a_k/a_{k_1}) N^{-(k-k_1)\alpha} - o\left(N^{-(k-k_1)\alpha}\right) \right) \right\}^2}.$$

It is clear that the sum in the denominator $\to 0$ as $N \to \infty$ since it is a sum of a fixed number of negative powers of N. In the numerator the exponent $(k_1 - 1)\alpha - 1$ is negative if $\alpha \le 1/k_1$. Hence the whole expression $\to 0$ as $N \to \infty$ if $\alpha \le 1/k_1$, so that $\sum_{t \ge 1} \sum_{(t)} p^{|A \cup B|} \to 0$ as $N \to \infty$.

From the results of the previous two paragraphs and the equation (3), we see that if $E(F_A) \neq 0$ (which certainly holds if $\alpha \leq \alpha_c$) then

$$\frac{E(F_A^2)}{E(F_A)^2} - 1 \to \frac{1}{E(F_A)} \quad \text{as } N \to \infty. \tag{8}$$

We can now wrap up (ii) and (iii). If $\alpha \leq \alpha_c$ then it is easily seen from (6) that $E(F_A) \to \infty$ as $N \to \infty$. Hence we have

$$Pr(F_A = 0) \leq \frac{E(F_A^2)}{E(F_A)^2} - 1 \to \frac{1}{E(F_A)} \to 0$$

as $N \to \infty$. This proves (ii).

Finally suppose $\alpha = \alpha_c$. From (6) we see that $E(F_A)$ has the term α_{k_1} and other terms in negative powers of N. These other terms then all $\to 0$ as $N \to \infty$, and as the number of these terms is fixed it follows that $E(F_A) \to a_{k_1}$ as $N \to \infty$. (iii) follows by (8). $\square$

In the case $\alpha = \alpha_c$, observe that if $a_{k_1} = 1$ then we have not really proved anything. However it is possible in this case to choose $K > 0$ such that if $p = K/N^{\alpha_c}$ then the probability that there are no atomic fail patterns is bounded above by the same constant < 1. This is easily verified.

8.2 Proof of Theorem 6.1

We only give the proof here for A_2. The other proofs follow exactly the same line.

It is easy to show that $\gamma_{A_1,N} = \Theta(N^{5/2})$, so $\beta = 5/2$ and by the above analysis if $\alpha > 5/8$ then almost certainly there is no failed A-pattern. This establishes part (i) of the transition property.

If $\alpha < 5/8$ then, as $E(F_A) = aN^{\frac{5}{2}-4\alpha}$ (4), clearly $E(F_A) \to \infty$ as $N \to \infty$.

Now it is an elementary matter to show that almost all pairs (A, B) of A_1-patterns satisfy $A \cap B = \emptyset$; more precisely, the number of such pairs is $a^2 N^5 - o(N^5)$. Hence

$$\sum_{(0)} p^{|A|+|B|} = \left(a^2 N^5 - o(N^5)\right) p^8 = E(F_A)^2 - o(E(F_A)^2).$$

Therefore

$$\frac{\sum_{(0)} p^{|A|+|B|}}{E(F_A)^2} = 1 - o(1) \to 1 \quad \text{as } N \to \infty.$$

It remains to determine $\sum_{(t)} p^{|A \cup B|}$ for each t, $1 \le t \le 3$. It is easy to show that for each $\mathcal{A}_1$-pattern A in G, $S_1(A) = O(n^3)$, $S_2(A) = O(n^2)$, $S_3(A) = O(n)$. Hence

$$\sum_{t=1}^{3} \sum_{(t)} p^{|A \cup B|} = O(N^{5/2} \cdot n^3 \cdot p^7 + N^{5/2} \cdot n^2 \cdot p^6 + N^{5/2} \cdot n \cdot p^5)$$

$$= O(N^{4-7\alpha} + N^{\frac{7}{2}-6\alpha} + N^{3-5\alpha}).$$

Therefore using (4) we have (provided $E(F_A) \ne 0$, which is true if $\alpha \le 5/8$)

$$\frac{1}{E(F_A)^2} \sum_{t=1}^{3} \sum_{(t)} p^{|A \cup B|} = O(N^{\alpha-1} + N^{2\alpha-3/2} + N^{3\alpha-2}).$$

If $\alpha \le 5/8$ then this quantity $\to 0$ as $N \to \infty$, so

$$\frac{E(F_A^2)}{E(F_A)^2} - 1 \le \frac{1}{E(F_A)} + o(1).$$

If $\alpha < 5/8$ then $E(F_A) \to \infty$ as $N \to \infty$ so the right hand side $\to 0$ as $N \to \infty$. Hence in this case $Pr(F_A = 0) \to 0$ as $N \to \infty$. This proves part (ii) of the transition property.

Finally if $\alpha = 5/8$ then $E(F_A) \to a$ as $N \to \infty$ so $Pr(F_A = 0) \le 1/a + o(1)$. If $p = K/N^{5/8}$ then the constant K could be chosen so that $Pr(F_A = 0)$ is bounded above by some constant < 1. This is part (iii) of the transition property, and the proof is complete. $\square$

References

[1] Erdös, Spencer, "Probabilistic methods in combinatorics", Academic Press, New York, 1974

[2] Feller, W., "An Introduction to Probability Theory and its Applications", Vol. I, Wiley, New York, 1958

[3] Kunde, M., Lang, H.W., Schimmler, M., Schmeck, H., Schröder, H., "The Instruction Systolic Array and its Relation to Other Models of Parallel Computers", in: M. Feilmeier, G. Joubert und U. Schendels (eds.): Parallel Computing '85, North Holland, pp. 491-497, 1986

148

[4] Kung, H.T., "Why Systolic Architectures?", Computer Magazine 15, pp 37-46, 1982

[5] Lang, H.W., "The Instruction Systolic Array, a Parallel Architecture for VLSI", Integration, The VLSI Journal, 4, 65-74, 1986

[6] Negrini, R., Sami, M., Stefanelli, R., "Fault Tolerance for Array Structures Used in Supercomputing", IEEE Computer, pp. 78-87, February 1986

[7] Palmer, E.M., "Graphical Evolution", Wiley, New York, 1985

[8] Sami, M., Stefanelli, R., "Reconfigurable Architectures for VLSI Processing Arrays", Proceedings of the IEEE, Vol. 74, No. 5, pp. 712-722, May 1986

[9] Schröder, H., "The Instruction Systolic Array -A Tradeoff between Flexibility and Speed", to appear in Computer Systems.

Parallel Testing of Parametric Faults in a DRAM

Pinaki Mazumder
Department of Electrical Engineering and Computer Science
University of Michigan
Ann Arbor, MI 48109

Janak H. Patel
Coordinated Science Laboratory
University of Illinois
Urbana-Champaign, IL 61801

This paper presents a testable design of DRAM architecture which allows one to access multiple cells in a word line simultaneously. The technique utilizes the two-dimensional organization of the DRAM and the resulting speed up of the conventional algorithms is considerable. This paper specifically investigates the failure mechanisms in the DRAM with trench-type capacitor. As opposed to the earlier approaches for testing parametric faults that employed sliding diagonal type tests with $O(n^{3/2})$ complexity, the algorithms discussed in this paper are different and have $O(\sqrt{n})$ complexity. These algorithms can be applied externally from the chip and also they can be easily generated for Built-In Self-Test (BIST) applications.

1. Introduction

Semiconductor Dynamic Random-Access Memory (DRAM) is the highest beneficiary of the rapid growth of VLSI technology. As the device feature width is decreasing every year, the DRAM size is quadrupling in every two to three years. Recently Nippon Telegraph and Telephone Company has announced the development of 16M-bits DRAM and by the turn of the decade, it is expected that several manufacturers will fabricate 64M-bits employing 0.5μ technology [3]. This enormous prospect of DRAM development cannot be economically exploited unless cost-efficient testing strategies are evolved to arrest the linear growth of testing cost with the increasing DRAM size. You and Hayes [8] have proposed the concept of parallel testing by partitioning an n-bit memory array into two or more subarrays and concurrently testing them. They have reconfigured the memory subarray of size s bits into an s-bits cyclic shift register where the data recirculate whenever a read operation is made. The reconfiguration has been done by introducing pass transistors on the bit lines which deteriorate both the sensitivity of the sense amplifiers by V_T (the threshold voltage of the MOS devices) and the access time of the DRAM in normal mode of operation. Also, their technique tests two subarrays simultaneously and detects the occurrence of a fault by comparing two cells which are simultaneously read. Thus if both the cells are identically faulty, it fails to detect the fault. Their test procedure is generated within the chip, involving a high overhead. Moreover, the test procedure detects only parametric faults and the proposed design of parallel testing cannot be adapted for a large class of functional faults, like coupling, and static-and-dynamic pattern-sensitive faults [1].

The intent of this paper is to employ a novel, general-purpose testable design which allows one to test multiple cells in a word line in parallel. The technique is not constrained to any specific test procedure and the test vectors can be applied externally [5] or generated by Built-In Self-Testing (BIST) circuit [6]. It can speedup any existing test procedure of $O(n)$ test length by a factor of $O(\sqrt{n})$. The paper investigates the parametric faults in a DRAM and proposes $O(\sqrt{n})$ test procedures to test the different parametric faults. The proposed design-for-testability technique employs very little overhead (only $2\sqrt{n} + 2\log_2 n + 11$ transistors) and needs only one transistor to fit within the pitch width of the memory so that it fits within 3λ inter-celler pitch width of the vertically-integrated, high-density DRAM.

The rest of the paper has been organized as follows. Section 2 enumerates the different faults which occur due to variation in processing technology in a DRAM, Section 3 proposes a new design-for-testability memory architecture, and the algorithms for testing these parametric faults are given in Section 4. The main contribution of this work is to propose a design-for-testability technique and to demonstrate how the parametric faults in a DRAM using trench-type memory cells can be tested in parallel.

2. Design for Testability

The organization of the testable RAM with augmented hardware is shown in Figure 1. The memory is organized as a $b \times w = n$ matrix, where b is the number of bit lines and w is the number of word lines. The normal 1-out-of-b decoder is modified to select multiple bit lines during test mode. In test mode, it divides the b bit lines into g groups such that the bit line i belongs to group j, where $j = i \pmod{g}$. Thus, a write operation in test mode results in writing the content of the data-in buffer on all cells at the crosspoints of the selected word line and the bit lines in group j. In read mode, the content of the cells located at the crosspoints of the selected word line and the selected bit lines group (say j) are read in parallel. Thus a 0 or 1 is entered in data-out buffer if all the multiple-accessed cells contain 0 or 1, respectively. If the content of all the cells are not identical, the data-out buffer may store a 0 or 1. It may be noted that it is not correct to assume that the resulting operation will be a wired-OR or wired-AND. On the contrary, it depends on the number of 0's and 1's in the multiple-accessed cells. If almost all the multiple-accessed cells contain 1's except a very few which contain 0's, then a 1 will be entered in data-out buffer when the cells are read in parallel. On the contrary, a 0 would have entered if almost all the cells contained 0's and a few cells contained 1's. To circumvent this problem, the content of all the cells in a group are compared by a parallel comparator. In the event that all the cells do not have identical contents, the parallel comparator triggers an error latch to indicate that a fault has been detected by the test.

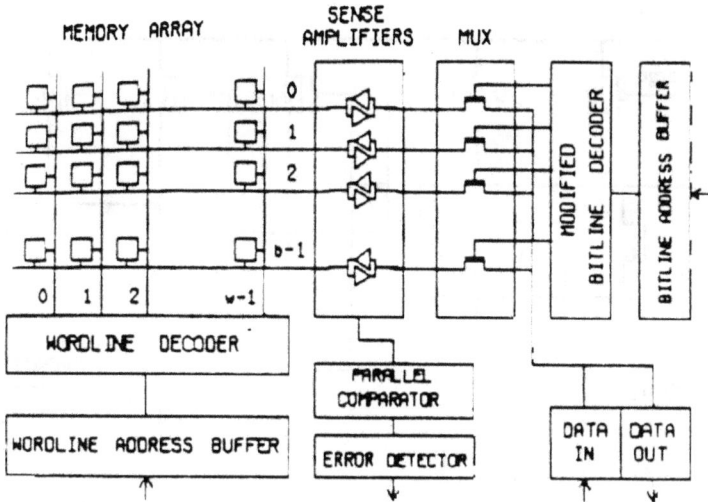

Figure 1: Testable RAM Organization

In contrast to the bit-line decoder, the word-line decoder is not modified and word lines are accessed one at a time. A parallel-word read operation is not meaningful because two or more cells will be sensed by the same sense amplifier resulting in a wired-OR or a wired-AND operation. A multiple-cells write through multiple word lines would require the sense amplifier to drive many cells at a time. For moderate size DRAM, this introduces high write cycle time delay. By increasing the physical size of the sense amplifier driver, delay can be improved to a certain extent. However, this increases power consumption and because of its large gate capacitance, the sense-amplifier slew rate decreases, and thereby the memory cycle time degrades for both read and write operations.

2.1 Modified DRAM Circuit

The modified CMOS decoder circuit is shown in Figure 2. Transistors $Q_1, \cdots, Q_7$ with the transmission gate constitute a normal decoder circuit. In the clock-phase ϕ_P, the transistor Q_1 turns on to precharge the common line connected to the address decoding transistor. If all the address bits, $a_0, \cdots, a_{k-1}$ are zeroes, transistor Q_6 pulls up the OUT to ONE, and the corresponding bit line is selected. The signal ϕ_{EN} enables the transmission gate so that the decoder selects the bit line only after all address lines have changed. Transistors Q_8 and Q_9 have been added so that in the test mode the decoder output can be selected by applying SELECT=0 independent of the input address. In the normal mode of operation, SELECT=1 and the decoder output is selected by the address input $a_0, \cdots, a_{k-1}$. The modified decoder is simulated using SPICE and the degradation in decoding time due to addition of the extra transistors have been found to be approximately 0.1 nsec.

152

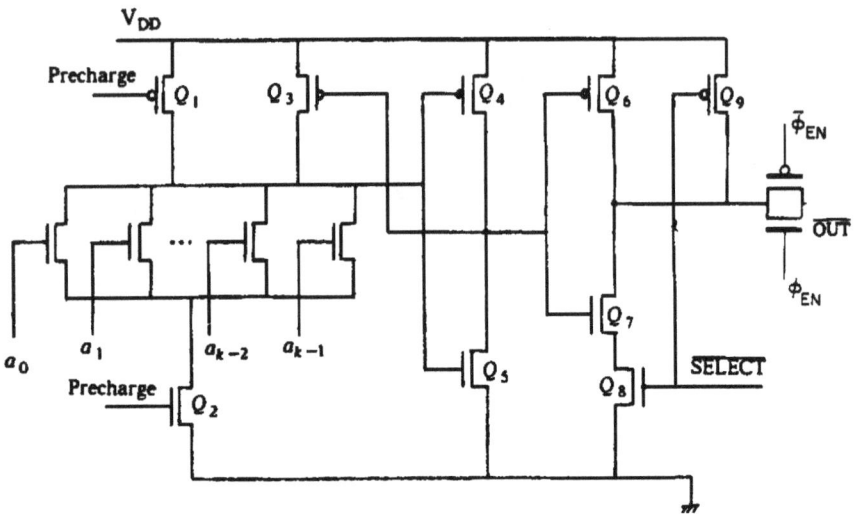

Figure 2: Modified Decoder Circuit

The parallel comparator and error detector monitors the output of sense amplifiers connected to bit lines which are selected in parallel and detects the concurrent occurrence of either 0's or 1's. If a selected bit line is different from the others it triggers the error latch indicating the occurrence of a fault. Figure 3 shows the parallel comparator and error detector. The p-channel transistors $T_1, \cdots, T_{m-1}$ are connected in parallel and detect concurrent occurrence of 1 in the bit lines. The n-channel transistors $P_1, \cdots, P_{m-1}$ are also connected in parallel and detect concurrent occurrence of 0 in the bit lines. Transistors T_0 and P_0 are the precharge transistors while transistor P_m is the discharge transistor which remains cutoff during precharge phase and turns on during discharge clock phase $\phi 2$. Since the bit lines are divided into $g=2$ classes, pass transistors are introduced to allow only the odd or even bit lines being compared simultaneously. Signals L_1 and L_2 select these bit lines. Transistors S_0, S_1 and S_2 form a coincidence detector. If all the selected bit lines are 0 or 1, then either S_1 or S_2 conducts and the output of the detector is 0. The output of the coincidence detector is connected to the error latch consisting of transistors $V_0, \cdots, V_3$. The error latch output is ERROR=0, when the selected bit lines are identical. If the bit lines are not identical, then both S_1 and S_2 remain cutoff and the detector output is 1. This triggers the error latch setting its output to ERROR=1. During the write phase and normal mode of operation, the error latch is clamped to zero by V_4. The error detector is inhibited by the discharge transistor P_m during the start of the read phase when the sense amplifiers outputs are not identical because of sluggish changes in some of the sense amplifiers.

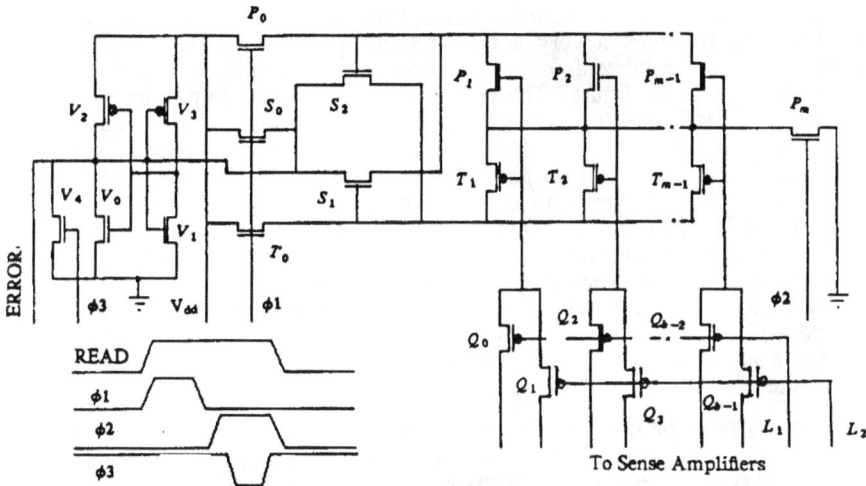

Figure 3: Parallel Comparator with Error Detector ($g=2$)

3. Parametric Faults in a DRAM

A typical configuration of the trench-type memory cells [7] with p^+ sidewall doping is shown in Figure 4. The access transistor is a PMOS transistor located within an n-well diffused over the p^+ substrate. A deep trench capacitor extends from the planarized surface through the n-well into the p^+ substrate. A conducting strap connects the p^+ doped polysilicon storage electrode inside the trench to the p^+ source region of the access transistor. With a thin composite insulator separating the polysilicon from the bulk silicon surrounding the trench, the storage capacitance comes primarily from the portions of the four trench sidewalls in the p^+ substrate region and the trench bottom. Some additional capacitance

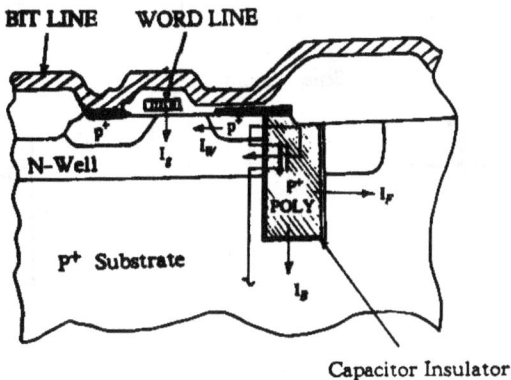

Figure 4: Memory Cell with Trench-Type Storage Capacitor

results from the four trench walls intersecting the n-well. The grounded p^+ substrate provides a very solid reference potential to the capacitor plate. The leakage currents due to process parameter variation have also been shown in the figure. These currents are divided into four components: Weak Inversion Current (I_W) from the storage area to the bit line, Field Inversion Current (I_F) between the two adjacent cells, Gate Leakage Current (I_G) due to pin holes defect in the gate-oxide, and the Dark Current (I_B) between the storage area and the p-type substrate. The weak-inversion current can degrade a stored ZERO by flow of minority carriers from the trench capacitor to the positively biased bit lines. Dark current which flows from the trench capacitor to the p^+ substrate can degrade stored ONE. It may also be observed that the cell forms a vertical parasitic FET device which occurs between the storage node and the substrate along the trench wall, gated by the node polysilicon as shown in Figure 4.

The effects of the leakage currents result in parametric faults such as the bit-line voltage imbalance and the bit-line to word-line crosstalk. The other types of parametric faults emanate due to a wide variation of timing signals in the decoding, address buffer, and peripheral circuits, such as the sense amplifiers. Incorrect timing between decoder enable, precharge clock, and decoder address signals may cause multiple address selection [4]. The above parametric faults are described below.

3.1 Bit-Line Voltage Imbalance

A typical memory array organization utilizes the differential amplifiers for sensing the signal partitioning of each array into two identical subarrays (called left and right in this paper) as shown in Figure 5. When most of the cells in the left-half of the memory subarray contain one type data (say ONE) and most of the cells in the right-half contain the opposite type data (ZERO), during a read cycle the precharge voltage on the two halves of the bit lines will be different. This is illustrated in Figure 5, where all the cells connected to the bit line B^L_i contain ONE and all the cells connected to the bit line B^R_i, except one also connected to the word line W_j, contain ZERO. If the cell containing ONE in the right-half of the memory is read, bit lines B^L_i and B^R_i will be

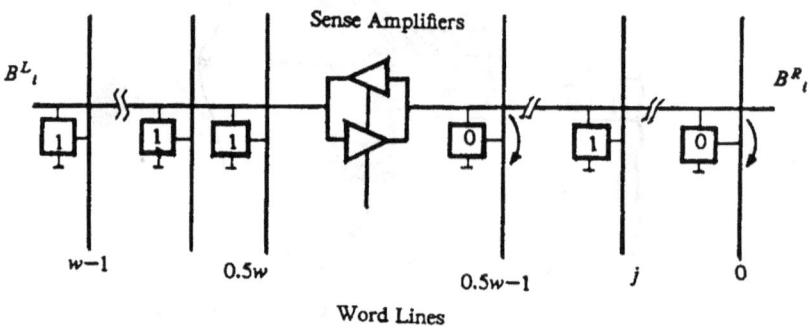

Figure 5: Bit-Line Voltage Imbalance

precharged to a voltage V_p. But due to the weak-inversion currents in the right-half cells, the precharged level will be degraded to $V_p - \sum_{j=0}^{3w-1} I_{W_j} \tau$, where I_{W_j} is the weak inversion current in C_{ij} and τ is the time interval for precharge and read. Consequently, when the word line W_j is selected, even though the selected cell contains ONE, it will be read as ZERO by the sense amplifier (because the ratio of bit-line capacitance and cell capacitance is usually greater than 15).

3.2 Bit-Line to Word-Line Crosstalk

From Figure 6 it can be seen that there is an overlap between the bit line and the word line, since they are orthogonal to each other. This overlapping forms a coupling between the bit lines and the word lines so that when the bit lines voltages change due to precharging and restoring operations during a read cycle, the unselected word lines may be inadvertently turned on. This coupling is maximum if all the cells in the selected word line contain ONE and if some of the cells of the coupled unselected word line contain ZERO, which will be degraded due to the weak inversion current of the access transistors. In the precharging phase, at first the bit lines voltage increases from ZERO to V_p which is coupled as a noise voltage V_{n_j} to a word line W_j. If all the cells in the word line W_i contain ONE, then if W_i is selected, its voltage will increase coupling a noise voltage V_{v_j} to W_j. The effect of these superimposed noise voltages may generate a sufficient weak inversion current such that a stored ZERO in a cell on W_j may be degraded.

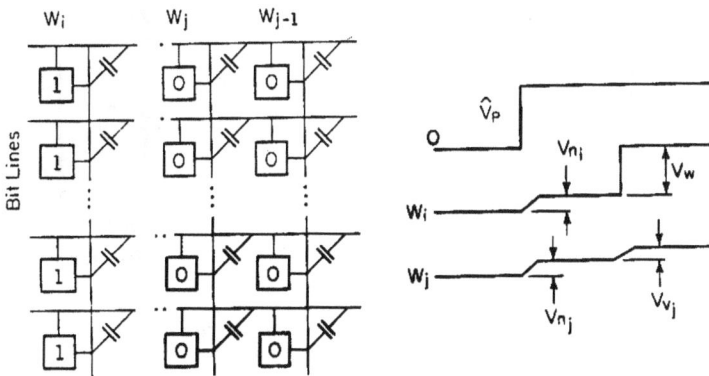

Figure 6: Bit-Line to Word-Line Crosstalk

3.3 Single-Ended Write

In a DRAM employing single-ended write technique, a single I/O line is used to write into the bit lines. In Figure 7 it can be seen that writing on the right bit line $B^R{}_i$ is controlled by the I/O line, while writing on right bit line $B^L{}_i$ is controlled by the sense amplifier. The ZERO level on B_i is determined by the input driver, but the ZERO level of B_i is determined by the sense amplifiers. Thus the level of ZERO in the two halves may be different.

3.4 Multiple Selection

In the decoder circuit in Figure 3, the precharge clock, ϕ_p, and the decoder_enable clock, ϕ_{EN}, should be nonoverlapping in the sense that always ϕ_p AND $\phi_{EN} = 0$. If due to incorrect timing they overlap, multiple selection may occur.

3.5 Transmission-Line Effect

In two-layer interconnect technology, either the bit lines or the word lines are made of metal and the other polysilicon or diffusion. Usually, the poly and diffusion lines have quadratic signal propagation delay. Because of the high resistivity in these interconnects the cells at the periphery of the chips, away from the sense amplifiers, are delivered a weak signal and thereby may fail. By inserting the repeaters at suitable intervals, the signal strength and delay may be improved, but it adds complexity to the layout. In order to check that all the cells in the array satisfy the limits of the stipulated memory cycle time, the transmission line effect should be tested.

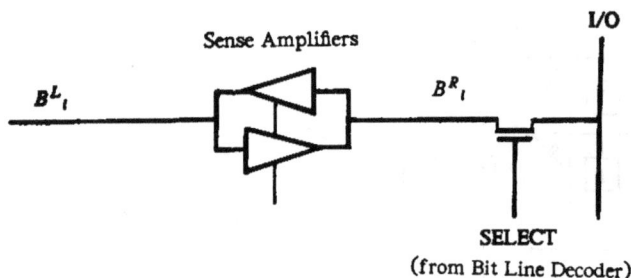

Figure 7: Single-Ended Write

4. Testing Strategy and Algorithms

In order to test all of the above faults, it is necessary to identify the circumstances in which each of them is likely to be maximum. It can be easily noted that the field inversion current (I_F) which occurs between two adjoining storage cells is maximum if the four adjacent cells of a base cell contain opposite data to that of the base cell, i.e., a checker board type pattern can test the effect of a field inversion current. This has been illustrated in Figure 8, where the base cell contains a ZERO and its four adjoining cells contain ONE. Similarly, the effects of a dark current and gate short can be tested by the checker board pattern, because the presence of these leakage currents manifests in the form of cell stuck-at ZERO or ONE. The following algorithm is a parallel version of the checker-board test.

The effect of a weak inversion current is maximum when all but one cell connected to a bit line contain ZERO. If the cell which contains ONE is attempted to read, the weak inversion currents in other cells will tend to degrade the precharge level of the bit line and thereby the cell containing ONE will be sensed as ZERO by the sense amplifier. Since the bit line capacitance is typically 10-20 times the capacitance of an individual cell, the stored ONE may not be sufficient to replenish the degraded precharge level. The testing strategy needs to test each memory cell so that when it is ONE all its bit lines neighbors will contain ZERO.

In order to test the bit-line voltage imbalance, it is necessary to write ZERO (ONE) on the cells at the bit line on the left-half of the subarray and write ONE (ZERO) on the cells at the bit line on the right-half of the subarray. Thus the test to detect the weak-inversion current can be utilized to test the bit-line voltage imbalance by testing the left and right subarrays with opposite background data. It may be noted that the test also detects faults due to single-ended write. Algorithm 2 tests all the above faults.

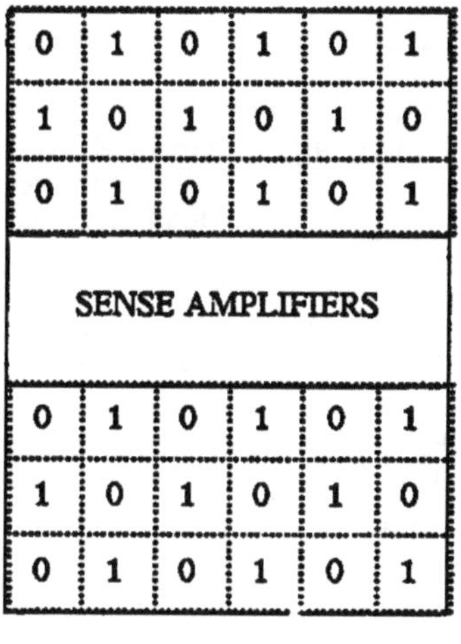

0	1	0	1	0	1
1	0	1	0	1	0
0	1	0	1	0	1
SENSE AMPLIFIERS					
0	1	0	1	0	1
1	0	1	0	1	0
0	1	0	1	0	1

Figure 8: Checker-Board Pattern

Algorithm 1: Parametric Checker-Board Test

(1)   Use complementing address sequence from word line W_0 until all word lines are scanned, write in two steps a pattern of $(01)^*$ if the word line is even and write in two steps a pattern of $(10)^*$ if the word line is odd.

(2)   Freeze the clock for the entire refresh interval, τ_F for testing *static refresh*.

(3)   Use a complementing address sequence from word line W_0 until all word lines are scanned; compare in parallel all even and odd bit lines to check ERRROR = 0.

(4)   Read continuously any arbitrary word line for the entire refresh interval, τ_F, to check ERROR = 0. This test checks the effect of temperature rise and tests the *dynamic refresh*.

(5)   Use the complementing address sequence from word line W_0 until all word lines are scanned; compare in parallel all even and odd bit lines to check ERRROR = 0.

(6)   Read continuously another distinct word line for the entire refresh interval, τ_F, to check ERROR = 0. This test checks the effect of temperature rise and tests the dynamic refresh.

(7)   Use the complementing address sequence from word line W_0 until all word lines are scanned; compare in parallel all even and odd bit lines to check ERRROR = 0.

(8)   Repeat Steps 1-7, with opposite data.

Algorithm 3 which runs a marching pattern of ONE at the background of ZERO and a marching pattern of ZERO at the background of ONE in each word-line will detect the multiple-access faults in the word-line decoder by comparing the read data with the expected data. Since the algorithm employs parallel writing by accessing all the even bit lines and odd bit lines in a single memory cycle, multiple access in the bit-line decoder will not be tested by the Algorithm 2. A separate algorithm is needed to test the bit-line decoders and is given in Algorithm 3.

It can be seen that the Algorithm 1 takes altogether $10w\,\tau_A + 6\tau_R$ time to complete all the steps, where τ_A is the average memory cycle time. Algorithm 2 takes $20w\,\tau_A + 2\tau_R$ time to test the entire RAM. Finally, the Algorithm 3 takes $4(w+b+1)\tau_A$ time to test the multiple-access faults in the decoder logic. Hence altogether $(34w+4b+4)\tau_A + 8\tau_R$ time is needed to test all the parametric faults in the DRAM. Table 1 depicts all types of parametric faults and how they are covered by these algorithms.

Algorithm 2: Parallel Parametric Walking Test

(1) Initialize the entire memory writing ZERO in all locations.

(2) Select two arbitrary word lines W_i and W_j and read them alternately for one refresh interval.

(3) From word line W_0 until all word lines are scanned, compare in parallel all even and odd bit lines to check ERRROR = 0.

(4) Initialize the entire memory, writing ONE in all locations.

(5) Select two arbitrary word lines W_p and W_q and read them alternately for one refresh interval.

(6) From word line W_0 until all word lines are scanned, compare in parallel all even and odd bit lines to check ERRROR = 0.

(7) Initialize the memory such that the left subarray contains ZERO in all locations and right subarray contains ONE in all locations.

(8) For all word lines starting from W_0, write a pattern of $(01)^*$ in the selected word line, parallel compare and check if ERROR = 0; initialize all the cells in the selected word line to ZERO if it is on left-half, otherwise to ONE.

(9) For all word lines starting from W_0, write a pattern of $(10)^*$ in the selected word line, parallel compare and check if ERROR = 0; initialize all the cells in the selected word line to ZERO if it is on left-half, otherwise to ONE.

(10) Repeat steps 7-9 with complementary bit patterns.

Table 1: Algorithms and Their Coverage of Parametric Faults

Fault Type	Algorithm 1	Algorithm 2	Algorithm 3
Weak Inversion Current	No	Yes	Yes
Field Inversion Current	Yes	No	No
Dark Current	Yes	No	No
Gate Short	Yes	Yes	Yes
Multiple Selection	No	No	Yes
Single-Ended Write	Yes	Yes	Yes
Bit-Line Voltage Imbalance	No	Yes	No
Bit-Line to Word-Line Crosstalk	No	Yes	No
Transmission-Line Effect	Yes	No	No

Algorithm 3: Bit-Line and Word-Line Decoder Test

/* Bit-Line Decoder Multiple-Access Test */

(1) Write in parallel ZERO in all cells on the word line W_j.

(2) Read and compare in parallel all the cells on W_j.

(3) Starting from the cell at the cross point of B_0 and W_j, for each cell on W_j, at first write ONE and read the cell (one cell at a time in ascending order of the word line).

(4) Starting from the cell at the cross point of $B_{\sqrt{n}-1}$ and W_j, for each cell on W_j, at first write ONE and read the cell (one cell at a time in descending order of the bit line).

/* Word-Line Decoder Multiple-Access Test */

(5) Write in parallel ZERO in all cells on the bit line B_i.

(6) Read and compare in parallel all the cells on B_i.

(7) Starting from the cell at the cross point of W_0 and B_i, for each cell on B_i, at first write ONE and read the cell (one cell at a time in ascending order of the word line).

(8) Starting from the cell at the crosspoint of $W_{\sqrt{n}-1}$ and B_i, for each cell on B_i, at first write ONE and read the cell (one cell at a time in descending order of the word line).

---

5. Conclusions

The main objective of this paper is to propose a modified DRAM architecture which enhances the speed for testing the occurrence of parametric faults in the DRAM employing trench-type capacitor. A novel testable architecture is proposed to test multiple cells in a word line simultaneously resulting in a speed up by a factor of $O(\sqrt{n})$. The testable design has been proposed to fit within the interceller pitch width of 3λ in the single-transistor DRAM. It employs only an additional $2\sqrt{n}+2\log_2 n+11$ transistors and has very low overhaed. Above all, the technique is not tailored to test only one specific test procedure and it can speedup the conventional test algorithms for functional faults [5,6].

A number of parametric faults that manifest due to a variation of processing parameters and also due to a critical circuit design, are enumerated in this paper. Leakage currents in a DRAM have been identified, and test procedures have been designed to test in parallel the faults due to these leakage currents. In addition to the faults due to leakage currents, circuit related design weaknesses are tested by the test procedures described here. Unlike the earlier approaches [2] which employ diagonal tests that have $O(n^{3/2})$ complexity, the algorithms described here have $O(\sqrt{n})$ complexity, and thereby a dramatic improvement by a factor proportional to the size of the DRAM can be achieved. For a 16M-bit DRAM organized into sixteen quadrants, the diagonal test algorithm will need more than a few hours as opposed to the proposed algorithms which will need only a few milliseconds. Moreover, these algorithms can be easily generated for Built-In Self-Test (BIST) applications.

Acknowledgements

This research has been partially supported by the Semiconductor Research Corporation under the contract number SRC 86-12-109.

References

1. M. S. Abadir and H. K. Reghbati, "Functional Testing of Semiconductor Random Access Memories," *ACM Computing Surveys* **15**(3) pp. 175-198 (September 1983).

2. M. A. Breuer and A. D. Friedman, in *Diagnosis and Reliable Design of Digital Systems*, Woodland Hills, Los Angeles (1976).

3. L. L. Lewyn and J. D. Meindl, "Physical Limits of VLSI dRAM's," *IEEE Journal of Solid-State Circuits* **SC-20**(1) pp. 231-241 (February 1985).

4. T. C. Lo and M. R. Guidry, "An Integrated Test Concept for Switched-Capacitor Dynamic MOS RAM's," *IEEE Journal of Solid-State Circuits* **SC-12**(6) pp. 693-703 (December 1977).

5. P. Mazumder, J. H. Patel, and W.K. Fuchs, "Design and Algorithms for Parallel Testing of Random Access and Content Addressable Memories," *Design Automation Conference* **24** pp. 688-694 (July 1987).

6. P. Mazumder and J. Patel, "An Efficient Built-In Self-Testing of Random Access Memory," *Proceedings of International Test Conference*, (August 1987).

7. A. H. Shah and et al., "A 4-Mbit DRAM with Trench-Transistor Cell," *IEEE Journal of Solid-State Circuits* **SC-21**(5) pp. 618-627 (October 1986).

8. Y. You and J. P. Hayes, "A Self-Testing Dynamic RAM Chip," *IEEE Journal of Solid-State Circuits* **SC-20**(1) pp. 428-435 (February 1985).

Meshes with Reconfigurable Buses

Russ Miller,[1] V. K. Prasanna-Kumar,[2] Dionisios Reisis,[3]

and Quentin F. Stout[4]

State University of New York at Buffalo.
University of Southern California.
University of Michigan Ann Arbor.

Abstract

This paper considers a mesh with reconfigurable bus (reconfigurable mesh), that consists of a VLSI array of processors overlaid with a reconfigurable bus system. The N PEs are laid out as a square mesh in $O(N)$ VLSI area. In addition to the 4 near neighbor mesh connections between processors, the reconfigurable bus system can be used to dynamically obtain various interconnection patterns amongst the PEs. In fact, the reconfigurable mesh can be used as a universal chip capable of simulating any $O(N)$ area organization without loss in time. The reconfiguration scheme also supports several parallel techniques developed on the CRCW PRAM model, leading to asymptotically superior solution times compared to those on the mesh with multiple broadcast buses, the mesh-of-trees, and the pyramid computer. These features are illustrated by presenting efficient reconfigurable mesh algorithms to solve a variety of problems involving graphs and digitized images.

1   Introduction

VLSI technology offers an environment for constructing parallel processing systems which consist of thousands of processors. A very attractive interconnection scheme is the *two-dimensional mesh-connected computer (mesh)* because of its simplicity, regularity, and the fact that the interconnect wires occupy only a fixed fraction of the area no matter how large the mesh.

[1] Department of Computer Science State University of New York at Buffalo, Buffalo, New York 14260.

[2] Department of Electrical Engineering-Systems SAL 344, University of Southern California Los Angeles, CA 90089-0781.

[3] Department of Electrical Engineering-Systems, University of Southern California Los Angeles, CA 90089-0781.

[4] Department of Electrical Engineering and Computer Science University of Michigan Ann Arbor, Michigan 48109.

However, since a *mesh of size* N is configured as an $N^{1/2} \times N^{1/2}$ grid of processors, the *communication diameter* (maximum of the minimum distance between any two processors in the network) is $\Theta(N^{1/2})$. Therefore, a lower bound on the time to solve problems that involve combining data residing in processors far apart on a mesh of size N is $\Omega(N^{1/2})$. In order to achieve faster solutions to problems, researchers have studied related organizations which add additional communication links to the mesh, while trying to keep the wire area small. Such organizations include the pyramid computer [7,16,27,28], the mesh-of-trees [12,19], and meshes with broadcast buses [3,4,20,25,26]. These organizations are static in nature, in that the communication patterns between processors cannot be altered during the execution of an algorithm.

For many algorithms though, it is desirable that more than one interconnection scheme be present during their execution. In this paper we consider a reconfigurable VLSI array that consists of a mesh-connected array of processors connected to a reconfigurable bus system. We show that this *reconfigurable mesh* can provide more efficient solutions than other mesh-based architectures for some graph and image problems. We also show that the reconfigurable mesh can act as a universal chip, simulating other VLSI organizations with equivalent area without a loss of time, and that the reconfigurable mesh can be used to simulate certain fundamental techniques that have been developed for the *concurrent read, concurrent write (CRCW) parallel random access machine (P-RAM)*.

In Section 2, we define the reconfigurable mesh and discuss related models. The algorithms presented in this paper rely on fundamental abstract data movement operations. Section 3 illustrates the power of the reconfiguration scheme in basic operations on data as well as in sparse data movement by giving efficient implementations for operations such as random access read/write, data reduction, and parallel prefix. Section 4 is devoted to embeddings of parallel organizations and the use of fundamental data movement operations and the reconfiguration scheme to develop efficient parallel solutions to several graph problems. Section 5 concludes the paper with some remarks.

Due to page limitations, a number of algorithms and simulation results have been omitted, and proofs have either been sketched or omitted. All algorithms, results, and details of all proofs appear in the final (journal) version of this paper [15].

2 Mesh with Reconfigurable Bus

A *mesh with reconfigurable bus (reconfigurable mesh) of size* N is a reconfigurable VLSI array that consists of N processors configured as a mesh of size N, with each processor connected to the reconfigurable broadcast bus. The bus, like the mesh, is constructed as an $N^{1/2} \times N^{1/2}$ grid. The N processors are situated at the grid intersection points of the bus. Each bus link between processors has a *switch* embedded in it, where the two processors at each end of the link can control the switch. These switches allow the broadcast bus to be divided into subbuses, where each subbus can function as a smaller reconfigurable mesh. Other than the buses and switches, the reconfigurable mesh is similar to the standard mesh in that it operates in SIMD (single instruction stream, multiple data stream) mode and has $O(N)$ area, under the assumption that processors, switches, and single links have constant size.

Subbuses are typically shared by multiple processors. The way in which simultaneous writes by multiple processors to a subbus are handled by the architecture has an affect on the algorithms that we design, and the running times of the such algorithms.

We consider that at most one processor is allowed to broadcast on a subbus. As it will be shown, the reconfigurable bus system can simulate without loss in time a bus system with the following properties:

1. Multiple identical values may be broadcast simultaneously to the bus.

2. Multiple (not necessarily identical) values may be broadcast simultaneously to the bus where the logical OR of these values is taken.

It should be noted that a value is assumed to consist of $O(\log N)$ bits. We also consider computational models that differ in the assumption regarding the delay that a broadcast requires. The *unit-time delay model* will assume that all broadcasts take $\Theta(1)$ time, as is the assumption in [3,20,22,25,26] for models that assume various broadcasting strategies. We will also consider the *log-time delay model* in which it is assumed that each broadcast takes $O(\log s)$ time to reach all processors connected to its subbus, where s is the maximum number of switches in a minimum switch path between two processors connected on the bus.

Major advantages of the reconfigurable mesh are as follows.

1. Buses can be used to speed up parallel arithmetic and logic operations among data stored in different processors. The reconfiguration scheme supports several techniques on the CRCW P-RAM model, leading to the same time performance as in the P-RAM model having N processors.

2. The reconfigurable mesh provides an environment for efficient sparse data movement operations.

3. A significant asymptotic improvement can be achieved in the running times of algorithms that solve several problems on the reconfigurable mesh compared to efficient algorithms for the mesh-of-trees, pyramid, and mesh with static broadcast buses.

4. The reconfigurable mesh can act as a universal chip, in that VLSI

organizations with equivalent area can be simulated without loss in time.

Many of the algorithms introduced in this paper will continually reconfigure the system by setting the switches to give the desired substructures. A number of algorithms have previously been derived that exploit row and column broadcasts [20,22,25,26]. Many of these algorithms use row (column) broadcasts simultaneously within every row (column). This technique can be accomplished on the reconfigurable mesh by having each processor set its switches to disconnect its column (row) links, creating $N^{1/2}$ separate row (column) buses. That is, processor $P_{i,j}$ of the reconfigurable bus has access the broadcast buses of row i and column j. Notice that by setting the switches properly, sub-row (column) buses can be created within each row (column), sub-reconfigurable meshes can be created, a global broadcast bus can be created, a bus can be created within sets of contiguously labeled processors, and so forth.

The reconfigurable mesh is conceptually similar to the *bus automaton* [5], which consists of a cellular automaton augmented with a locally switchable global communication network. Problems studied for the bus automaton involve parallel parsing of formal languages, parallel processable cryptographic methods, and immediate recognition and computation of certain properties. The reconfigurable mesh is also similar to the *polymorphic-torus network* [13], except that in the polymorphic-torus network there are arbitrary crossbars at each processor to control connections between the north, south, east, and west bus ports. Therefore, the more complex logic of the switches creates more powerful, yet slower, processors on the polymorphic-torus network. Finally, the reconfigurable mesh is similar to the latest version of the *Content Addressable Array Parallel Processor (CAAPP)* [30], which consists of a mesh, augmented with a global some/none signal that computes the logical OR of a single bit of all processors after each instruction, and the *Corterie network*, which is a reconfigurable mesh bus, as defined in this paper, that takes the logical OR of any values simultaneously written to a subbus. To the best of our knowledge, this version of the CAAPP was developed independently of the reconfigurable mesh.

3   Data Movement Operations

Data movement operations form the foundation of numerous algorithms for machines constructed as an interconnection of processors. In fact, algorithms that are designed in terms of fundamental *abstract data movement operations (ADMOs)* [14,18] provide the possibility of portability to architecturally related machines. This notion can be viewed as the parallel analogue to designing serial algorithms in terms of *abstract data types (ADTs)*. Therefore, the algorithms given in this paper will be described in terms of ADMOs. In this section, a variety of these operations are given for the reconfigurable mesh.

A powerful data movement operation that has been exploited recently on a number of architectures is *parallel prefix*, which can be used to sum values, broadcast data, solve problems in image processing, solve graph problems, and so forth. Assume processor P_i, $0 \leq i \leq N - 1$, initially contains data element a_i. Then the *parallel prefix problem* requires every processor P_i, $0 \leq i \leq N - 1$, to determine the ith initial prefix $a_0 \otimes a_1 \otimes \cdots \otimes a_i$, where $\otimes$ is a binary associate operator. The algorithm given below assumes a row-major indexing scheme of the processors of a reconfigurable mesh of size N.

Proposition 3.1 *Given a set $S = \{a_i\}$ of N values, distributed one per processor on a reconfigurable mesh of size N so that processor P_i contains a_i, $0 \leq i \leq N - 1$, and a unit-time binary associative operation $\otimes$, in $\Theta(\log N)$ time using the unit-time delay model, and in $\Theta(\log^2 N)$ time using the log-time delay model, the parallel prefix problem can be solved so that each processor P_i knows $a_0 \otimes a_1 \otimes \ldots \otimes a_i$.*

Proof: The basic idea of the algorithm is to perform parallel prefix within each row of the reconfigurable mesh so that each processor knows the initial prefix of those values restricted to its row. Next, in the last column of the reconfigurable mesh perform parallel prefix to determine row-wise prefix solutions. Finally, within each row, broadcast the prefix of the previous rows so that all processors can update their entry appropriately. The details of the algorithm are now given.

The first step is to perform parallel prefix in every row, by an algorithm that is based on being able to set switches appropriately at each iteration of this step so as to group processors into disjoint linear *strings* (1-dimensional meshes) of processors. During the ith iteration of this step, each row is grouped into disjoint strings of size 2^i, $1 \leq i \leq \log_2 N^{1/2}$. Within every string of size $k = 2^i$, assume the processors are $P_1, P_2, \ldots, P_k$. Within every string, processor $P_{k/2}$ broadcasts its value on its string bus, and all processors P_j, $k/2 \leq j \leq k$, perform the operation $a_j \leftarrow a_j \otimes a_{k/2}$. After $\Theta(\log N)$ iterations, every processor will know the initial prefix of the values restricted to its row, and in particular every processor $P_{(r+1)N^{1/2}-1}$, $0 \leq r \leq N^{1/2} - 1$, will know $a_{rN^{1/2}} \otimes a_{rN^{1/2}+1} \otimes \cdots \otimes a_{(r+1)N^{1/2}-1}$, the final initial prefix restricted to row r.

Next, perform the parallel prefix, as just described, on the values in column $N^{1/2} - 1$ of the mesh. After this is complete, every processor $P_{(r+1)N^{1/2}-1}$, $0 \leq r \leq N^{1/2} - 1$, will know its initial prefix with respect to S. That is processor $P_{(r+1)N^{1/2}-1}$, $0 \leq r \leq N^{1/2} - 1$, now knows $a_0 \otimes a_1 \otimes \ldots \otimes a_{(r+1)N^{1/2}-1}$. To finish the algorithm, every processor in column $N^{1/2} - 1$ gets the initial prefix just computed from the processor to its north, and broadcasts this value to all processors in its row, which update their current prefix value to determine the final answer.

The running time of the algorithm is dominated by the time it takes for $\Theta(\log N)$ communication operations and internal computations. $\square$

A technique called *bus splitting*, shows how the processors can exploit the ability to locally control the effective size of subbuses. We also show how the splitting mechanism can be used to simulate bus systems with different properties. In propositions 3.2 and 3.4 we show that the reconfigurable bus can perform as a bus on which if more than one values are broadcast, the logical OR of these values is taken. To illustrate how the reconfigurable bus can perform as a bus on which broadcast of multiple identical values is allowed, we consider the algorithm in proposition 3.3.

The next algorithm is used to compute the logical OR of a single bit of data stored one per processor in the ith row of the reconfigurable mesh and store the result of the operation in processor $P_{i,0}$, for all $0 \leq i \leq N^{1/2} - 1$.

Proposition 3.2 *Given a reconfigurable mesh of size N, in which each processor stores a bit of data, the logical OR of the data in each row (column), or the entire reconfigurable mesh, can be determined in $\Theta(1)$ time using the unit-time delay model, and in $\Theta(\log N)$ time using the log-time delay model.*

Proof: Each processor $P_{i,j}$ that has a 1 as its data value uses *bus splitting* to divide its row bus. This is performed by having each such processor set the bus control switch to its right to disconnect its row bus. Next, each processor $P_{i,j}$ that has a 1 as its data value broadcasts the 1 on its subbus. Processor $P_{i,0}$, for all $0 \leq i \leq N^{1/2} - 1$, will read from the row subbus that it is connected to the leftmost 1 in its row, if such a value exists. If it is desired to find the logical OR of the data stored in all processors, then first the OR of each row is determined, and then the OR of these values in the first column is determined. $\square$

In [8] has been shown that the exclusive OR function cannot be computed in $O(1)$ time using polynomial number of processors. Further analysis of the exclusive OR computation on the reconfigurable mesh, as well as related results, are presented in [15].

Many algorithms are designed to reduce data at intermediate stages of the algorithm. It is, therefore, often useful to be able to efficiently perform fundamental operations on reduced sets of data. The maximum (or minimum) of $N^{1/2}$ values initially stored in a row of a reconfigurable mesh of size N can be determined as follows. First, column broadcasts can be used so that every processor $P_{i,j}$ contains entry x_j. Next, a row broadcast is used within every row i so that processor $P_{i,i}$ informs all processors $P_{i,j}$ as to the value of x_i, $0 \leq i,j \leq N^{1/2} - 1$. At this point, every processor $P_{i,j}$ contains data values x_i and x_j. Now, all processors in column j use a logical OR, as described in Proposition 3.2, to decide whether x_j is the maximum (minimum) and the result is stored in the top processor of column j. Note that there may be more than one processors having the maximum (minimum) value. This value has to be broadcast to all other processors. A bus system on which broadcast of multiple identical values is allowed, can perform the above in one step. Here, concidering the above situation as an example, we will show how broadcast of multiple identical values can be simulated by the reconfigurable bus in $O(1)$ number of steps. The bus splitting mechanism can resolve the situation as follows. First, the row bus of the first row is isolated by disconnecting the column switches between the first and the second row. Next, each processor on the first row having the maximum (minimum) value disconnects the switch to its right and broadcasts the maximum (minimum) value. Thus, only one value is broadcast on each subbus and $PE_{0,0}$ reads from the leftmost subbus the maximum (minimum) value. Then the switches are formed to set a single global bus and $PE_{0,0}$ broadcasts the maximum (minimum) to all the other processors.

Proposition 3.3 *Given a reconfigurable mesh of size N, in which no more than one processor in each column stores a data value, the minimum (maximum) of these $O(N^{1/2})$ data items can be determined in $\Theta(1)$ time using the unit-time delay model, and in $\Theta(\log N)$ time using the log-time delay model.* $\square$

Note that considering a set of values, stored one value per processor in contiguous processors, we can create a reconfigurable subbus system dedicated to perform any kind of operations on these values. Assuming that among these values there exist some with a specific property, then using the bus splitting mechanism all the processors in the set will read this property from the subbus system in $O(1)$ time in the unit delay model, and in $O(\log N)$ time in the log-time delay model. The proof of the above and proposition 3.4 are in [15].

Proposition 3.4 *Assume that a set of data items S is stored in contiguous processors, one item per processor and a subset of S has a specific value. Then, all processors having items in S can read this value in $O(1)$ time using the unit delay model, and in $O(\log N)$ time using the log-time delay model.* $\square$

By a somewhat more complicated sequence, Valiant's P-RAM algorithm for finding maximum [29] can be simulated on a reconfigurable mesh to find the maximum of all N values, assuming they are stored one value per processor.

Proposition 3.5 *Given a set of data items S of size N stored one per processor on a reconfigurable mesh of size N, the minimum value of S can be determined in $O(\log \log N)$ time using the unit-time delay model, and in $O(\log N \log \log N)$ time using the log-time delay model.* $\square$

It is often desirable to model P-RAM algorithms on other machines. The CRCW P-RAM consists of a set of processors and a shared memory in which concurrent reads from the same memory location are allowed, as are concurrent writes to the same location. In the case of concurrent writes, a predefined scheme is used to decide which value succeeds. In order to do efficiently simulate the CRCW P-RAM, one must be able to efficiently simulate the concurrent read and concurrent write properties. Define a *Random Access Read (RAR)* to be a data movement operation that models a concurrent read, in which each processor knows the index of another processor from which it wants to read data. Similarly, a *Random Access Write (RAW)* will model a concurrent write in that each processor knows the index of a processor that it wishes to write to. In the case of multiple writes to the same processor, a tie-breaking scheme is used, such as minimum or maximum data value, or arbitrarily letting one value succeed.

Proposition 3.6 *Given a reconfigurable mesh of size N, in $O(k^{1/2} + \log N)$ time using the unit-time delay model, and in $O(k^{1/2} \log N + \log^2 N)$ time using the log-time delay model, k data items may be moved in a RAR or RAW, where $k \leq N$.*

Proof: This can be accomplished by distributing the data such that the rows of the reconfigurable mesh are partitioned into $k^{1/2}$ sets, with each set having at least $k^{1/2}$ data items and no more than $2k^{1/2}$ such items. Once such a partition of the rows is accomplished, the data in can be moved into the upper left block of size $\Theta(k)$ in constant time. The reconfigurable buses play a crucial role in this operation. In the block of size $\Theta(k)$, the data is sorted using a mesh computer sort, according to their destination index. The final step of the algorithm is to use the broadcast buses to move the data to their destinations. The details of partitioning the rows and moving the k data items into the upper left submesh of size $\Theta(k)$ follow.

1. Each processor that has a data item to be moved marks itself *active*.

2. Using parallel prefix with $\otimes$ as addition $(+)$, and active processors using a data value of 1 and inactive processors using a data value of 0, all active processors can know their position in their row with respect to the other active processors in their row, their position in the mesh with respect to the other active processors in the mesh, and the total number of data items k.

3. Bus splitting is performed by each active processor that is numbered $ck^{1/2}$, c an integer, with respect to the other active processors in its row. This divides the active processors in each row into disjoint strings of processors, where the number of active processors in a string is $k^{1/2}$. Let R_i denote the *remaining elements* in the ith row, $0 \leq i \leq N^{1/2} - 1$. That is, the remaining elements are those elements in a row that do not belong to a string of size $k^{1/2}$. Note that $|R_i| < k^{1/2}, 0 \leq i \leq N^{1/2} - 1$.

4. Partition the rows into disjoint sets such that each set has at least $k^{1/2}$ remaining elements and at most $2k^{1/2}$ remaining elements. Simulta-

neously within each partition, in $O(k^{1/2})$ time, move the *remaining elements* to a contiguous set of processors in a single row, forming a string.

5. At this time each row has a set of strings. Notice that in rows containing remaining elements, it is possible for the string of remaining elements to overlap other strings in the row. Define S_i to be the ith string in the reconfigurable mesh. Using parallel prefix, each processor in the reconfigurable mesh that contains an active record can know the value of i corresponding to the S_i that it is a member of. Notice that there are no more than $k^{1/2}$ S_i's.

6. Starting with $i = 0$, and continuing until $i = k^{1/2} - 1$, move S_i to row i. This requires $O(k^{1/2})$ column broadcasts.

7. In each row i, $0 \leq i \leq k^{1/2} - 1$, compress the active records to the leftmost set of processors, one item per processor. This requires $O(k^{1/2})$ row broadcasts.

Parallel prefix takes $O(\log N)$ time in the unit-time delay model, and $O(\log^2 N)$ time in the log-time delay model. The data movements involving row and column broadcasts take $O(k^{1/2})$ time in the unit-time delay model, and $O(k^{1/2} \log N)$ time in the log-time delay model. Now we show how the records in the submesh of size $\Theta(k)$ can be sent to their final destination.

1. Using a standard $\Theta(k^{1/2})$ time mesh sorting algorithm, order the records in the submesh of size k by their row destination (primary key), with ties broken by their column destination (secondary key).

2. Within every row of the submesh of size k, use a mesh row rotation to mark the processors that have records with the same primary key as other processors in its row. This takes $O(k^{1/2})$ time and does not use the bus.

3. For each of the $\Theta(k^{1/2})$ rows, one row at a time, use a column broadcast followed by a row broadcast to send the data in the marked processors of the given row of the submesh of size $\Theta(k)$ to the appropriate destination processor in the reconfigurable mesh.

4. For each of the $\Theta(k^{1/2})$ columns, one column at a time, the remaining records in the submesh of size $\Theta(k)$, i.e., the unmarked processors that contain distinct row destinations within the row of the submesh, use a row broadcast followed by a column broadcast to send their record to its destination.

The second step of the algorithm requires $\Theta(k^{1/2})$ uses of row and column buses. Therefore, total time for the second step is $O(k^{1/2})$ in the unit-time delay model, and $O(k^{1/2}\log N)$ time in the log-time delay model. Hence, the running time of the RAR and RAW is as claimed. $\square$

More efficient data movement can be performed if the distribution of the source processors, i.e., those processors sending data, as well as the destination processors, i.e., those processors receiving data, is uniform over the array.

Proposition 3.7 *If the number of source and destination processors within any block of size k^2 is $O(k)$, $1 \leq k \leq N^{1/2}$ then RAR and RAW can be performed in $O(\log N)$ time under the unit-time delay model, and in $O(\log^2 N)$ time under the log-time delay model.* $\square$

Another fundamental operation that involves data movement is data reduction. Assume that each processor has at most one record having a *key* field and a *data* field. *Data reduction* will perform an associative binary operation on the data of records having the same key. At the end of the data reduction operation, each processor with key k will have the result of the binary operation performed over all data with key k.

Proposition 3.8 *Given a binary associative operator $\otimes$, data reduction can be performed on k distinct keys in $O(k^{1/2} + \log N)$ time on a reconfigurable mesh of size N under the unit-time delay model, and in $O(k^{1/2}\log N + \log^2 N)$ time under the log-time delay model, so that each processor knows the result of applying $\otimes$ over all data items with its key.* $\square$

Lemma 3.1 *Given a reconfigurable mesh of size N with k distinct keys randomly distributed one key per processor, the number of distinct keys can be determined in $O(k^{1/2} + \log N)$ time under the unit-time delay model, and in $O(k^{1/2}\log N + \log^2 N)$ time under the log-time delay model.* $\square$

4 Applications

In this section, we illustrate the performance of the reconfigurable mesh by giving parallel algorithms to solve several problems and by giving simulations of other low wire area organizations. In the first part of this section, we show that the numerous communication patterns provided by the reconfigurable bus system can be exploited to allow the reconfigurable mesh to efficiently simulate two hierarchical organizations. Step by step simulations of the mesh-of-trees and pyramid are given.

A *mesh-of-trees (MOT)* of base size N, where N is an integral power of 4, has a total of $3N - 2N^{1/2}$ processors. N of these are base processors arranged as a mesh of size N. Above each row and above each column of the mesh is a perfect binary tree of processors. Each row (column) tree has as its leaves an entire row (column) of base processors. All row trees are disjoint, as are all column trees. Every row has exactly one leaf processor in common with each column tree. Each base processor is connected to 6 other processors (assuming they exist): 4 neighbors in the base, a parent in its row tree, and a parent in its column tree. Each processor in a row or column tree that is neither a leaf nor a root is connected to exactly 3 other processors in its tree: a parent and 2 children. Each root in a row or column tree is connected to its 2 children. Notice that in the MOT the processors in each row and in each column can be looked upon as placed at levels $0, 1, \ldots, k$ where $N^{1/2} = 2^k$.

A *pyramid computer (pyramid) of size N* is a machine that can be viewed as a full, rooted, 4-ary tree of height $\log_4 N$, with additional horizontal links so that each horizontal *level* is a mesh. It is often convenient to view the pyramid as a tapering array of meshes. A pyramid of size N has at its base a mesh of size N, and a total of $\frac{4}{3}N - \frac{1}{3}$ PEs. The levels are numbered so that the base is level 0 and the apex is level $\log_4 N$. A processor at level i is connected via bidirectional unit-time communication links to its 9 neighbors (assuming they exist): 4 siblings at level i, 4 children at level $i - 1$, and a parent at level $i + 1$.

The basic steps of any computation on the reconfigurable mesh that need to be considered when using embeddings for simulation are

1. computations within the processors,

2. communications between processors using the local (mesh) connections, and

3. communication between processors via the reconfigurable bus.

Define a c-embedding of a hierarchical organization onto the reconfigurable mesh to have the following properties.

1. A constant number of processors of the hierarchical organization are mapped to each processor of the reconfigurable mesh.

2. The number of edges between levels l and $l+1$, $0 \leq l \leq k-1$, incident on any row or column bus segment is $\leq c$.

Define a class of algorithms to be *normalized algorithms* if the following hold.

1. During a computation step of a hierarchical algorithm, all data operated on are located at the same level of the hierarchical organization.

2. All communication steps are performed between at most two adjacent levels.

This leads to the following results.

Proposition 4.1 *Any normalized algorithm running in $T(N)$ time on a mesh-of-trees of base size N can be simulated on a reconfigurable mesh of size N to finish in $O(T(N))$ time, under the unit-time delay model, and in $O(T(N) \log N)$ time, under the log-time delay model.* $\square$

Proposition 4.2 *Any algorithm running in $T(N)$ time on a mesh-of-trees of base size N, can be simulated on a reconfigurable mesh of size N to finish in $O(T(N) \log N)$ time under the unit-time delay model, and in $O(T(N) \log^2 N)$ time under the log-time delay model. Further this time is optimal.* □

Proposition 4.3 *Any algorithm running in time $T(N)$ on a pyramid of size N can be simulated on a reconfigurable mesh of size N in $O(T(N))$ time under the unit-time delay model, and in $O(T(N) \log N)$ time under the log-time delay model.* □

In fact, it is possible to show the following.

Theorem 4.1 *Any architecture that can be laid out in an $N^{1/2} \times N^{1/2}$ grid, (assuming wires have unit width) can be simulated on the reconfigurable mesh in constant time per unit time of the target architecture.* □

The remainder of this section concentrates on developing efficient parallel algorithms to solve fundamental problems on the reconfigurable mesh. As discussed earlier, the reconfigurable mesh can provide efficient communication between processors when the amount of data is sparse. Such fast communication of data is useful in efficient parallel solutions to many problems involving graphs and digitized images. Due to space limitations, only a few algorithms for graphs will be sketched.

The first problem considered is that of computing the connected components of an undirected graph with $N^{1/2}$ vertices, given as an adjacency matrix. The $(i, j)^{th}$ entry of the adjacency matrix of the graph is initially stored in processor $P_{i,j}$ of the reconfigurable mesh. The algorithm that we use is based on the $O(\log N)$ time algorithm presented for the CRCW P-RAM model [24]. It should be noted that the algorithm presented in [24] assumes a less restrictive unordered edge input and requires $O(\log N)$ iterations. During each iteration two operations are performed, which are now briefly described.

1. The first operation is called a *shortcut*, which consists of each vertex 'connecting' itself to its grandparent. This operation can be implemented on the reconfigurable mesh by the following steps.

   (a) First, perform bus splitting and broadcasts within every row i, so that all processors in row i know the parent of vertex i.

   (b) Next, using column broadcasts from every diagonal processor $P_{i,i}$, allow every processor $P_{j,k}$ to know the parent of vertex k.

   (c) Finally, a broadcast within every row allows all processors $P_{i,j}$ to know the grandparent of vertex i.

2. The second operation is *hooking*, where each vertex i pointing to a root j tries to hook the root to a non leaf node p such that $label(p) < label(j)$. This is accomplished on the reconfigurable mesh as follows.

(a) Every processor $P_{i,p}$ in row i compares $label(j)$ with $label(p)$, for all p such that (i,p) is an edge in the graph.

(b) Using bus splitting and row broadcasts, processor $P_{i,j}$ will know $label(p)$, if there exists a p such that (i,p) is an edge in the graph and $label(p) < label(j)$.

(c) Using bus splitting and a column broadcast, processor $P_{j,i}$ will know its new label, if such a label exists.

(d) Using a row broadcast, all processors in row j will know their new label, if such a label exists.

Notice that the implementation of both operations requires a fixed number of reconfigurable bus operations. Therefore, the following is obtained.

Theorem 4.2 *Given the adjacency matrix of a undirected graph with $N^{1/2}$ vertices distributed so that the $(i,j)^{th}$ element of the matrix is stored in processor $P_{i,j}$ of a reconfigurable mesh of size N, the connected components of the graph can be determined in $O(\log N)$ time under the unit-time delay model, and in $O(\log^2 N)$ time under the log-time delay model.* $\square$

The reconfigurable mesh can also be used to provide efficient solutions to some graph problems that assume unordered edges as input. As an illustration, consider computing the connected components of a V vertex graph with N edges on a reconfigurable mesh of size N. The general component labeling algorithm that we follow has been used for other architectures, and can be found in a variety of sources (c.f., [18]). Initially, each processor $P_{p,q}$ has an edge (i,j). At the end of the algorithm, processor $P_{p,q}$ has the label of the component to which vertices i and j belong. The algorithm iterates $\Theta(\log V)$ times, where each iteration consists of two operations.

1. In the first operation, each component chooses the component with the minimum label to which it is connected and keeps a pointer to this.

   - On the reconfigurable mesh, this is implemented by the key reduction operation using as keys the labels of the components, and using minimum as the binary associative operator.

2. In the second operation, the labels of the vertices are updated according to the labels just chosen.

   (a) On the reconfigurable mesh, collect the l active labels to the upper leftmost block of size l. In this block, a pointer update procedure finishing in $O(l^{1/2})$ time allows each active label to know the minimum index of the vertices it is connected to with respect to the iteration of the algorithm that is in progress.

   (b) The vertices are then relabeled according to the updated labels.

Therefore, the following result holds.

Theorem 4.3 *The connected components of a V vertex graph with N edges given in unordered edge input format, can be computed in $O(V^{1/2}+\log V \log N)$ time on the reconfigurable mesh of size N under the unit-time delay model, and in $O(V^{1/2} \log N + \log V \log^2 N)$ time under the log-time delay model.* $\square$

The strong similarities between component labeling algorithms and minimal spanning forest algorithms for parallel models of computation are well known. In particular, others have noted that small changes to a component labeling algorithm for a parallel computer can give a minimal spanning forest algorithm for the same computer [6,9,23]. The changes mainly consist of choosing edges with minimal index at each stage of the algorithm, rather than edges incident on vertices of minimal label.

Corollary 4.1 *A minimal spanning forest of a V vertex graph with N edges given in unordered edge input format, can be computed in $O(V^{1/2} + \log V \log N)$ time on the reconfigurable mesh of size N under the unit-time delay model, and in $O(V^{1/2} \log N + \log V \log^2 N)$ time under the log-time delay model.* $\square$

Several graph properties can be deduced once a spanning tree of the graph is determined [2]. Using Corollary 4.1 and the data movement operations presented in Section 3, the following results can be obtained.

Corollary 4.2 *Given N edges of a graph G with V vertices distributed one vertex per processor in a reconfigurable mesh of size N, in $O(V^{1/2} + \log V \log N)$ time under the unit-time delay model, and in $O(V^{1/2} \log N + \log V \log^2 N)$ time under the log-time delay model, one can*

    a) check if G is bipartite,

    b) compute the cyclic index of G, and

    c) compute the articulation points of G. $\square$

5   Further Remarks

This paper has introduced the reconfigurable mesh. We have presented efficient implementations of fundamental data movement operations for the reconfigurable mesh and shown that it can be used as a universal chip, in that the reconfigurable mesh is capable of simulating any organization of processors occupying the same area without loss of time. We have also presented algorithms that show how the reconfigurable mesh can efficiently solve a number of graph problems using the fundamental data movement operations.

Due to space limitations, a number of simulations, algorithms, and proofs were omitted. Efficient algorithms for the reconfigurable mesh are given in [15] to solve image problems involving connectivity, convexity, proximity, and other fundamental properties involving figures. [15] also provides details of all algorithms and additional simulation results, including embeddings that show how related architectures, such as the mesh-of-trees and pyramid, can be used to simulate the reconfigurable mesh.

6   Acknowledgements

The work of Russ Miller was supported in part by the National Science Foundation under grant DCR-8608640.

The work of V. K. Prasanna Kumar was supported in part by the National Science Foundation under grant IRI-8710863.

The work of Dionisios Reisis was supported in part by DARPA under contract F 33615-84-K-1404 monitored by the Air Force Wright Aeronautical Laboratory.

The work of Quentin F. Stout was supported in part by the National Science Foundation undergrant DCR-8507851, and by an Incentives for Excellence award from Digital Equipment Corporation.

We would like to thank Mehrnoosh Mary Eshaghian for many helpful suggestions, and Mr. Janto Wongso for his technical help in preparing this paper.

References

[1] Hussein Alnuweiri and V. K. Prasanna Kumar *Efficient image computations on VLSI arrays with reduced hardware.* In proceedings of IEEE workshop on PAMI, 1987.

[2] M. Atallah and R. Kosaraju. *Graph problems on a mesh connected processor array.* JACM, 1983.

[3] S. H. Bokhari. *Finding Maximum on an Array Processor with a Global Bus.* IEEE Transactions on Computers, Vol. C-33, No. 2, February 1984, pp 133-139.

[4] D. Carlson. *The mesh with a Global mesh: A flexible high speed organization for parallel computation.* Tech. Report, Electrical and Computer Engineering Department, University of Massachusetts, 1985.

[5] D. M. Champion and J. Rothstein, *Immediate parallel solution of the longest common subsequence problem,* 1987 International Conference on Parallel Processing, pp. 70-77.

[6] F.Y. Chin, J. Lam and I.-N. Chen, Efficient parallel algorithms for some graph problems, *Communications of the ACM*, 25 (1982), 659-665.

[7] C. R. Dyer. *A VLSI pyramid machine for hierarchical parallel image processing.* Proc. IEEE conference on Pattern Recognition and Image Processing, 1981

[8] Furst, Saxe and Sipser. *Parity, Circuits and Polynomial Time Hierarchy.* Proc. IEEE Foundations on Computer Science, pp. 260-270, 1981.

[9] S.E. Hambrusch and J. Simon, Solving undirected graph problems on VLSI, Tech. rep. CS-81-23, Computer Science, Penn. State Univ., 1981.

[10] J. Ja'Ja' and V. K. Prasanna Kumar. *Information Transfer in Distributed Computing with Applications to VLSI.* Journal of ACM, Jan. 1984.

J. Ja'Ja', V. K. Prasanna Kumar and J. Simon. *Information transfer under different sets of protocols.* SIAM Journal on Computing, 1984.

[12] F. T. Leighton. *Parallel computations using Mesh of Trees.* Technical Report, MIT, 1982.

[13] H. Li and M. Maresca. *Polymorphic-Torus Network.* Proc. International Conference on Parallel Processing, 1987.

[14] R. Miller, *ADLSIMD: An abstract description language for the SIMD family,* Proc. of the 1984 IEEE Southern Tier Tech. Conf., pp. 89-94.

[15] R. Miller, V.K. Prasanna Kumar, D. Reisis, and Q.F. Stout, *Meshes with reconfigurable buses,* submitted.

[16] R. Miller and Q. F. Stout. *Data Movement Techniques for the Pyramid Computer.* SIAM Journal on Computing, Vol. 16, No. 1, February 1987.

[17] R. Miller and Q. F. Stout. *Some graph and image processing algorithms for the hypercube.* SIAM Conference on Hypercube Multiprocessor, 1987.

[18] R. Miller and Q. F. Stout. *Parallel Algorithms for Regular Architectures.* The MIT Press, 1988.

[19] D. Nath, F. N. Maheshwari, P. C. P. Bhatt. *Efficient VLSI networks for parallel processing based on orthogonal trees.* IEEE Transactions on Computers, 1983.

[20] V. K. Prasanna Kumar and C. S. Raghavendra. *Array Processor with Multiple Broadcasting.* Proceedings of the 1985 Annual Symposium on Computer Architecture, June 1985.

[21] V. K. Prasanna Kumar and M. Eshaghian. *Parallel Geometric algorithms for Digitized pictures on the Mesh of Trees organization.* International Conference on Parallel Processing, 1986.

[22] V. K. Prasanna Kumar and D. Reisis *Parallel Image Processing on Enhanced Arrays .* International Conference on Parallel Processing, 1987.

[23] C. Savage and J. Ja'Ja, Fast, efficient parallel algorithms for some graph problems, *SIAM Journal on Computing,* 10 (1981), 682-691.

[24] Y. Shiloach and U. Vishkin. *A $O(\log N)$ Parallel Connectivity Algorithm.* Journal of Algorithms 3, 1982.

[25] Q. F. Stout. *Mesh Connected Computers with Broadcasting.* IEEE Trans. on Computers C-32, pp. 826-830, 1983.

[26] Q. F. Stout. *Meshes with Multiple Buses.* Proc. 27th IEEE Symposium on the Foundations of Computer Science, 1986, pp. 264-273.

[27] S. L. Tanimoto. *A Pyramidal Approach to Parallel Processing.* Proc. 1983 International Symposium on Computer Architecture.

[28] L. Uhr. *Algorithm-Structured Computer Arrays and Networks.* Academic Press, 1984.

[29] L. G. Valiant. *Parallelism in comparison problems.* SIAM J. on Computing 3, 1975.

[30] C.C. Weems, S.P. Levitan, A.R. Hanson, E.M. Riseman, J.G. Nash, D.B. Shu, *The image understanding architecture,* COINS Tech. Rept. 87-76, University of Massachusetts at Amherst.

Embedding Large Tree Machines into Small Ones
(Extended Abstract)

Ajay K. Gupta and *Susanne E. Hambrusch*

Department of Computer Sciences
Purdue University
West Lafayette, IN 47907, USA

The problem of embedding an n-processor architecture T into an m-processor architecture H for $n > m$ arises when algorithms designed for architectures of an ideal size are simulated on existing architectures which are of a fixed size. We consider the case when both architectures are complete binary trees and present embeddings that minimize various cost functions and maximize processor utilization. One of our embeddings achieves a dilation of 2, a level-load of 4, a load of $min\{\lceil \frac{n}{m} \rceil, 2\log n\}$, and a utilization of $\lceil \frac{n}{m} \rceil$. A second embedding assigns to every processor of H the same number of leaf and interior processors of T. It achieves a dilation of $2\log\log m + 1$, a level-load of $O(\log\log m)$, and a load of $O(\log n)$.

1. Introduction

Parallel algorithms are generally designed for parallel machines or VLSI architectures of an ideal size; i.e., n processors are available for a problem of input size n [4,13,14]. However, in practice, parallel machines or VLSI architectures are of a fixed size m [2,8]. In the realistic case when $n > m$, the problem of how to efficiently implement an algorithm designed for an n-processor network T on an m-processor host network H has to be solved. One possible solution is to partition the problem of input size n into subproblems that fit the size of the host and to obtain the final solution by combining the solutions to the subproblems [6,9]. Another, inherently different solution is to simulate the algorithm on the host by having one processor of H take over the function of a number of processors of T. This approach, which is easily modeled as a graph embedding problem, is the

---

This work was supported by the Office of Naval Research under Contracts N00014-84-K-0502 and N00014-86-K-0689, and by the National Science Foundation under Grant DMC-84-13496.

one studied in this paper. We present new and efficient embeddings for
the case when both networks are complete binary trees. Our embeddings
consider a number of new cost measures. One such measure is the level-
load, which is motivated by the data movement characteristic to functions
commonly executed on tree machines. Another consideration is the processor
utilization. We call a utilization balanced if every processor of H simulates
at most $\lceil \frac{n}{m} \rceil$ processors of T. Furthermore, we call a utilization balanced
with respect to leaf and interior processors if every processor of H simulates
the same number of leaf and interior processors of T.

When both networks are viewed as graphs, an embedding $< f, g >$ of
T into H is defined by an injective mapping f from the processors of T
to the processors of H together with a mapping g that maps every edge
$e = (v, w)$ of T onto a path $g(e)$ connecting $f(v)$ and $f(w)$. We refer to f as
the assignment. Two commonly and extensively studied cost measures of an
embedding are the dilation and the load [1,3,10,12]. The *dilation* δ is defined
as the maximum distance in H between two adjacent processors in T, and
the *load* λ is defined as the maximum number of paths containing an edge in
H, where every path represents an edge in T. A new cost measure considered
by our embeddings is the level-load which captures the data movement characteristic to broadcasting and reduction functions (e.g., computing min and
max). In these functions only one level of the tree T is active at any time.
Hence, the *level-load* λ_l is defined as the maximum load over any edge of H
when only one level of T is active. Another new cost measure we consider is
the utilization μ which is defined as the maximum number of processors of
T assigned to any processor of H. As already stated, in a balanced utilization we have $\mu = \lceil \frac{n}{m} \rceil$. This requirement is of practical importance, since
it makes every processor of H share an equal load - with respect to time,
as well as memory utilization. In a tree network not all processors may be
of the same type [5], and we consider the situation when the leaf processors
of T have different processing and/or memory capabilities from the interior
processors. In this case an embedding that achieves a balanced utilization
with respect to leaf and interior processors of T is desired, i.e., we require a
leaf utilization μ_l of $\lceil \frac{n+1}{2m} \rceil$ and an interior utilization μ_i of $\lceil \frac{n-1}{2m} \rceil$.

The main results of this paper are two embeddings, the EP (Even Processor) and the EL (Even Leaf) embedding. The EP embedding achieves

a dilation of 2, a level-load of 4, a load of $\min\{\lceil\frac{n}{m}\rceil, 2\log n\}$, and a utilization of $\lceil\frac{n}{m}\rceil$. However, it has an unbalanced leaf and interior utilization for some values of n and m. The EL embedding, which achieves a balanced leaf and interior utilization, has a dilation of $2\log\log m + 1$, a level-load of $O(\log\log m)$, and a load of $O(\log n)$. Both embeddings can easily be computed by the host tree H in $O(\lceil\frac{n}{m}\rceil + \log m)$ time. Our embeddings make use of a number of interesting and new embedding strategies. In order to keep the level-load and load small, we assign subtrees of T to the processors of H. However, for a small level-load, these subtrees have to come from different levels of T. This in turn results in embedding strategies that, at first, appear unnatural for trees since they are not symmetric. Obtaining a level-load of 4 in the EP embedding is a non-trivial task. The processors of T assigned to a processor of H belong to different levels in T and/or form a subtree of T which is a complete binary tree and whose leaves are the leaves of T. We allow at most a constant number of violation of the above two conditions at every processor of H. Obtaining a balanced leaf and interior utilization and keeping the cost measures small is considerably harder than just obtaining a balanced utilization. In fact, we have shown that a dilation of 2 cannot be achieved in the case when leaf and interior utilization is required to be balanced and $n = 2m+1$ [7]. We next describe a number of basic embedding strategies that will be used in the EP and EL embeddings.

2. Preliminaries

In this section we first describe two embeddings, the isomorphic and the shrink embedding. Let T_{h_1} (resp. T_{h_2}) be a complete binary tree of height h_1 (resp. h_2). Let $t_{h_1}(i,j)$ denote the jth leftmost node on level i in T_{h_1}. For clarity reasons, we will refer to nodes of T_{h_1} as PEs and nodes of T_{h_2} simply as nodes. The *isomorphic embedding* of T_{h_1} into T_{h_2} is defined for $h_1 \le h_2$. In this embedding PE $t_{h_1}(i,j)$ is assigned to node $t_{h_2}(i,j)$, for $0 \le i \le h_1 - 1$ and $0 \le j \le 2^i - 1$, and no PEs are assigned to nodes in T_{h_2} at levels $\ge h_1$. The *shrink embedding* is defined for $h_1 > h_2$. For levels 0 through $h_2 - 2$ the shrink embedding is same as the isomorphic embedding and level $h_2 - 1$ is handled as follows. Every leaf $t_{h_2}(h_2 - 1, j)$ of T_{h_2} is assigned the $2^{h_1 - h_2 + 1} - 1$ PEs of T_{h_1} which represent the PEs in the subtree rooted at $t_{h_1}(h_2 - 1, j)$ in T_{h_1}. Observe that both embeddings achieve $\delta = 1$,

$\lambda_l = 1$, and $\lambda = 1$. Furthermore, the shrink embedding gives an embedding of T into H with $\delta = \lambda_l = \lambda = 1$, and in which one half of the nodes of H are assigned 1 and the other half are assigned $2\frac{n+1}{m+1} - 1$ processors of T. This embedding has a rather unbalanced utilization.

We next describe the DLS (Different Level Subtree) embedding which embeds a $2m + 1$-PE tree T' into an m-node tree H. Let $m = 2^l - 1$. The DLS embedding achieves a dilation and level-load of unity, a load of $l - 1$, and it distributes the leaves of T' evenly (i.e., $\mu_l = 2$). In order to achieve this, the tree T' is partitioned at PEs on different levels and the subtrees rooted at these PEs are embedded by an isomorphic embedding as follows. Let $T'_1, T'_2, \ldots, T'_l$ be l subtrees of T', where T'_i has height i and is rooted at PE $t'(l - i + 1, 1)$, $1 \leq i \leq l$. We embed T'_i into H using the isomorphic embedding and then assign all the PEs on the path from $t'(0,0)$ to $t'(l,0)$ to the root $h(0,0)$. The DLS embedding assigns thus $l + (l-1) = 2l - 1$ interior PEs of T' to the root of H and $l - r - 1$ interior PEs of T' to nodes on level $r \geq 1$. Furthermore, the root is assigned 2 leaves of T' and every other node of H is assigned 1 leaf each. Figure 1 shows this embedding for $l = 4$.

We now outline briefly the general strategies used in the EP and the EL embeddings. Throughout, let $n = 2^k - 1, m = 2^l - 1$, and $c = 2^{k-l}$. In both the embeddings we first embed the n-PE tree T into a $2m + 1$-PE tree T' by a shrink embedding. Next, an initial assignment is obtained by embedding T' into H by a DLS embedding. This initial assignment has a rather unbalanced utilization. We obtain a balanced utilization in 2 steps: a balancing step and a recursive step. The recursive step recursively embeds one of the two leaves of T' assigned to the root into H. Each such leaf represents a subtree of height $\log c$ in T. The balancing step, in which we reassign PEs to obtain a balanced utilization, is the most delicate part of the embeddings. The balancing done in the EP embedding is of quite a different nature from the one done in the EL embedding. In the EL embedding only interior PEs are reassigned, since the initial assignment and the recursive step achieve a balanced leaf utilization. In the EP embedding we reassign both interior and leaf PEs to obtain a balanced utilization. We keep the dilation at 2 and the level-load at 4 by carefully reassigning PEs that are in the subtrees representing the leaves of T'. As will be described later, when $k \leq l + \lceil \log l \rceil$ a different strategy is needed for the EP embedding.

3. The EP Embedding

The main result of this section is the EP embedding. In order to achieve a balanced utilization the embedding strategy repeatedly performs balancing steps which refine current assignments of processors to nodes. Section 3.1 presents two refinement lemmas used in the EP embedding. As mentioned in Section 2, the EP embedding requires two different strategies. The first one is used when $k > l + \lceil \log l \rceil$ and it is based on shrinking T, followed by a DLS embedding, a balancing and a recursive step. An alternate strategy is used when $k \leq l + \lceil \log l \rceil$ and it requires a forest-in-tree (FT) embedding. This FT embedding, which is interesting in its own right, is described in Section 3.3.

3.1. The Refinement Lemmas

In this section we prove two lemmas. Each lemma changes an initial assignment into a refined assignment. If the initial assignment has a dilation of 1 and a level-load of at most 2, then the refined assignment obtained by Lemma 3.1 has dilation and level-load at most 2.

Lemma 3.1: *Let H be an m-node complete binary tree of height l in which every node at level i has $y + l - i - 1$ PEs assigned to it, $0 \leq i \leq l - 1$ and $y = 2^\alpha - 1$. Assume that, out of the $y + l - i - 1$ PEs at a node, y PEs form a complete binary tree of height α. Then, there exists a refined assignment of PEs to the nodes such that*

(i) the root of H is assigned $y - l + 1$ and every other node of H is assigned $y + 1$ PEs, and

(ii) if a PE is at node v in the refined assignment, then it was either at node v or at v's parent in the initial assignment.

Proof: It is easy to see that node $h(i,j)$ has to reassign $l - i - 1$ PEs to each of its children, $0 \leq i \leq l - 2$. The reassigned PEs will come from the y PEs assigned to $h(i,j)$ representing a complete binary tree of height α. Let T_α be this tree. Depending on the value of α we distinguish two cases.

Case 1 : $l - 1 < \alpha$. Let L (resp. R) be the leftmost (resp. rightmost) path of length $l - i - 1$, starting at $t_\alpha(1,0)$ (resp. $t_\alpha(1,1)$) and ending at $t_\alpha(l - i - 1, 0)$ (resp. $t_\alpha(l - i - 1, 2^{l-i-1} - 1)$). We reassign the PEs on path L to node $h(i+1, 2j)$ and the PEs on path R to node $h(i+1, 2j+1)$.

Case 2 : $l - 1 \geq \alpha$. Node $h(i,j)$ reassigns $l - i - 1$ PEs of the subtree rooted at $t_\alpha(1,0)$ to its left child and $l - i - 1$ PEs of the subtree rooted at $t_\alpha(1,1)$ to its right child. Here we only describe the reassignment process from $h(i,j)$ to its left child, since the reassignment process to the right child is symmetric. Let $l - i - 1 = b_s 2^s + b_{s-1} 2^{s-1} + ... + b_0 2^0$, where b_r is either 0 or 1 and $s = \lfloor \log(l - i - 1) \rfloor$. For every index r such that $b_r = 1$, $0 \leq r \leq s$, let T_r be the subtree rooted at $t_\alpha(\alpha - r, 1)$. We then reassign $t_\alpha(\alpha - r - 1, 0)$ and the PEs in T_r to node $h(i+1, 2j)$. Since T_r contains $2^r - 1$ PEs, a total of $l - i - 1$ PEs are reassigned to $h(i+1, 2j)$.

Since the PEs reassigned to a node v in Case 1 and 2 come from the subtree T_α initially assigned to the parent of v, condition (ii) is satisfied. ∎

Observe that the reassignment strategy given for Case 2 of the above lemma (i.e., when $l - 1 \geq \alpha$) could also be used for Case 1 (i.e., $l - 1 < \alpha$). However, we do need to make this distinction since the reassignment strategy for Case 1 has certain properties needed by the EP embedding. We next state a second refinement lemma which is used by the EP embedding. The proof of the lemma will appear in the full version of this paper.

Lemma 3.2: Let H be an m-node complete binary tree of height l in which the root is assigned $2y - 1$ PEs, every other node is assigned y PEs, and $y - 1 \leq l$. Then, there exists a refined assignment such that

(i) every node on the path from $h(0,0)$ to $h(y - 2, 0)$ is assigned $y + 1$ PEs, and

(ii) if a PE is at node v in the refined assignment, then it was either at node v or at v's parent in the initial assignment.

3.2. The Embedding Strategy

In this section we describe the embedding strategy for the EP embedding which achieves a dilation δ of 2, a level-load λ_l of 4, a load λ of $\min\{\lceil \frac{n}{m} \rceil, 2 \log n\}$, and a utilization μ of $\lceil \frac{n}{m} \rceil$. We point out that it is easy to obtain an embedding that achieves $\delta = 1$, $\lambda_l = \lambda = c$ and in which $2c - 1$ PEs are assigned to the root of H and c PEs are assigned to every other node of H. This embedding assigns to only one node of H roughly twice the number of PEs assigned to every other node of H, but it has a high level-load and load. Details will be given in the full version of this paper.

As described in Section 2, when $k > l + \lceil \log l \rceil$, the initial assignment for the EP embedding is provided by the DLS embedding after a shrink embedding. The DLS embedding assigns 2 leaves of the "shrunk" tree T' to the root of H and 1 leaf of T' to every other node of H, where every leaf represents a subtree of T of height $\log c$. In the balancing step of the EP embedding we reassign PEs from these subtrees. However, if the size of these subtrees is small, i.e., $c \leq l$, we do not have enough PEs for a reassignment to achieve a balanced utilization. Furthermore, we can not reassign PEs at levels 0 through $l-1$ in T (i.e., the interior PEs of T') since it would result in a high dilation. For this reason we need a different embedding strategy for $k \leq l + \lceil \log l \rceil$.

The EP embedding strategy is recursive and encompasses two base cases, referred to as Base-Case 1 and Base-Case 2. Should neither Base-Case hold, the embedding strategy proceeds as described above.

Base-Case 1: $k \leq l$. In this case T is smaller than H, and we use the isomorphic embedding to embed T into H with $\delta = \lambda_l = \lambda = \mu = 1$.

Base-Case 2: $l < k \leq l + \lceil \log l \rceil$. Would we perform a shrink embedding of T into T', the leaves of T' would not be large enough to obtain a balanced utilization after the DLS embedding. Hence, a different embedding strategy tailored towards this range of k is needed. The embedding starts with an LLRR (Left-child-to-Left-child, Right-child-to-Right-child) embedding which assigns sets of PEs on different levels in T to nodes in H. Once the LLRR embedding strategy reaches a PE on level $k-1$ in T, we switch to a FT embedding strategy. We briefly describe both strategies. The details of the FT embedding are given in Section 3.3.

We start with the assignments made for $h(0,0)$, $h(1,0)$ and $h(1,1)$ in the LLRR embedding strategy. Let L be the set of PEs on the path from $t(1,0)$ to $t(c-1,0)$ and let R be the set of PEs on the path from $t(1,1)$ to $t(c-1,2^{c-1}-1)$. Then, $h(0,0)$ is assigned set $A(0,0) = L \cup R \cup \{t(0,0)\}$, a total of $2c - 1$ PEs. Next, the right children of the PEs in set L and PE $t(c,0)$ are assigned to $h(1,0)$. The left children of the PEs in set R and PE $t(c,2^c-1)$ are assigned to $h(1,1)$. Observe that at this point level-load is already 2 and the load is c. Figure 2 shows these first embedding steps for the case when $c = 3$. Assuming that set $A(i-1,j)$ has been determined, sets $A(i,2j)$ and $A(i,2j+1)$ are formed as follows. Set $A(i-1,j)$ contains c PEs, two of which are on the same level in T and the remaining $c-2$ PEs are on different levels. We assign to node $h(i, 2j)$ the left children of the PEs in $A(i-1, j)$ and to $h(i, 2j+1)$ the right children of the PEs in $A(i-1, j)$. Hence, c PEs, 2 of which are at the same level in T, namely level $c+i-1$, are assigned to each of $h(i, 2j)$ and $h(i, 2j+1)$.

The LLRR embedding strategy terminates when $i = k - c$, at which point leaves of T are included into sets $A(i, j)$ for the first time. Every node $h(k-c, j)$ is the root of a subtree of height $l + c - k = c - \log c$. Let H' be such a tree (we omit an indexing depending on j). Consider the PEs assigned to $h(k-c, j)$. Set $A(k-c, j)$ consists of 2 PEs on level $k-1$, one PE on level $k-2$, one PE on level $k-3$, ..., and one PE on level $k-c+1$. Let $T_1', T_1, T_2, ..., T_{c-1}$ be the subtrees of T rooted at the PEs in $A(k-c, j)$, where the subscript indicates the height. In order to assign the non-root PEs of the T_i's to the non-root nodes of H', we use the FT embedding strategy whose details are given in Section 3.3. It embeds the trees $T_1, T_2, ..., T_{c-1}$ forming forest F into tree H', which is of height $c - \log c$, so that the root of H' gets $c - 1$ PEs and every other node gets c PEs assigned to it. The FT embedding achieves $\delta = 1$, $\lambda_l = 1$, and $\lambda = c$. It places the root of every T_i of F at the root of H' and thus our original assignment $A(k-c, j)$, which includes also the single PE representing T_1', does not change.

After the LLRR and FT embeddings have been performed, the root $h(0, 0)$ has $2c - 1$ PEs assigned to it and every other node of H has c PEs assigned to it. The embedding obtained so far has unit dilation, a level-load of 2 (which is caused by the LLRR embedding) and a load of c. In order to achieve a utilization of $\lceil \frac{n}{m} \rceil = c + 1$ we use Lemma 3.2 with $y = c$. We next describe which of the PEs to reassign when using Lemma 3.2.

For $i = 0, 1, ..., c-1$, node $h(i, 0)$ needs to reassign $c - i - 2$ PEs to its left child $h(i+1, 0)$. In order to keep the level-load at 2, the PEs reassigned to a node belong to different levels in T. Recall that, among the c PEs assigned to a node on level j with $j \leq k - c$ during the LLRR embedding, $c - 1$ PEs come from different levels of T. If $k \leq l + \lceil \log l \rceil - 1$, we have $k - c > c$ and the reassignment done by Lemma 3.2 terminates before level $k - c$ in H. In this case the $c - i - 2$ PEs to be reassigned are chosen from the $c - 1$ PEs assigned to a node by the LLRR embedding. If $k = l + \lceil \log l \rceil$, we have $k - c < c$ and the PEs for Lemma 3.2 are selected as follows. Nodes on levels $0, 1, ..., k - c$ reassign PEs assigned to the node by the LLRR embedding as described above. For nodes on level $k - c + 1$ and higher, we reassign PEs which were assigned to a node by the FT embedding. As will be shown in Section 3.3, among the c PEs assigned to a node $h(i, j)$ during the FT embedding, at least $l - i - 1$ PE come from different levels in T, for $k - c < i \leq l - 1$. The number of PEs which need to be reassigned from $h(i, 0)$ to $h(i + 1, 0)$ is $c - i - 2$ for $k - c < i \leq c - 3$. Now, since $c - i - 2 < l - i - 1$, Lemma 3.2 reassigns $c - i - 2$ PEs out of the $l - i - 1$ PEs which are assigned to node $h(i, 0)$ and are on different levels.

It is easy to see that the reassignment done by Lemma 3.2 causes dilation 2 and load $c + 1 = \lceil \frac{n}{m} \rceil < 2 \log n$. The level-load is kept at 2 since the PEs reassigned to a node belong to different levels in T. Now, considering the level-load of the entire embedding, a level-load of 2 is achieved when PEs at level $c + i - 1$ communicate with the PEs at level $c + i - 2$ of T, $1 \leq i \leq k - c$, and a level-load of 1 or 0 is achieved when PEs at level j communicate with PEs at level $j - 1$ for $1 \leq j \leq c - 1$. This completes the embedding of T into H with dilation and level-load 2, load $\lceil \frac{n}{m} \rceil$, and utilization $\lceil \frac{n}{m} \rceil$ for the case when $l < k \leq \lceil \log l \rceil$.

Recursive EP Step: If neither of the two base cases holds (i.e. $k > l + \lceil \log l \rceil$), we embed T into H as outlined earlier: Embed T into a $2m + 1$-node tree T' by the shrink embedding and then perform a DLS embedding. We then obtain the final embedding in two steps: a balancing step and a recursive step. The DLS embedding assigns $2l - 1$ interior and 2 leaf nodes of T' to the root $h(0, 0)$, and $l - i - 1$ interior and 1 leaf node to every other node of H at level i. Since each leaf of T' represents a subtree of T of height $\log c$ containing $c - 1$ PEs, $2l + 2c - 3$ PEs are assigned to the root $h(0, 0)$ and $l + c - i - 2$ PEs are assigned to every other node of H at level i. So far the dilation is 1, the level-load is 1, and the load is $l - 1$. The balancing step reassigns the PEs so that the root of H is assigned $2c - 1$ PEs and every other node is assigned c PEs of T. This is done by using Refinement Lemma 3.1 with Case 1 when $\alpha = k - l > l$, Case 2 when $k - l \leq l$, and with $y = c - 1$. Note that none of the $c - 1$ PEs representing one of the subtrees assigned to the root of H participate in the reassignment of Lemma 3.1. As stated earlier in Section 3.1, as far as Lemma 3.1 is concerned, Case 1 of this lemma does not appear to be necessary. However, if the balancing step would always use Lemma 3.1 with Case 2 (and not use Case 1, when possible), we would end up with a level-load of $O(\frac{\log n}{\log m})$, instead of 4.

When Case 1 of the Lemma 3.1 is used, the embedding achieves a level-load of 2, since PE $t(2l-i-1, j')$ is reassigned to the left child of $h(i, j)$ while PEs $t(2l-i, 2j')$ and $t(2l-i, 2j'+1)$ remain assigned to $h(i, j)$. The remaining PEs assigned to a node $h(i, j)$ after the balancing step either belong to different levels of T (which results in a level-load of 1), or, if two siblings are assigned to $h(i, j)$, then their common parent is also assigned to $h(i, j)$ (which results in a level-load of zero). When Case 2 of Refinement Lemma 3.1 is used, the embedding achieves a level-load of 1. In both of the cases the dilation is kept at unity. The load increases to at most $2(l - 1)$. Observe that after applying Lemma 3.1, the root of T is assigned to the root of H and $c - 1$ of the $2c - 1$ PEs assigned to node $h(0, 0)$ form a complete binary tree of height $k - l = \log c$. Let T^* be this tree. The last step recursively embeds T^* into H. The embedding of T^* achieves dilation 2, level-load at most 4, load at most $\min\{\lceil \frac{c-1}{m} \rceil, 2\log n - 2\log m\}$, and utilization $\lceil \frac{c-1}{m} \rceil$. This completes the embedding of T into H for the case when $k \geq l + \lceil \log l \rceil$.

Theorem 3.3: *An n-PE complete binary tree T can be embedded into an m-node complete binary tree H so that the dilation δ is 2, the level-load λ_l is 4, the load λ is $\min\{\lceil \frac{n}{m} \rceil, 2\log n\}$, and the utilization μ is $\lceil \frac{n}{m} \rceil$.*

Proof: We only need to prove that the embedding of T^* into H, together with the embedding of $T - T^*$ into H, achieves the claimed bounds.

It is easy to see that the total number of PEs assigned to every node of H is at most $c + \lceil \frac{c-1}{m} \rceil = \lceil \frac{n}{m} \rceil$. The embeddings of T^* and $T - T^*$ into H have dilation at most 2, respectively, and since the root of T^* is assigned to $h(0, 0)$, the embedding of T has dilation at most 2. That combining the embeddings of $T - T^*$ and T^* results in a level-load of at most 4 is shown by a careful case-by-case analysis. Assume first that $k \leq 2l$. In this situation the balancing step uses Case 2 of Lemma 3.1. Since the height of T^* is less than or equal to l, T^* is embedded using Base-Case 1. The embedding of $T - T^*$ yields a level-load of 1 and the isomorphic embedding of T^* also yields a level-load of 1. Combining the embeddings of $T - T^*$ and T^* yields a level-load of precisely 2. Assume now that $k > 2l$. In this situation the balancing step in the embedding of $T - T^*$ used Case 1 of Lemma 3.1. It is easy to see that this yields a level-load of 2. If the height of T^* is less than or equal to $l + \lceil \log l \rceil$ then T^* is embedded using Base-Case 2. It can be shown that this results in a level-load of 2. Hence, combining the embeddings of $T - T^*$ and T^* yields a level-load of at most 4. Finally, consider the situation where $k > 2l + \lceil \log l \rceil$. The embedding of T^* has a level-load of at most 4 and that of $T - T^*$ has level-load of 2. The proof that the combined level-load is at most 4 is omitted in this abstract.

Showing that combining the embeddings of $T - T^*$ and T^* results in a load of at most $2 \log n$ is straightforward and hence Theorem 3.3 follows. ∎

We conclude this section by describing the leaf and interior utilization μ_l and μ_i, respectively, achieved by the EP embedding. For some values of k and l the EP embedding achieves a balanced or close to a balanced leaf utilization , while for others the leaf utilization is rather unbalanced. If k is a multiple of l, then it is easy to see that the EP embedding achieves $\mu_l = \lceil \frac{n+1}{m} \rceil$ and $\mu_i = \lceil \frac{n-1}{m} \rceil$. For $dl < k < (d+1)l$ with $d \geq 2$, it achieves a balanced μ_l within a constant factor and an exactly balanced μ_i as d approaches infinity. The most unbalanced μ_l occurs when $l < k < 2l$. In this case some nodes of H have no leaf PE assigned to them, while others have $\Theta(\frac{n+1}{m})$ leaf PEs assigned to them. A natural question that arises is whether there exist other embedding strategies for the range $l < k < 2l$ that achieve a better leaf utilization. Achieving a balanced leaf and interior utilization and keeping the cost measures small, appears to be most difficult for the case when $k = l + 1$. In fact, we have recently shown that a dilation of 2 can not be achieved together with a balanced leaf and interior utilization when $k = l + 1$ [7]. In Section 4 we describe an embedding strategy that achieves a balanced leaf and interior utilization; i.e., $\mu_l = \lceil \frac{n+1}{2m} \rceil$ and $\mu_i = \lceil \frac{n-1}{2m} \rceil$.

3.3 Forest-in-Tree (FT) Embedding

In this section we describe the Forest-in-Tree embedding needed by Base-Case 2 of the EP embedding. We are given a forest F consisting of the complete binary trees $T_1, T_2, ..., T_{c-1}$, where T_i is of height i, and a complete binary tree H' of height $c - \log c$. The embedding of F into H' assigns $c - 1$ PEs of F, namely the roots of the T_i's, to the root of H' and c PEs of the forest to every other node of H'. Note that F contains a total of $2^c - c - 1$ PEs and H' contains a total of $2^{c - \log c} - 1$ nodes. The FT embedding achieves unit dilation, unit level-load and a load of c. The level-load is measured between the levels of a tree T that contains trees $T_1, T_2, ..., T_{c-1}$ as subtrees (as in Base-Case 2). The main idea in the FT embedding is to "shrink" the trees of the forest F at particular different levels so that the PEs can be reassigned from the subtrees embedded at every node of H' to achieve a balanced utilization.

The FT embedding is recursive. The base-case of the FT embedding embeds a forest F' consisting of trees $T_{c/2}, ..., T_{c-1}$ into H' and then a recursive call embeds the remaining trees of F into H'. The base-case assigns $\frac{c}{2}$ PEs to every node at level j of H' with $j \leq \frac{c}{2} - \log c$ and c PEs to every other node of H'. The embedding strategy in the base-case is as follows.

FT Base-Case: We first use the shrink embedding to embed every tree T_i of F' into a tree of height $i - \log c + 1$. This gives us a forest F''' consisting of trees T_i'', $i = \frac{c}{2} - \log c + 1, ..., c - \log c$, where every leaf of T_i'' contains $c - 1$ PEs of the original tree $T_{i + \log c - 1}$. For example, when $c = 8$ the forest F' consists of trees T_4, T_5, T_6, T_7 and H' is of height 5. Forest F''' obtained by the shrink embedding consists of trees $T_2'', T_3'', T_4'', T_5''$. We then embed the forest F''' into H' by embedding every tree T_i'' into H' using the isomorphic embedding. After every tree of F''' has been embedded, every node of H' at level j with $j \leq \frac{c}{2} - \log c - 1$ has $\frac{c}{2}$ PEs of F''' assigned to it. Every node at level j with $j \geq \frac{c}{2} - \log c$ has $c - \log c - j$ PEs of F''' assigned to it, out of which exactly 1 PE corresponds to a leaf PE of F'''. Thus, the number of PEs of F' assigned to every node at a level in H' is:

Levels	No. of PEs assigned	
	before balancing	after balancing
$j \leq \frac{c}{2} - \log c - 1$	$\frac{c}{2}$	$\frac{c}{2}$ (no change)
$j = \frac{c}{2} - \log c$	$\frac{3c}{2} - 2$	$\frac{c}{2}$
$j \geq \frac{c}{2} - \log c + 1$	$2c - \log c - j - 2$	c

Next, we reassign PEs in a balancing step so that the final assignment shown in the above table is obtained. Let H'' be a subtree of height $\frac{c}{2}$ rooted at a node at level $\frac{c}{2} - \log c$ in H'. We apply Case 2 of the Refinement Lemma 3.1 with parameters $l = \frac{c}{2}$ and $y = c - 1$ to every subtree H''. Recall that Lemma 3.1 requires a complete binary tree consisting of y PEs at every node in H''. The y-PE subtrees are the subtrees assigned to nodes when performing the shrink embedding. It is easy to see that every H'' satisfies the conditions needed by Lemma 3.1. After the reassignment, the root of H'' has $\frac{c}{2}$ PEs assigned to it, and every other node of H'' has c PEs assigned to it. Now, considering this in H', our claim about the final assignment of PEs after the balancing step follows. This completes the description of the base-case.

Recursive FT Step: Let F^* be the forest consisting of trees $T_1, T_2, ..., T_{\frac{c}{2}-1}$ of F. Let H^* be the subtree of height $\frac{c}{2} - \log c + 1$ rooted at $h'(0,0)$. We recursively embed F^* into H^*. This embedding assigns $\frac{c}{2} - 1$ PEs to the root of H^* and $\frac{c}{2}$ PEs to every other node of H^*. Observe that when $c = 2$, the forest F^* is empty and thus the recursive step does not assign any PEs to the root of H^*.

It is easy to see that when combining the embeddings of F' into H' and F^* into H^* (and thus H'), the root of H' is assigned $c-1$ PEs of F and every other node of H' is assigned c PEs of F. Since the shrink and isomorphic embeddings are used, and since the roots of trees T_i of F are assigned to the root of H', the dilation is unity. We have thus the following result.

Theorem 3.4: *A binary forest F consisting of complete binary trees $T_1, T_2, ..., T_{c-1}$, where T_i is of height i, can be embedded into a complete binary tree H' of height $c - \log c$ with unit dilation such that the root of H' is assigned $c - 1$ PEs and every other node of H' is assigned c PEs.*

We next show that the FT embedding achieves a unit level-load and a load of c when applied to Base-Case 2 of the EP embedding. Observe that forest F consists of binary trees $T_1, T_2, ..., T_{c-1}$, where each tree T_i corresponds to a subtree of the tree T whose leaves are also the leaves in T. Thus, level j of the tree T_i corresponds to level $k - i + j$ of the tree T, where $0 \leq j \leq i - 1$ and $1 \leq i \leq c - 1$. Since the root of T_i is assigned to the root of H' and T_i is embedded into H' using a shrink embedding, all the PEs assigned to a node of H' either belong to different levels of T or, if two sibling PEs are assigned to a node of H', then their common parent is also assigned to the same node. Thus, before the application of Case 2 of Refinement Lemma 3.1 the level-load is unity. Since applying Case 2 of Lemma 3.1 does not increase the level-load, the embedding of F into H' yields a level-load of 1 with respect to the tree T. It is easy to see that the load is at most c, since at most c PEs are assigned to a node of H and Case

2 of Lemma 3.1 is used.

4. The EL Embedding

We now describe the EL embedding which embeds an n-PE complete binary tree T into an m-node complete binary tree H with balanced leaf and interior utilization (i.e., $\mu_l = \lceil \frac{n+1}{2m} \rceil$ and $\mu_i = \lceil \frac{n-1}{2m} \rceil$). The dilation δ of this embedding is $2 \log \log m + 1$, the level-load λ_l is $O(\log \log m)$, and the load λ is $O(\log n)$. As stated in Section 2, in the EL embedding we first use the shrink embedding to embed T into T', then embed T' into H using the DLS embedding and a balancing step, and finally use a recursive step. The recursive step embeds one of the two subtrees representing leaves of T' and assigned to the root of H into H. The most delicate part of the EL embedding is the balancing step after the DLS embedding of T' into H. We thus start by describing how to embed an $2m+1$-PE tree T' into an m-node tree H. Recall that $m = 2^l - 1$. The embedding of T' into H achieves a dilation of $2 \log l + 1$, a level-load of $O(\log l)$, a load of $O(l)$, and a balanced leaf and interior utilization. In this utilization exactly one node of H, namely the root, is assigned 2 leaf PEs and 1 interior PE of T', and all the other nodes of H are assigned 1 leaf and 1 interior PE each.

We assume w.l.o.g. that l is a power of 2 (the bounds change by an additive constant otherwise). The embedding of T' into H consists of 2 steps: a DLS embedding step and a balancing step. The DLS embedding assigns $2l - 1$ interior and 2 leaf PEs of T' to the root of H, and $l - r - 1$ interior and 1 leaf PE of T' to every other node of H at level r. Furthermore, it achieves unit dilation, unit level-load, and a load of $l - 1$. Recall that in the balancing step of the EP embedding PEs from the subtrees of T representing leaves of T' were reassigned in order to achieve a balanced utilization. But doing so did not achieve a balanced leaf and interior utilization. After the DLS embedding of T' into H, we already have a balanced leaf utilization. The balancing step described next reassigns only interior PEs of T' in order to achieve a balanced interior utilization. This reassignment causes the dilation to increase to $2 \log l + 1$ and the level-load to $O(\log l)$. The load remains $O(l)$. Note that, given the DLS embedding, one could not achieve a smaller dilation, since we assigned $2l - 2$ additional interior PEs to the root of H and their reassignment has to result in a dilation of $\Theta(\log l)$.

In the balancing step the left and right subtree of H are handled separately. Within each such subtree H' every interior PE is pushed onto a higher level. If a PE is assigned to node v at level l_v in H', then it will be pushed to a node that is a descendent of v on level $l_v + r$, $r \leq \log l$. The balancing is done in a bottom-up fashion in $\log l - 1$ stages. The ith stage considers subtrees H'' of height $2^i + i$ rooted at nodes on level $l - 2^{i+1}$. At the beginning of the ith stage the situation within each subtree H'' is as shown in Figure 3(a). The bottom i levels of H'' have no interior PEs assigned to them, and the top 2^i levels have the assignment of interior PEs as generated by the DLS embedding. During the ith stage the interior PEs are pushed down so that at the end the top $i+1$ levels of H'' have no interior PEs assigned to them, and every node on another level has exactly 1 interior PE assigned to it. Furthermore, a PE at a node v in H'' is reassigned to a descendent of v at most $i + 1$ levels below the level of v. Figure 3(b) shows the result of stage i on H''. We next give an informal outline of the balancing step.

procedure BAL
Input: a tree H of height l and the assignment of interior PEs generated by the DLS embedding.
Output: the final assignment of the embedding of T' into H.
1. for $p = 0$ to 1 do
2.   Let H' be the subtree of height $l - 1$ rooted at $h(1,p)$.
3.   for $i = 1$ to $\log l - 1$ do (* execute i^{th} stage *)
4.   for every node v in H' on level $l - 2^{i+1}$ do
            Let H'' be the subtree of height $2^i + i$ rooted at v;
5.   RE_INT (H'', i).
   endfor (* 1, 3 and 4 *)
   (* the assignment of $2l - 1$ interior PEs to $h(0,0)$ has not changed, nodes on levels 1 to $\log l$ have no interior PEs assigned to them, and nodes on level r with $r > \log l$ have 1 interior PE assigned to them *)
6. Arbitrarily reassign $2l - 2$ interior PEs from $h(0,0)$ to nodes on levels 1 through $\log l$ so that every node on these levels gets 1 interior PE assigned to it.
end BAL.

The heart of the balancing step is the procedure RE_INT which does the actual reassignment of interior PEs. At the ith stage procedure RE_INT works with subtrees H'' of height $2^i + i$. Recall the assignment of interior PEs to nodes of H'' from Figure 3(a). A crucial observation is that the total number of interior PEs assigned to nodes at the top 2^i levels of H'' is same as the total number of nodes in the bottom $2^i - 1$ levels. Procedure RE_INT reassigns interior PEs to obtain the assignment of Figure 3(b). This is done in a bottom-up fashion by starting to assign interior PEs to nodes at the bottom level of H'' and then marching upwards in H'', level by level. We keep the dilation at $2(i + 1) + 1$ by considering subtrees S of height $i + 2$. Initially, the leaves of S are the leaves of H''. We reassign interior PEs assigned to the root and its left and right child of every such subtree S to the leaves of S. Note that if a PE u is initially assigned to node v in H'', then the parent of u is assigned to the parent of v. Let PE u be assigned to the root of S by the DLS embedding. Since PE u is reassigned by RE_INT to a leaf of S and even if the parent of u is not reassigned to a node in S (i.e., the parent of u is reassigned to a leaf of the subtree of height $i + 1$ rooted at the sibling of the root of S in H''), the dilation can not be more than $2(i+1)+1$. The precise process of the reassignment in the rth iteration is described next.

Assume by induction hypothesis that we have reassigned interior PEs in $(r - 1)$ iterations so that the assignment to nodes at level j in H'' with $j \leq 2^i - r - 1$ did not change (i.e., every node has $2^{i+1} - j - 2$ interior PEs assigned to it), and every node at level $2^i - r$ has $2^i - r$ interior PEs assigned to it. Every node at level j with $2^i - r < j \leq 2^i + i - r$ has no interior PE assigned to it, and every node at level j with $j > 2^i + i - r$ has 1 interior PE assigned to it. Observe that for $r = 1$ our hypothesis trivially holds. The rth iteration considers subtrees S of H'' of height $i + 2$ rooted at nodes on level $2^i - r - 1$ in H''. In every such subtree S, consider the subtree SL (resp. the subtree SR) of height $i + 1$ rooted at the left child (resp. right child) of $s(0,0)$. There are 2^i leaves in SL. We first reassign all interior PEs assigned to $s(1,0)$ to the leaves of SL. Since there are only $2^i - r$ interior PEs assigned to $s(1,0)$ after $(r-1)$ iterations, we reassign r interior PEs assigned to $s(0,0)$ to the leaves in SL that did not get a interior PE from $s(1,0)$. A similar reassignment is done for the leaves of SR. Once the reassignment is done for all the subtrees in the rth iteration, every node at level $2^i - r$ in H'' has no interior PE assigned to it, and every node at level $2^i - r - 1$ has

$(2^{i+1} - (2^i - r - 1) - 2) - 2r = 2^i - r - 1$ interior PEs assigned to it. The assignment to nodes at level j with $j \leq 2^i - r - 2$ has not changed. Since the leaves of S correspond to nodes at level $2^i + i - r$ in H'', every node at this level has 1 interior PE assigned to it. There are $2^i - 1$ iterations in RE_INT and after the $(2^i - 1)^{st}$ iteration, every node on level j in H'' with $j \leq i$ has no PE assigned to it and every node on level j in H'' with $j > i$ has 1 PE assigned to it. This completes the description of procedure RE_INT.

We now return to procedure BAL. It follows easily from RE_INT that after all $\log l - 1$ stages in BAL have been completed, the assignment of $2l - 1$ interior PEs to the root $h(0,0)$ did not change, every node on level j in H with $1 \leq j \leq \log l$ has no PE assigned to it, and every node on level j in H with $j > \log l$ has 1 PE assigned to it. There are $2l - 2$ nodes at levels 1 through $\log l$ in H. In Step (6) of procedure BAL $2l - 2$ of the $2l - 1$ interior PEs at the root $h(0,0)$ are reassigned arbitrarily to these nodes so that every one of them gets 1 interior PE assigned. This completes the reassignment in the balancing step. We now can state the following result.

Lemma 4.1: *A $2m + 1$-PE complete binary tree T' can be embedded into an m-node complete binary tree H so that the dilation is $2 \log \log m + 1$, the level-load is $O(\log \log m)$, the load is $O(\log m)$, the leaf utilization is 2, and the interior utilization is 1.*

Proof: Recall that $m = 2^l - 1$. It is easy to see that the dilation achieved in the reassignment of interior PEs by the balancing step is $2 \log l + 1$. In order to prove that the level-load is $O(\log l)$, consider an edge (v_1, v_2) in H where v_1 is the parent of v_2. Let i be the stage of procedure BAL so that at the end of stage i nodes v_1 and v_2 are reassigned 1 interior PE each. Let H_q be the subtree of height q rooted at v_2, where $q = i + 2$. Assume that the PEs on level x of T give the maximum level-load over the edge (v_1, v_2), $1 \leq x \leq l - 1$. The level-load over edge (v_1, v_2) is determined by the number of interior PEs on level x which are reassigned to nodes in H_q but whose parent PEs are not reassigned to nodes in H_q. In the reassignment done in Step (5), interior PEs assigned to at most $q - 1$ ancestors of v_1 are reassigned to nodes in H_q. Now, since at most 1 interior PE on level x is assigned to a node by the DLS embedding, at most $q - 1$ interior PEs on level x are reassigned to nodes in H_q. Their parent PE however, may not be reassigned to a node in H_q. Thus, the total level-load of edge (v_1, v_2) is at most $O(q - 1)$. Since q is at most $\log l + 1$, the level-load before Step (6) in BAL is $O(\log l)$. Similarly, we can show that the load is $O(l)$. It is clear that the reassignment from $h(0, 0)$ in Step (6) results in a dilation of $2 \log l + 1$, a level-load of at most 2, and a load of at most $O(\log l)$.

Finally, consider the assignment of leaf PEs done by the DLS embedding and the reassignment of their parent PEs by the balancing step. It is easy to see that the dilation is $\log l$, the level-load is $\log l + 1$, and the load is $\log l + 1$. Combining this with the bounds on dilation, level-load and load after the balancing step yields the claimed bounds and hence the lemma follows. ∎

By using ideas similar to the ones of the recursive step of the EP embedding and the balancing step of the embedding of T' into H, we can establish the following result about the EL embedding. The details of the EL embedding will be given in the full version of the paper.

Theorem 4.2: *An n-PE complete binary tree T can be embedded into an m-node complete binary tree H with dilation δ of $2 \log \log m + 1$, level-load λ_l of $O(\log \log m)$ and load λ of $O(\log n)$ so that the leaf utilization μ_l is $\lceil \frac{n+1}{2m} \rceil$ and the interior utilization μ_i is $\lceil \frac{n-1}{2m} \rceil$.*

5. References

[1] R. Aleliunas, A. Rosenberg, "On Embedding Rectangular Grids into Square Grids," *IEEE Trans. on Comp.*, Vol. V-31, pp. 907-913, 1982.

[2] K. E. Batcher, "Design of Massively Parallel Processor," *IEEE Trans. on Computers*, Vol. C-29, 1980.

[3] S. Bhatt, F. Chung, F. T. Leighton, A. Rosenberg, "Optimal Simulations of Tree Machines," *Proc. of 27th FOCS*, pp. 274-282, 1986.

[4] A. Borodin and J. E. Hopcroft, "Routing, Merging, and Sorting on Parallel Models of Computation," *JCSS* 30, pp. 130-145, 1985.

[5] S. A. Browning, "The Tree Machine: A Highly Concurrent Computing Environment," *Ph.D. Thesis*, Computer Science Dept., Caltech., 1980.

[6] L. Guibas, H. T. Kung, and C. D. Thompson, "Direct VLSI Implementation of Combinatorial Algorithms," *Caltech Conf. on VLSI*, 1979, pp.509-525.

[7] A. K. Gupta and S. E. Hambrusch, *manuscript*, 1987.

[8] W. D. Hillis, "The Connection Machine," MIT Press, 1985.

[9] K. Hwang and Y. H. Chung, "Partitioned Algorithms and VLSI Structures for Large-scale Matrix Computation," *Proc. of the 5th Symposium on Computer Arithmetic*, May 1981, pp. 222-232.

[10] H. Jai-Wei, K. Mehlhorn, A. Rosenberg, "Cost Trade-offs in Graph

Embeddings, with Applications," *JACM*, pp. 709-728, 1983.

[11] S. R. Kosaraju, M. Atallah, "Optimal Simulations between Mesh Connected Array of Processors," *18th ACM SIGACT Conf.*, pp.264-271, 1986.

[12] A. L. Rosenberg, "Preserving Proximity in Arrays," *SIAM J. on Computing*, pp. 443-460, 1979.

[13] J. T. Schwartz, "Ultracomputers," *ACM Trans. on Prog. Lang. and Systems*, Vol. 2, No. 4, Oct. 1980, pp. 484-521.

[14] J. D. Ullman, "Computational Aspects of VLSI," CSP, 1984.

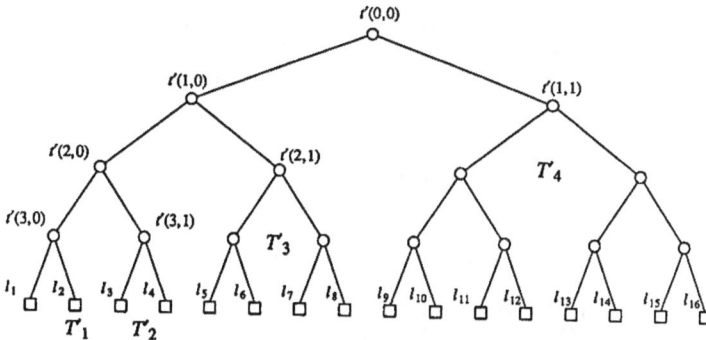

(a) Tree T' used in the DLS embedding.

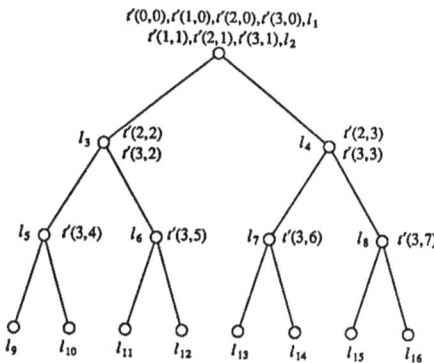

(b) Tree H' in the DLS embedding showing assignment of T' when $l=4$.

Figure 1.

T :

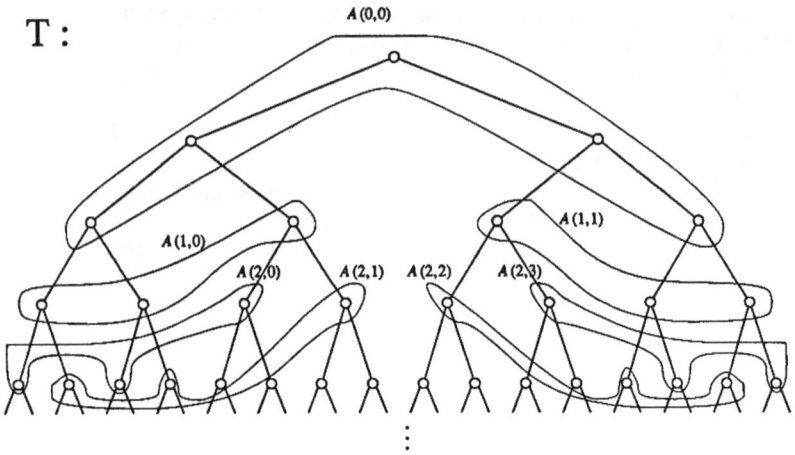

Figure 2: Some assignments of the LLRR embedding for $c=3$.

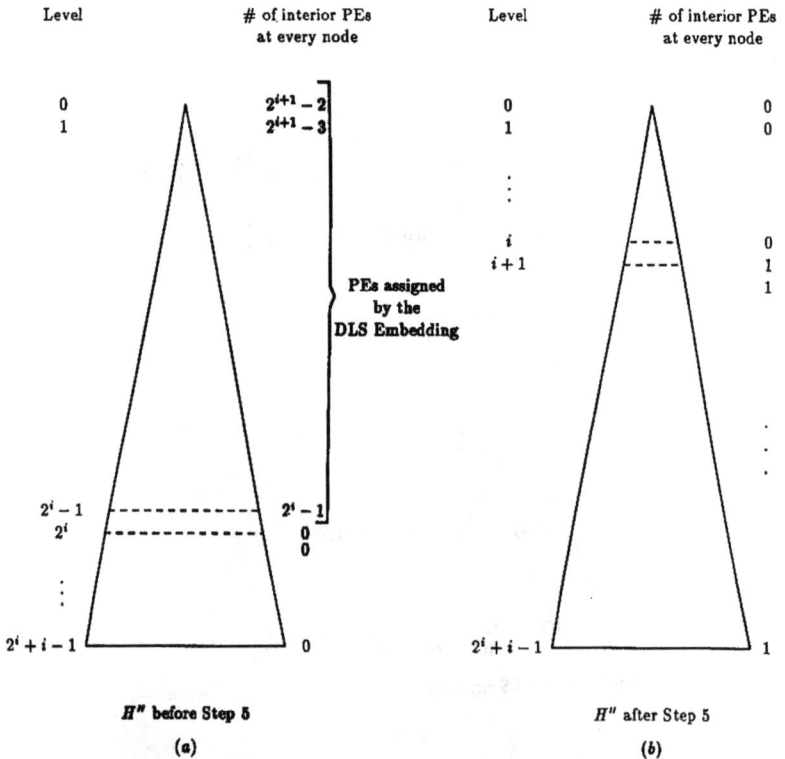

Figure 3: Schematic Process in the ith stage of procedure BAL.

Invited Talk

Algorithms for Multi-Level Logic Synthesis

R. K. Brayton

Department of Electrical Engineering and Computer Science
University of California Berkeley, CA

This talk will give an overview of the algorithmic approach to multi-level logic synthesis that has been developed over the last few years. The algorithms are divided into two groups; technology independent methods, and technology mapping. The technology independent approach depends on two important abstractions, the concept of a Boolean network, and the concept of the factored form representation of a logic function. The objective of multi-level logic synthesis is to take a description of multi-level logic, typically as extracted from some high level language, and optimize it for minimum area subject to timing constraints. We do this in three stages. The first is to synthesize and minimize the logic for area, using the number of literals in the factored forms as the area measure. The second stage is to map this into some target library of gates, again using minimum area as the criterion. The third stage involves restructuring the logic to fit the timing constraints, which are inferred from the system's timing requirements and estimates of the timing for the systems components, and remapping into the library, now using timing as the criterion.

A system, MIS, has been developed which embodies these ideas. We will give results using MIS which illustrate the effectiveness of algorithms.

A Global Approach to Circuit Size Reduction
Extended Abstract

Leonard Berman and Louise Trevillyan

IBM T. J. Watson Research Center
Yorktown Heights, New York 10598

This paper is concerned with reducing the size of an arbitrary combinational circuit. We use techniques of data flow analysis to summarize a given circuit in a manner which appears to be useful across a range of logic design applications. We characterize a class of legal (i.e. function preserving) circuit transformations in terms of this summary information and describe a new algorithm from this class that uses approximate summary information to map the problem of reducing circuit size to that of finding a small cut in an associated graph. This algorithm has been implemented and forms part of an automatic design system in use within IBM. The algorithm we describe is both more general and more efficient than previous methods.

I. Overview

The problem of constructing small circuits for a function is important throughout computer science. From a theoretical standpoint it is significant since the size of a minimal circuit is a natural measure of the complexity of a function. From a practical standpoint, it is important since the cost of building a circuit for a function is closely related to the size of the circuit.

This paper is concerned with both aspects of producing small circuits. We describe a new theoretical approach to optimizing circuits based on the idea of maintaining a sequence of easily computed global invariants of the circuit. This approach yields several polynomial time algorithms for reducing the number of connections in a circuit. In this abstract we describe one that has been implemented and forms part of an automatic design system in use within IBM.

Our algorithm uses data flow analysis to approximate the effect which a wire has in a circuit. Using this approximation, we reduce the problem of finding an equivalent small set of connections of that wire to the problem of finding a small cut in a weighted directed graph. We believe the algorithm is the first to use global flow analysis, (or, in fact, any type of global analysis other than some flavor of exhaustive search) in the actual design of combinational networks rather than the design of data paths [7,12]. It is also the first to apply techniques from network flow to the problems of combinational logic design (as opposed to problems of physical design [10] or retiming [9]).

The algorithm is based on a number of new ideas. Given a circuit, $\mathscr{C}$, composed of NOR gates we define a function, $\mathscr{F}$, which associates with every wire, w, in the circuit a set of gates, $\mathscr{F}(w)$. We show that if we rearrange the connection points of w in such a way that $\mathscr{F}(w)$ is not changed, then the function computed by the new circuit (i.e. $\mathscr{C}$ with the connection points of w rearranged) is identical to the function computed by $\mathscr{C}$. We then show, using techniques of dataflow analysis, how to compute a quick approximation to $\mathscr{F}(w)$ that has the same property. Finally, we construct a weighted, directed graph with the property that every cut through the graph corresponds to a set of connection points for w that guarantees that $\mathscr{F}(w)$ is unchanged, and such that the cost of the cut equals the number of connection points of w in the new circuit. A straightforward application of the MIN/CUT [6] algorithm yields the new set of connections of w.

II. Background and Related Work

Automatic logic design has been an important area of both theoretical and practical research since the early days of digital design [13,14]. In the early days, the conceptual basis for representing boolean functions was as subsets of the vertices of the hypercube. This representation leads naturally to questions concerning minimal disjunctive form representations [3,11]. These investigations, in turn, lead naturally to efficient algorithms for the automatic design of PLAs.

Unfortunately, the representation of boolean functions as a sum of cubes did not lead naturally to efficient algorithms for designing multi-level circuits. There was work on this problem [8]; however, no practical algorithms were developed. Over the last few years work has revived in this style of logic minimization, and there have been significant advances in the problem of factoring single output logic from the sum of cubes representation [2]. These new algorithms have made clever use of the similarity between logic and polynomial algebra; however, they have not been successful in dealing with multi-output logic directly and have therefore had limited practical impact. Our method deals with both multiple output and single output circuits with equal facility.

Recently a new approach to the specification of boolean functions has been explored by logic designers [5]. In this style, a boolean function is represented as a straight-line program. Given this spec-ification as a starting point and using the techniques of compiler op-timization, researchers have begun to develop practical algorithms for synthesis of multi-level circuits. These *logic compilers* [4,16] would be familiar to most compiler writers. They function primarily as peep-hole optimizers, using local pattern matching to improve a circuit. The algorithm which we present has been applied in this context.

In this paper, we begin with a circuit. Using flow analysis, we derive a representation which seems to be a natural framework within which to consider optimizations of multi-level circuits. In this framework, the local transforms [4], which appear arbitrary at first glance, are seen to maintain a simple invariant. The algorithm we present is a natural generalization of these local circuit re-writes when viewed as an operator which is maintaining the same invariant. The representation is a generalization of the sum of cubes, in that, given a 2-level implementation our representation is identical to the sum of cubes. Although we have not pursued this, it suggests that some of the work directed towards manipulating and optimizing disjunctive form may apply in our context.

III. Summary of Technical Material

Circuits are represented as DAGs. For simplicity, we assume the nodes of the DAG to be NORs, INPUTs, or OUTPUTs although it is easy to extend the work to include arbitrary primitives. We will use the terms gate, node, and box interchangeably to refer to the nodes of our DAG. We will use the the terms wire or signal to refer to the collection of edges beginning at a node. Since we assume each gate has a single output, we will also identify the wire with the node which is its source. The meaning of terms such as *sinks of a wire* will, we hope, be clear from context.

Our work makes use of the following summary information[15]. For each wire, x, we are interested in four sets of wires, called *forcing sets*, which are defined:

$$F_{ij}(x) \equiv \{s : \text{if } x = i \text{ then } s = j\} \qquad i,j \in \{0,1\}$$

The condition in the set definition should be interpreted to mean: if each consistent assignment which results in an 'i' on wire x also results in a 'j' on wire s then s is in the set. The above set as well as the solution to the recurrences given below, obviously depend on the circuit involved; however, in order to simplify notation, we suppress this. We trust that the dependencies will be clear from context.

The exact computation of these forcing sets is not feasible since it would involve solving the satisfiability problem. However, we note that this collection of sets is a fixed point of the following recurrence equations (where $X(s) = \{inputs \ of \ box \ s\}$):

- $C_{10}(x) = \{s|\exists y \in X(s)[y \in C_{11}(x)]\} \cup \{s|\exists y \in C_{11}(x)[s \in X(y)]\} \cup$

$$\{s|x \in C_{10}(s)\} \cup C_{10}(x)$$

- $C_{11}(x) = \{s|\exists y \in C_{10}(x) \forall t \in X(y)[t \neq s \Rightarrow t \in C_{10}(x)]\} \cup$

$$\{s|\forall y \in X(s)[y \in C_{10}(x)]\} \cup \{x\} \cup C_{11}(x)$$

- similarly for C_{00} and C_{01}

It is not particularly important that the reader study these equations, but it is interesting to note that one wire can force values on wires either upstream or downstream, and by iteration, wires not in either "cone of influence" can be forced. We mention in passing that more complicated sets of recurrence equations could be used. This would result in improved summary information which would in turn lead to more opportunities for optimizations. (Similarly, simpler equations could be used [15] if the goal is to improve running times at the cost of optimization quality.)

The fact that the forcing sets are fixed points for these equations suggests that it may be useful to approximate them with the least fixed point of these equations since this can be computed easily.

We now make a definition that is central to our approach to circuit rewriting. Given summary information, $\mathscr{S}$, for a circuit, $\mathscr{C}$, we define a set for each signal, i, in the circuit. The set is denoted $\mathscr{F}_{\mathscr{C}}(i,\mathscr{S})$ and is called the $\mathscr{S}$-*frontier of i*. (The subscript may be omitted when it is clear from context.) The elements of $\mathscr{F}_{\mathscr{C}}(i,\mathscr{S})$ are the nodes, j, for which the following hold:

- $j \in \mathscr{S}(i)$,
- there is a path $j \to j_1 \to j_2 ... \to OUTPUT$ such that for no j_l is $j_l \in \mathscr{S}(i)$,
- j is reachable in the circuit from i.

We are now ready to state our first theorem. This theorem has as a corollary a result which characterizes a class of legal circuit transformations in terms of the notion of frontier.

Theorem 1: Let $\mathscr{C}$ be a circuit, and s a wire in $\mathscr{C}$. Let C_{10} be the least fixed point of the above set of recurrence equations and let F_{10} be the forcing sets defined earlier. Assume $C_{10}(s) \leq D \leq F_{10}(s)$ (ordered by inclusion). If $\mathscr{C}'$ is identical to $\mathscr{C}$ except that in $\mathscr{C}'$ s is an input only to those nodes in $\mathscr{F}_{\mathscr{C}}(s,D)$, then $\mathscr{C} \equiv \mathscr{C}'$.

Two applications of the above theorem yield the following corollary.

Corollary 1: Let $\mathscr{C}$ and $\mathscr{C}'$ be two circuits which are identical except in the set of gates to which one wire, s, is an input. Let C_{10}, C_{10}', F_{10}, and F_{10}' be as defined above for $\mathscr{C}$ and $\mathscr{C}'$ respectively. Assume $C_{10} \leq D \leq F_{10}$ and $C_{10}' \leq D' \leq F_{10}'$. If $\mathscr{F}_{\mathscr{C}}(s,D) = \mathscr{F}_{\mathscr{C}'}(s,D')$ then $\mathscr{C} \equiv \mathscr{C}'$.

This corollary suggests a method for reducing the size of a circuit. If we rearrange the connections of a wire without changing its associated C -frontier, the circuit will compute the same function. The questions we must answer are how to do this efficiently and whether optimizations based on this idea will produce smaller circuits than previous methods.

From a practical standpoint, we see immediately that because of efficiency questions, we will be interested primarily in the C -frontier; however, we note that by extending C_{10} upwards (by adding elements in F_{10}) we can approximate the F-frontier. Theorem 1 holds for these enlarged fixed points, and, furthermore if the optimization which we obtain with the C-frontier is not adequate, we can attempt to improve it by one of these enlarged fixed points. We note also that the problem of finding the minimal number of connections of a signal which maintain the same C-frontier can be translated easily into integer linear programming. Because of the large size of the problems which occur we do not think this is a practical method for applying Theorem 1.

Our method of using the above corollary is to construct a DAG, G_s, which represents the effect that wire, s, has in $\mathscr{C}$. Intuitively, in G_s there will be a source node which represents s, and a sink node which represents the "outside". Paths in G_s will correspond to ways in which the effect of s travels through the circuit to the C-frontier of s and then on to the external world. Our construction of G_s will guarantee that cuts through G_s can be translated into sets of connections of s which leave the C-frontier of s invariant.

We construct a DAG, G_s, which we call the *derived graph of s* using the following procedure:

   0)Put a source node labelled s and a sink node labelled SINK in G_s.

   1)For each i, if $i \in C_{10}(s)$ and i is reachable from s then add node i to G_s.

   2)For each i added in step 1,
         for each $k \in \{inputs\ of\ i\}$
            if $k \neq s$ and $k \in C_{11}(s)$
                  then for each $l \in \{inputs\ of\ k\}$ add the edge (l,i) to G_s
            if $k = s$ then add an edge (s,i) to G_s

   3)Take the transitive reduction [1,p.219] of G_s and assign weight of 1 to the remaining edges.

   4)For each node, k, added in step 2 above, if $k \in C_{10}$-frontier of s then add an edge of infinite weight from k to the sink.

   5)Split nodes so that a cut of the graph will correspond to cutting nodes of this graph rather than edges.

We prove the following theorem concerning this graph.

Theorem 2: Let K be any cut separating the source and sink in G_s, and let $\mathscr{C}'$ be a circuit which is identical to $\mathscr{C}$ except that the sinks of signal s are precisely those nodes which are cut by K, then $\mathscr{F}_{\mathscr{C}}(s, C_{10}) = \mathscr{F}_{\mathscr{C}'}(s, C_{10})$.

This result combined with Theorem 1, shows how to rearrange the connections of signal s without changing the function computed by the circuit. The circuit rewriting is done as follows: Given a circuit and a wire, s, construct G_s, use the MIN/CUT algorithm [6] to find a cut-set of the derived graph, and rearrange the connections of s based on this cut-set.

IV. Practical Considerations & Experience

Many current practical logic compilers [4,16] use pattern matching, based on DeMorgan's laws, with an objective function oriented towards reducing the number of connections. These are implemented with 5-10 seemingly dissimilar graph re-writing transformations. (The accompanying figure illustrates two typical re-writes. A more complete list of these re-writes, as used in the Logic Synthesis System -LSS- is given in [4].)

The *ad hoc* nature of these rewrite rules has been a source of concern to those of us who have been trying to develop practical logic compilers. One early concern was whether the set of re-writes in use was powerful enough. In fact, during the early phases of the development of LSS, additional re-writes were added on a number of occaisions and there is no theoretical reason to think that no more will be needed in the future. Another concern involves the order of application of re-writes. The order of application of transformations is significant; however, there is no reason, other than experimentation, for preferring one order to another.

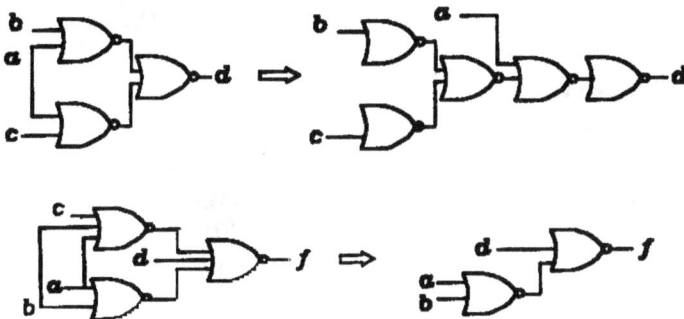

OBSOLETE LOCAL TRANSFORMS

Our new optimization results address both of these concerns. A close examination, of the rewrites shown in [4], shows that the first five of the re-writes, which form the main part of logic optimization of [4], produce changes to the circuit which Theorem 1 guarantees to be correct. This shows that, when viewed in the proper framework, these rewrites are actually different aspects of the same process and permits us to replace these five re-writes with one application of our optimization technique in the compilation process. Because the approximations which we make in our method, by using the C − frontier and the MIN/CUT based algorithm, are above board, we are much less concerned by the nature of our techniques than by the local transforms which they replace. (Of the other four re-writes shown in [4], three of them can be characterized conveniently in another way using the notion of forcing sets. We will describe this in a later paper.)

Algorithms similar to those described here have been implemented and form part of LSS. Before incorporating them into our system, experiments were run to evaluate their effectiveness. In these comparisons with earlier versions of LSS, which were deemed acceptable for production use, our methods yielded significant improvements in both cell and connection count. In addition, it reduced run time by 1/3.

Table 1 (see adjacent page) gives a comparison of the results obtained using these methods with those obtained using LSS with local transformations. The results presented compare the system using the methods described here, as well as the other usage of forcing information alluded to above, with the system using only local transformations. The test cases come from various IBM products that range from the low end to the high end of the product line. All of this logic was synthesized in its product form and no manual implementations are available. No timing correction was included in the LSS scenario. The first line in each is the result from applying local transformations, and the second line is the result from global transformations. The times given are for the total run time of the logic on an IBM 3081 processor, including processes such as parsing, factoring, fan-in correction, and repowering. The target technology was 4-way NORs with fanout constrained to a maximum of 8.

NAME	SIG	CON	BOX	S//C	A//W	CPU
CNTL	73	101	50	1.38	2.87/9	0:10
	71	96	48	1.35	2.87/9	0:06
RS	64	122	60	1.91	10/10	0:09
	65	122	61	1.81	10/10	0:06
VJ	76	149	69	1.96	7.66/9	1:11
	75	145	68	1.93	7.66/9	1:09
XK	296	470	238	1.59	13.8/16	0:34
	301	469	249	1.56	12.2/14	0:25
ADDS	874	1141	673	1.31	5.86/6	0:47
	873	1140	672	1.31	5.85/6	0:39
BMUX	529	956	393	1.8	6.11/7	0:54
	514	824	378	1.6	6.11/7	0:43
AMUX	738	1260	541	1.71	7.14/9	1:08
	717	1091	520	1.52	7.14/9	0:51
PACK	1452	3335	1321	2.30	15.5/18	12:22
	1386	2718	1255	1.96	17.1/21	9:48
S2M	2769	5977	2687	2.16	23.5/47	13:41
	2703	5670	2622	2.08	21.6/47	9:58
Z40	3167	6395	3016	2.02	23.4/44	13:32
	3123	5828	2973	1.87	21.4/42	7:13

S//C -- ratio of connections to signals
A//W -- average and worst number of levels in design

TABLE 1

In all, twenty examples were run. In ten of them, the results were identical with those of the local transformations, although the CPU times for the global method were about 75% of those for the local methods. The cases in which the results were identical tended to be those with small amounts of shallow logic. This

result is to be expected, since the broadening of the optimization window has less effect in shallow logic. This trend can also be seen in the results given below; as the logic becomes larger and deeper the global algorithms have more effect on the implementation.

From these experiments, it can be seen that the new algorithms generally produce logic that is no worse than the local methods and that they do it in less time.

In the cases that were not identical, both the NOR-gate count and the levels of logic were generally reduced. The exceptions to this are examples ADDS1, in which the gate count is smaller but the logic levels are the same, PACK, in which the gate count is smaller but the levels have increased, and example RS in which the gate count has gone up by 2% but the levels of logic have remained the same. In the case of ADDS1, the logic is marginally better than with the old method. With PACK, it is difficult to judge which implementation is better since a decrease of 66 gates is offset by an increase of 1.6 in the average number of levels of logic. In the case of RS, the new implementation is definitely somewhat worse.

V. Conclusion and Future Work

We have presented a new approach to practical circuit size reduction. Our methods are global in nature, yet efficient to implement. They look at information propagation rather than the precise connection of wires in a circuit. Our technique is based on the idea of summarizing a circuit and using the summary information to perform optimizations. We have found that our methods yield superior results and offer insight into connections among the apparently unrelated peep-hole optimizations which have been used in logic compilers.

Our approach suggests many areas for further work. We present a few of the more theoretical here.

- Is F_{10}-frontier (or C_{10}-frontier) a representation of the circuit, i.e. can two non-equivalent circuits have the same frontiers for all signals?

- The third clause in the definition of frontier was somewhat arbitrary. It could have been 'i not reachable from j' or it could have been omitted entirely. If the frontier does not represent the circuit, perhaps with the definition changed in one of these (or some other) ways it would be a representation. We think this would be worthwhile as any new representation for circuits is likely to lead to added insights.

- Real circuits are large and changing the connections of one signal may change the C_{10} sets and frontiers of other signals, even though its own frontier is not altered. How can we maintain the C_{10} information incrementally?

- Even if we limit ourselves to methods guaranteed correct through Theorem 1, our results are far from optimal. As described, G_s contains unneeded edges and may result in unnecessary connections. We can show that constructing an optimal derived graph is NP-hard. Are there alternative methods (i.e., other than constructing an analog of G_s and using MIN/CUT) for utilizing Corollary 1? For example, can the dominator's algorithm be extended to apply here?

- How can this approach be used to reduce the connections of all the wires simultaneously?

- Can algorithms which were developed to manipulate disjunctive form be generalized to this context?

- In our construction of the derived graph, each connection is given equal cost. Our weighting function could be adjusted to take other aspects of the design goal into account, e.g. timing constraints. How can this be done most effectively?

Acknowledgements

We would like to thank Larry Carter and Bill Joyner for many helpful discussions.

Bibliography

[1] A. V. Aho, J. E. Hopcroft, and J. D. Ullman, *The Design and Analysis of Computer Algorithms*, Addison-Wesley, 1974.
[2] R. K. Brayton, *Algorithms for Multi-Level Logic Synthesis and Optimization*, IBM Research Report RC 11938, June 1986.

214

[3] A. Cobham, R. Fridshal, and J. H. North, *An Application of Linear Programming to the Minimization of Boolean Functions*, Proc. 2nd Symposium of Switching Circuit Theory and Logical Design, Detroit, Oct. 16-20, 1961, pp.3-9.

[4] J. A. Darringer, W. H. Joyner, Jr., C. L. Berman, L. Trevillyan, *Logic Synthesis Through Local Transformation*, IBM Journal of Research and Development, v.25, no. 4, pp.272-280, July 1981.

[5] S. Dudani and E. Stabler,*Types of Hardware Description*, Proc. 6th Symp. on Computer Hardware Description Languages, Pittsburgh, Pa., May 23-25, 1983. pp.43-54.

[6] L. R. Ford, Jr., D. R. Fulkerson, *Flows In Networks*, Princeton University Press, Princeton, New Jersey, 1962.

[[7] S. C. Johnson, *Code Generation for Silicon*, Proc. 10th POPL, Austin, Texas, Jan. 24-26, 1983. pp.14-19.

[8] R. M. Karp, F. E. McFarlin, J. P. Roth, and J. R. Wilts, *A Computer Program for the Synthesis of Combination Switching Networks*, Proc. 2nd Symposium of Switching Circuit Theory and Logical Design, Detroit, Oct. 16-20, 1961, pp.182-194.

[9] C. E. Leiserson and J. B. Saxe, *Optimizing Synchronous Systems*, Proc. 22nd FOCS pp.23-36 (1981).

[10] A. S. LaPaugh, *Algorithms for Integrated Circuit Layout: An Analytic Approach*, Ph.D. dissertation, MIT, 1980.

[11] E. J. McCluskey, Jr. *Minimization of Boolean Functions*, Bell System Tech. Journal, v.35, no. 6, Nov. 1956.

[12] J. M. Siskind, J. R. Southard, K. W. Crouch, *Generating Custom, High-Performance VLSI Designs from Succinct Algorithmic Descriptions*, Proc. Conference on Advanced Research in VLSI, MIT, Jan.25-27, 1982. pp.28-39. Artech House, Inc. 1981.

[13] C. E. Shannon, *A symbolic Analysis of Relay and Switching Circuits*, Trans. A.I.E.E. v. 57, pp.713-723, 1938.

[14] C. E. Shannon, *The Synthesis of Two Terminal Switching Circuits*, Bell Systems Technical Journal, pp.59-98.

[15] L. Trevillyan, W. Joyner, L. Berman, *Global Flow Analysis in Automatic Logic Design*, IBM Research Report RC 10340, Jan. 1983.

[16] A. K. Vaidya, D. L. Dietmeyer, M. K. Engh, *WISLAN-- Technology Transformation and Optimization*, Proc. 6th Symp. on Computer Hardware Description Languages, Pittsburgh, Pa., May 23-25, 1983. pp.43-54.

Depth First Search and Dynamic Programming Algorithms for Efficient CMOS Cell Generation

Reuven Bar-Yehuda†, Jack A. Feldman, Ron Y. Pinter, and Shmuel Wimer

IBM Israel Scientific Center
Technion City
Haifa 32 000, ISRAEL

We describe a new algorithmic framework for mapping CMOS circuit diagrams into area-efficient, high-performance layouts in the style of one-dimensional transistor arrays. Using efficient search techniques and accurate evaluation methods. the huge solution space that is typical to such problems is traversed extremely fast, yielding hand-layout quality designs. In addition to generating circuits that meet prespecified layout constraints in the context of a fixed target image, on-the-fly optimizations are performed to meet secondary optimization criteria. A novel routing tecnnique is employed to accommodate the special conditions that arise in this context. Several instances of such algorithms have been implemented, and applied successfully (in a real-time setting) to an entire library of fairly large leaf cells.

1. Introduction

The automatic generation of high-performance integrated circuits in the layout style of a one-dimensional transistor array, as suggested by Uehara and vanCleemput [1], has been studied recently from a number of different angles. In some cases [2,3,4], the primary goal was to minimize the amount of diffusion required in the artwork, while other considerations, such as reducing internal wiring and accommodating performance constraints, were handled (if at all) as secondary issues.

In other cases [5,6], the order of importance was reversed: another criterion, such as low routing density or minimal wire length, was

---

† main affiliation: Computer Science Department, Technion, Haifa, Israel.

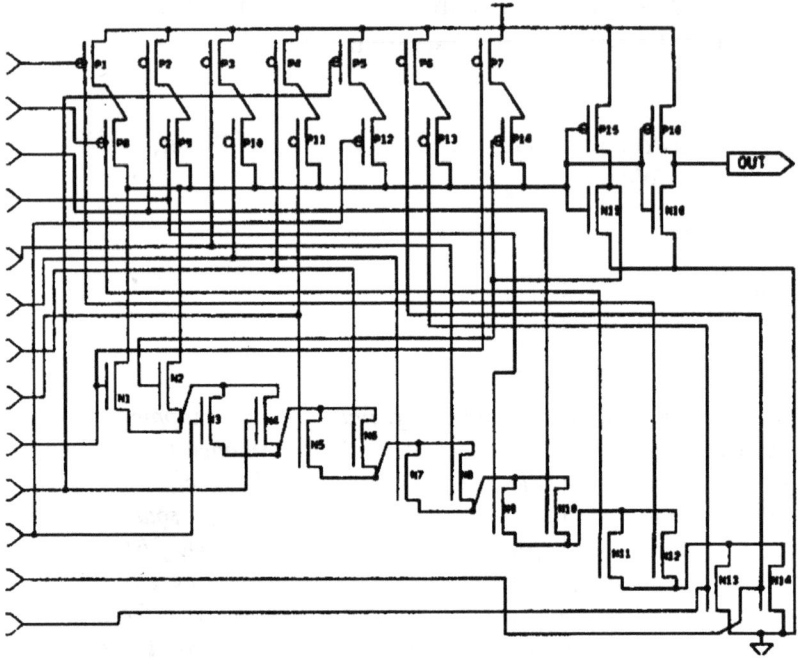

Figure 1. The circuit diagram for a CMOS latch

the driving goal, and then a considerable amount of time was spent minimizing diffusion gaps. In both cases, the primary goal was obtained reasonably efficiently, but accommodating other concerns

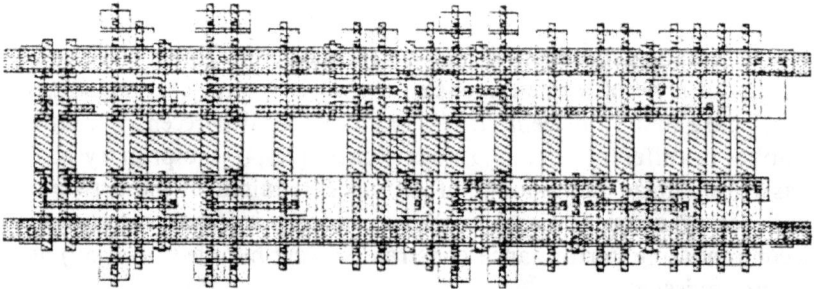

Figure 2. The layout of the CMOS latch from Figure 1. This result was obtained by running the algorithm allowing 2 routing tracks on each side, with minimal stretching of the diffusion runs.

came at a significant running-time cost (as the circuits' sizes grew larger).

In this paper, we propose an algorithmic[1] method in which the generation of the layout is driven by a set of optimization criteria (including diffusion adjacencies) and composition constraints at the same level of importance, allowing better control over the generation process. The solution space is expanded by a depth-first-search procedure using the constraints to effectively reduce the branching factors, and applying the optimization criteria to sharply bound and eliminate unnecessary expansions. Particular care has been taken in choosing the implementation method so as to provide truly optimal solutions when possible in linear-time: several incremental dynamic programming calculations are conducted on-the-fly, and the supporting data structures are maintained efficiently.

The results of our technique are layouts that fit tight intracell routing requirements, utilizing diffusion adjacencies wherever possible (but not at the cost of invalidating the routing constraints), and meet user specified performance constraints. In addition, the algorithms proposed here can handle *arbitrary* circuit graphs, as opposed to several restricted algorithms that can handle only series-parallel circuits [2,3] or only circuits having an equal number of p-type and n-type transistors. For example, the CMOS latch shown in Figure 1 was laid out as shown in Figure 2, using only four routing tracks (two over the p-transistors and two over the n-transistors), relatively short metal straps, and using as many diffusion adjacencies as possible. Our algorithms have been coded (in Pascal), and were applied to an entire library of fairly large leaf cells (containing several dozen transistors each) at substantial productivity gains.

The rest of this paper is organized as follows. Section 2 defines the target image of the layout, explaining some of the constraints. Then we describe our algorithmic framework in Section 3, and summarize our results in Section 4.

The functionality of such an algorithm is to map a circuit diagram (topology) into a layout (geometry).

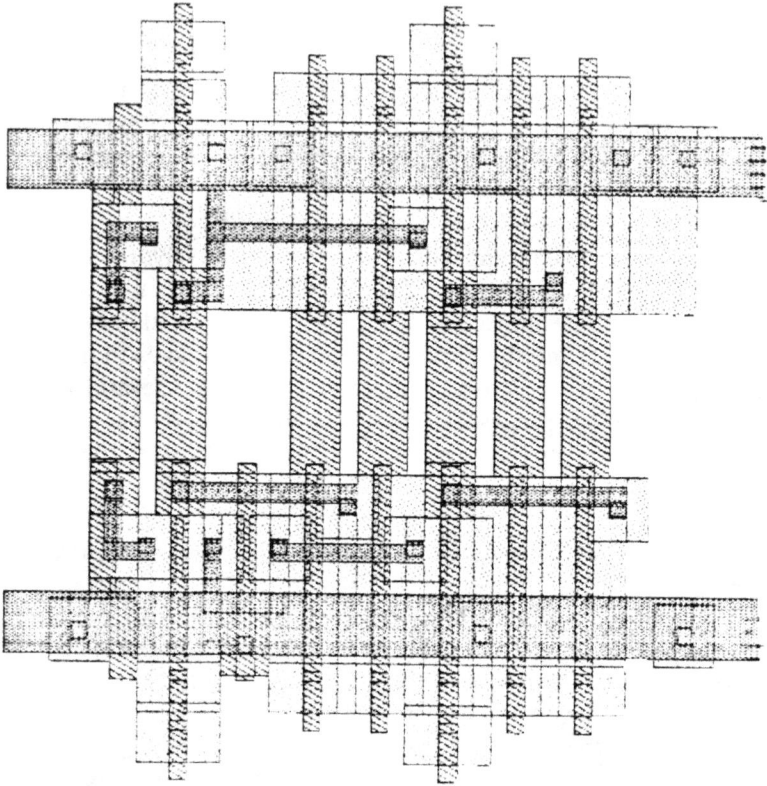

Figure 3. A small cell, exemplifying the fixed image. Notice the internal routing and the isolation device.

2. The layout image

The *image* of the target layout affects both the constraints and the optimization criteria that guide the generation algorithm. Here is one possible setting, which will also be used as the framework for the layout algorithm as described in the next section:

- the transistor are arranged in two horizontal, parallel rows - one for the p-type devices and the other for the n-type transistors
- the transistors can only be assigned to fixed horizontal locations, grid-fashion, at a fixed pitch

- diffusion is used to connect adjacent diffusion ports (source or drain) on each row, and vertical polysilicon lines connect horizontally aligned gates
- diffusion breaks (on either side) may be instantiated as gaps or as grounded devices [3,4]; either way, a break incurs a significant increase in width
- internal routing is performed in one layer of metal over each row of transistors, using a fixed number of wiring tracks; this (usually small) number is set as a parameter for the generation process, and is normally fixed for the whole cell library
- a reserved area is left between the rows for external routing, but it can be (and is, when possible) used for internal horizontal routing on poly between adjacent gates to reduce density
- contacts between diffusion and metal are placed anywhere outside the active area of a device, and contacts between poly and metal are allowed on top of gates; the contact spacing rules are somewhat complicated and are further explained in Section 3.3

This image is illustrated by the layout shown in Figure 3. Notice that only one layer is used for internal routing over the devices, aside from diffusion adjacencies and vertical polysilicon connections. Also, the diffusion break in the bottom row is implemented as an isolation device (a transistor whose gate is connected to power).

Working with a (relatively) fixed image as the one described here has a number of advantages, as argued in [7]:

- *control:* the algorithms used to generate cells have full control over all parameters of the layout and are not left to the mercy of a compactor which is used as a post process
- *complexity:* a fixed grid provides a succinct and clear level of abstraction of the physical layout, even more than the commonly used virtual grid (which suffers from lack of control and predictability), thereby making the layout algorithms easier to develop and maintain
- *composition:* when library cells are combined into a macro, much simpler layout tools than complex placement and routing techniques can be used in such a structured environment. The composition of cells into larger designs is smoother and - due to

the reduction in interconnect - most of the time also more space efficient

In addition, the utility of such an image is enhanced when having to formulate constraints and objectives other than just minimizing diffusion breaks or cell density, namely more accurate physical design measures such as the number and type of contacts, routing jogs, and wire lengths.

3. The layout algorithm

The layout algorithm has three main components, as follows:

1. The primary driver is a depth-first-search (DFS) [8] routine, expanding potential transistor *placements* (or orderings) along the generated array. This driver generates a virtual search tree on which a branch-and-bound procedure is performed. Only placements that can be subsequently routed are generated.
2. During the DFS expansion, the internal optimal orientation of the transistors in each potential placement is determined on the fly. The computation of these orientations is done in constant time for each new node that is expanded.
3. The internal, detailed *routing* of the best placement that was found by the DFS is then performed.

Each component is now described in greater detail in the following sections.

3.1. Depth-first-search expansion of transistors

Let $N_1,..., N_n$ and $P_1,..., P_p$ be the n-type and p-type transistors[2] of the input circuit, respectively. The set of all possible placements can be generated using an enumeration tree, where each node (except the root) is marked by an ordered pair $< P_i, N_j >$. If each transistor appears exactly once on the labelled path from the root to a leaf, then such a path represents a (say) left-to-right assignment of the transistors in the layout. A path from the root to an

---

[2]   If $n \neq p$ we introduce dummy devices, whose number and type resolve the difference.

internal node corresponds to a partial assignment. Note that each label stands for all four possible orientations of its associated P-N pair, each having a (possibly different) cost. We explain in Section 3.2 how the best of these orientations is picked; the cost of a partial layout depends on wiring measures and a number of other layout characteristics that can be computed dynamically, such as

- total or maximal wire length
- number of diffusion breaks
- wiring density
- alignment of transistors having the same signal for a gate (to facilitate vertical polysilicon connections)

These measures can be either combined into one single-valued function, using a weighting scheme, or put together in vectors that are then compared lexicographically. The first approach has the problem of mixing apples and oranges, whereas the second tends to bias the optimization towards the "primary" criterion. Our experience indicates that (with appropriate tuning) the second approach is quite reasonable.

A simplified variation of the depth first search (DFS) algorithm is given here (using pseudo-Pascal notation):

```
v:=root;
repeat
    while (there exists an acceptable orientation)
            and (v has an unvisited outgoing edge (v,u)) do
            begin
                for each orientation calculate its cost;
                v:=u; (* forward step *)
            end;
    if (v is a leaf) and (Cost(v) is acceptable)
        then begin
                record solution;
                update acceptability threshold;
            end;
    v := the father of v in the tree; (* backward step *)
until v=root;
output (minimal cost solution);
```

This DFS procedure is implemented efficiently, using the following branch-and-bound criteria:

1. the number of feasible edges that can be extended from each vertex is strongly bounded by wiring density considerations

2. backtracking occurs early due to the usage of a sharp objective function that prunes most of the suboptimal solutions
3. an efficient data structure is used to control the order in which the edges emanating from each vertex is visited

It is important to note that imposing constraints on the resulting layout - as opposed to minimizing a layout measure - can be gracefully incorporated into the DFS scheme. Moreover, it speeds up the DFS expansion. For example, there is no point in minimizing the length of a specific wire as long as it is far less then the allowable maximum - we might as well use this length as best we can.

3.2. Optimal orientation of devices

During placement, several possible orientations of the devices in each growing solution may be feasible at the same time, as long as the wiring density of the P and N sides does not exceed the given bound. We would like, however, to keep only the best solution, *i.e.* the one yielding the fewest diffusion breaks (which also eases the routing, reduces the number of contacts, and improves the cell performance).

This "optimal flip" problem was addressed in [5], and a branch-and-bound post-placement algorithm, which enumerates all the 2^{2n} possible orientations (for a circuit with $2n$ devices), was suggested. This problem was also addressed implicitly in [4] at the pairing stage, which is performed before placement. In this section we present a dynamic programming algorithm that solves this optimization problem in linear time. Moreover, our solution can be integrated into the DFS procedure, requiring constant time per node expansion.

For sake of presentation, we describe the scheme as if it were applied to a given sequence of pairs realizing the whole circuit, but it is easy to see how it can be combined with the DFS. The algorithm proceeds from left to right. At each step a new pair is processed and concatenated in its optimal orientation to the previously processed pairs. A device is in the "0" orientation if its left terminal is the drain and the right terminal is the source, or in the "1" orientation if the device is flipped. The orientation of a pair is denoted by the concatenation of the p- and the n-orientations, and the set

$\Psi = \{00,01,10,11\}$ denotes all possible orientations of a pair. Let P_i^σ, where $1 \le i \le n$ and $\sigma \in \Psi$, denote that the i-th P-N pair from the left is in the σ orientation (the superscript is omitted when the orientation is immaterial). Let c_i^σ be the cost of assigning the orientation σ to the i-th pair. Also, $d_{i,i+1}^{\sigma,\tau}$ denotes the cost resulting from concatenating P_i^σ to P_{i+1}^τ.

Finally, let $S_k^\sigma = (P_1, \ldots, P_{k-1}, P_k^\sigma)$, $1 \le k \le n$, $\sigma \in \Psi$ denote the k leftmost pairs whose orientation has been found so as to minimize the cost of their implementation, where the rightmost P-N pair is in orientation σ, and let C_k^σ denote the cost of S_k^σ.

It is easy to prove that given S_k^σ, the optimal orientation of P_{k+1} is independent of the orientations of all the P_i, $1 \le i \le k - 1$, and the cost of S_{k+1}^σ, $\sigma \in \Psi$, is given by

$$C_{k+1}^\sigma = \min_{\tau \in \Psi} \{C_k^\tau + c_{k+1}^\sigma + d_{k,k+1}^{\tau,\sigma}\}, \qquad \sigma \in \Psi. \tag{1}$$

The above observation yields a dynamic programming procedure to find the optimal orientation for P_{k+1}. After the n-th step is done, the optimal orientation is obtained by taking S_n^σ yielding the minimum among $C_n^{00}, \ldots, C_n^{11}$. From this final state we retrieve the optimal orientation by going backward from $k = n$ down to $k = 1$. Note that the calculation in (1) is performed four times for each pair, thus requiring $O(n)$ time altogether. Since S_k^σ is a sequence whose length does not exceed n, also only $O(n)$ memory is required.

3.3. Routing

At this stage, some of the internal connections were already taken care of by abutting adjacent diffusion ports, and others are handled trivially using vertical connections on poly between aligned gates. Also, all other connections between nets that reside on both rows (and these are rare) are realized by vertical poly lines that are inserted during placement. The routing step deals with the wiring of the remaining connections that must be performed in the given tracks such that as little stretching as possible is incurred (due to jogging and some special design rules that are explained shortly).

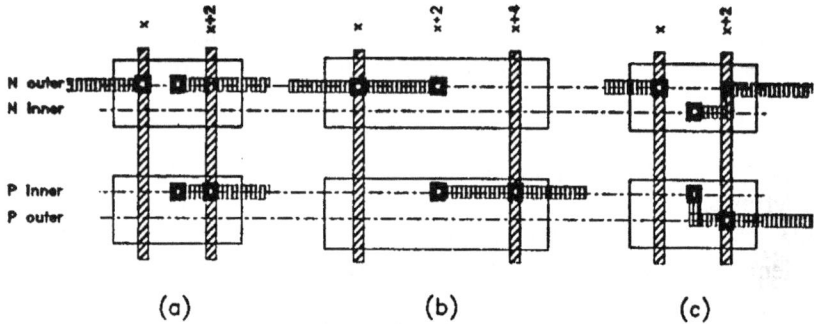

Figure 4. Accommodating the special spacing rules requires ragged intervals. (a) is the result of the placement (on the virtual grid), (b) is a naive solution (using straight intervals) for our image, and (c) is the more efficient solution, as supported by our algorithm.

Since routing is allowed over the two transistor rows in a prescribed number of tracks, the situation seems reminiscent of one-dimensional routing for which the track assignment problem can be solved using interval graph coloring [9]. Such a solution, however, restricts the metal to be instantiated as straight lines, but we have the additional flexibility (possibly at some cost) of stretching the layout horizontally to make room for jogs, affording better utilization of the available tracks (whose number is given and does not need to be minimized). Moreover, there are a few additional, special spacing rules that we must respect:

- contacts on the same track cannot be placed in consecutive columns (even if they belong to the same net)
- a metal wire segment and a contact that belong to different nets cannot reside in consecutive columns within one track
- a contact between poly and metal (that is placed on top of a transistor) is not allowed in the outside tracks of either row

One can model these additional constraints within the framework of interval graph coloring (*e.g.* by extending each interval on both ends), but such a simplistic solution would deteriorate the quality of the layout significantly. Figure 4 compares the solutions that would be obtained using straight lines to those achievable with less rigid techniques, using jogs and doglegs (notice the interaction between the two rows). This figure shows only the consequences of the first rule; similar effects result from the other two rules.

To handle this unique situation, we propose a dynamic programming algorithm which accomplishes the wiring whenever possible with minimal space insertion. The functionality of this algorithm is to assign (poly and diffusion) contacts and straight wiring segments to tracks, as well as vertical metal wires to connect between the tracks where necessary. We model contacts as 0-length intervals, and break up wiring segments into unit-length intervals. Initially, each such interval is assigned to its original x-coordinate as decided by the placement phase.

The algorithm propagates a wave of possible track assignments from left to right, keeping only the solutions that can potentially affect the optimum - dynamic programming fashion - according to the minimum width and other criteria. Two observations make this approach workable:

- the track population in the last 3 columns is all that matters even when the special spacing rules are taken into account
- the total layout cost (reflecting both the amount of stretching as well as wire-length) is *additive*

At every column, we first decide how to extend existing wires with the current unit-length intervals, and then decide on where to put the contacts, so as to minimize the cost. The set of potential solutions that can still be optimal is then updated. The size of this set might be exponential in the number of tracks, but since this number is small (4 for our image, and typically not much more) the algorithm performs reasonably fast.

This novel routing scheme can be viewed as a combination of maze routing [10] and column-by-column channel routing [11] (in one layer), offering new insight into wiring techniques. The proposed approach can be used for a wide class of problems, where the number of wiring tracks is given, and the objective is to minimize the width of the layout rather than its height.

226

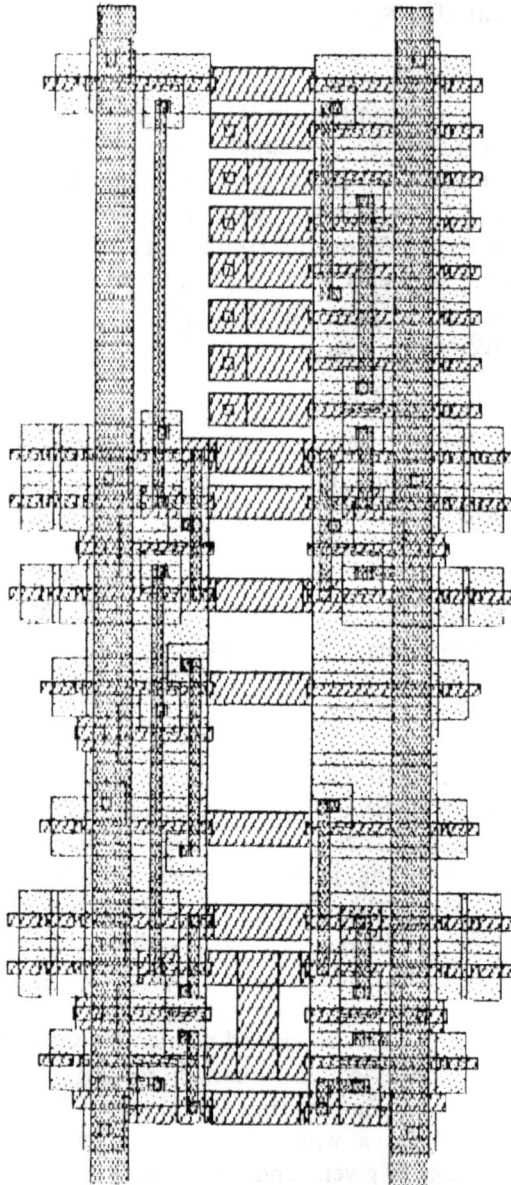

Figure 5. A large cell, with more n-type transistors than p-type devices

4. Discussion

We have described a new algorithmic method for laying out circuits from their schematics which has several advantages over other published algorithms, especially in terms of running times and the quality of the layouts produced. The algorithm has a clear formulation and at the same time is practical in the context of a cell generation tool. Our framework can accommodate additional optimization criteria as well as constraints that arise in certain design environments.

Our algorithms have been programmed into approx. 3000 lines of Pascal (not including interfaces to the front-end and back-end subsystems). Running times range from 5 to 50 CPU seconds on an IBM 4381 model 13 to generate one circuit with up to 60 transistors (depending on its structure), compared to several hours (or even days) per circuit by hand. To test the programs, an entire library (containing more than 300 cells) was generated automatically and the results were compared to an existing version that was done by hand. The whole process took one day (with one designer), compared to 3 months manually. In terms of quality, the automatically generated cells were never worse - and often better - than those created manually. Figures 3 and 5 exemplify two typical cells, displaying features that were traditionally considered out of the scope for automatic tools, such as routing jogs, dealing with an unequal number of p-type and n-type transistors, and polysilicon "bridges" between adjacent gates.

Acknowledgements. We would like to thank Israel Berger for his support and encouragement. We would also like to thank Emanuel Gofman for his contribution to the ideas that lead to the algorithm of Section 3.2, and to Ravi Nair of the IBM Watson Research Center at Yorktown Heights for making a layout back-end system available to us.

References

[1] Uehara T., and W. M. vanCleemput: Optimal Layout of CMOS Functional Arrays, *IEEE Transactions on Computers C-30, 5* (May 1981), 305-312.

[2] Nair, R., A. Bruss, and J. Reif: Linear Time Algorithms for Optimal CMOS Layout, in *VLSI: Algorithms and Architectures,* P. Bertolazzi and F. Luccio *(Eds),* Elsevier (North-Holland), 1985, 327-338.

[3] Müller, R., and T. Lengauer: Linear Algorithms for Two CMOS Layout Problems, *Proceedings of the Aegean Workshop on Computing, LNCS 227,* Springer-Verlag, July 1986.

[4] Wimer, S., R. Y. Pinter, and J. A. Feldman: Optimal Chaining of CMOS Transistors in a Functional Cell, *IEEE Transactions on Computer-Aided Design CAD-6, 5* (Sept. 1987), 795-801.

[5] Hill, D. D.: Sc2 - a Hybrid Automatic Layout System, *Proceedings of the ICCAD-85,* 172-174.

[6] Bhasker, J., and S. Sahni: Optimal Linear Arrangement of Circuit Components, *Journal of VLSI and Computer Systems, 1987,* 87-109.

[7] Nair, R.: MLG - a Case for Virtual Grid Symbolic Layout without Compaction, *Proceedings of the ICCAD-87,* 180-183.

[8] Even, S.: *Graph Algorithms,* Computer Science Press, Rockville, MD, 1979.

[9] Ohtsuki, T., H. Mori, E. S. Kuh, T. Kashiwabara, and T. Fujisawa: One-dimensional Logic Gate Assignment and Interval Graphs, *IEEE Trans. Circuits Syst., CAS-26* (Sept. 1979), 675-684.

[10] T. Ohtsuki: Maze-running and Line-Search Algorithms, in *Layout Design and Verification* (T. Ohtsuki, ed.), North-Holland, Amsterdam, 1986, 99-131.

[11] Burstein, M.: Channel Routing, in *Layout Design and Verification* (T. Ohtsuki, ed.), North-Holland, Amsterdam, 1986, 133-167.

Invited Talk

Structure in Parallel Processor Arrays

Arnold L. Rosenberg

Dept. of Computer & Information Science
Univ. of Massachusetts
Amherst, MA 01003

One can discover many computational properties of arrays of identical processing elements by studying the structures of the graphs that abstract the arrays' interprocessor communication networks. Such studies afford one a vehicle for comparing the abilities of competing architectures to handle a given pattern of algorithmic (data and control) dependencies, and to simulate arrays of other structures. Additionally, such studies allow one to uncover uniformities in the structure of an array, which one can exploit when devising algorithms to execute on the array. In this talk we describe and illustrate two mathematical approaches to studying the structure of processor arrays, one based on the theory of graph embeddings and one based on the theory of algebraically specified graphs. We survey recent results and sketch some proofs to suggest the nature of the modes of analysis.

Invited Talk

Artificial Neural Networks

J.J. Hopfield

California Institute of Technology
Pasadena, CA
AT&T Bell Labs
Murray Hill, NJ

Artificial neural networks are electrical (or optical) circuits designed
as a metaphor for the computational system of neurobiology. They
differ from most conventional VLSI in emphasizing large connectivity
of active devices; use of device analog properties in computation; use
of time evolution and time delays as a dynamical computational
system rather than simply as a transition between logical states; use of
feedback in the computational structure; and in the extensive use of
learning at the hardware level. Such differences presumably account
for the effectiveness of brains in difficult and computationally
intensive problems such as speech and vision. The mode of
computation by such networks will be described. A minimization
principle is of great use in understanding the stability and
programmability of these networks. A neural network approach to a
problem of word recognition in continuous speech for a small
vocabulary will be used to illustrate some of the origins of the
computational effectiveness of such networks. The status of
conventional and unconventional approaches to large scale neural
network hardware will be briefly reviewed.

A Dynamically Configurable Architecture For Prototyping Analog Circuits

Massimo A. Sivilotti

Department of Computer Science
California Institute of Technology
Pasadena, California 91125

A system intended to facilitate analog circuit design is presented. At its heart is a VLSI chip that is electrically configured in the field by selectively connecting predesigned elements to form a desired circuit, which is then tested electrically. An architecture built around a hierarchy of interconnect switches has been developed to accomodate the wiring requirements of large classes of systems. A polynomial-time programming algorithm for embedding the desired circuit graph onto the prefabricated routing resources is presented. Prototypes capable of synthesizing circuits of approximately 1000 transistors have been constructed; test results are presented.

1 Introduction

Even with fast turnaround fabrication, a considerable fraction of prototyping lead-time is consumed by the VLSI fabrication step. This delay is particularly expensive when prototyping subcircuits, or exploring new circuit structures and logic forms as a prelude to a large design project. The fabrication step is beyond the control of the designer, raising technical issues of yield and parametric behavior, as well as the logistic issues of cost, time, and the need to interface with an external vendor.

This paper presents the design of a VLSI prototyping platform, in which the Proto-chip, a field-reconfigurable IC, is used to actually wire up the circuit to be tested. The Proto-chip is particularly intended for synthesis and testing of analog neural-network architectures [10], which generally require large numbers of active elements, but usually with relaxed precision requirements. These demands make neural-networks a poor match for conventional software simulators. In addition to the obvious advantages of a hardware accelerator and simulator for fast prototyping, this system provides the first opportunity for a truly integrated design–fabricate–test environment.

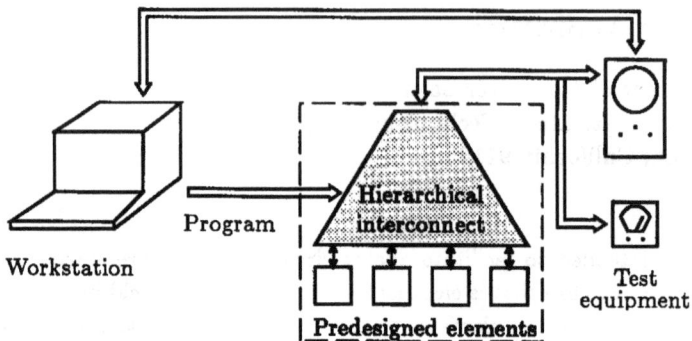

Figure 1 The Proto-chip environment. The Proto-chip is programmed by interconnecting the elements into a circuit that is then tested electrically.

The basic structure of the Proto-chip is a set of circuit elements arranged within a hierarchy of interconnect switches, facilitating the synthesis of the desired circuit. This programming process is controlled in the field by a workstation (Figure 1). Because this chip can effectively be considered an extension of a design-tool suite, similar performance criteria are applicable:

1. **Universality.** Because the circuits to be prototyped generally are designed *after* the Proto-chip has been fabricated, can the user be reasonably assured the system is capable of testing a usefully wide range of circuits?

2. **Functionality and performance.** Do the prototyped circuits function correctly? How useful is the tool as a simulator?

3. **Programmability.** How directly can the Proto-chip be configured into the desired circuit for testing?

4. **Scalability.** Can the prototyping-chip architecture exploit the additional capacity of improving VLSI processes? This concern raise issues of hierarchical architectures and modular programming algorithms.

5. **Flexibility.** Can the design of the Proto-chip itself be usefully adapted for synthesizing particular classes of circuit networks? The system can be customized at *two* levels: the basic circuit elements and interconnect structure are bound at fabricate time, and the actual Proto-chip is configured electrically in the field at run-time.

This paper concentrates on three aspects of the Proto-chip's design. First, an analysis of a general model for circuit connectivity is used to explore the requirements on the interconnect matrix. The case for a hierarchical architecture is presented, based on a desire to minimize these wiring resources, as well as some general observations regarding the organization of large systems. Second, an algorithm is developed for embedding a circuit graph onto the Proto-chip's hierarchical interconnect. Third, the structure and design of the fabricated Proto-chip is discussed, and test results presented.

2 Connectivity of Circuits

The general form of the prototyping chip is a collection of functional elements interconnected by a programmable switch matrix. This matrix embodies both the great flexibility of the system *and* its most severe constraint. Because the design of the switch matrix is bound when the Proto-chip is fabricated, it needs to accommodate unanticipated circuits, while not incurring a prohibitive area penalty.

To this end, it is highly instructive to examine some general interconnect properties of circuits. This section uses Rent's rule as a connectivity model to investigate various wiring strategies. The ultimate objective is the development of a hierarchical interconnect matrix.

It is important to distinguish between topology and structure; the former deals exclusively with connectivity, the latter adds placement and layout constraints. The high cost of wiring in VLSI circuits makes consideration of structure imperative, as do several objectives for particular networks (for example, an imaging array should be regularly placed); discussion of structure is, however, deferred to Section 5.2.

The least restrictive model of circuit connectivity requires every element to be connectable to any other. Defining N to be the number of elements permitted on the chip, and P_L to be the number of ports (pins) on each of these leaf elements, the required number of switches is

$$N_S = P_L^2 N^2$$

Of course, the requirement that any pin connect arbitrarily to any other pin (and possibly to any number of other pins) requires the full generality of a crossbar switch, with its associated quadratic overhead. We define a useful measure of circuit wiring complexity to be

$$\beta = \frac{N_S}{P_L^2 N}$$

which can be thought of as the number of other *elements* to which a particular element must fan out. For the crossbar,

$$\beta_{\text{crossbar}} = N$$

A circuit may be considered as a graph the vertices of which represent elements, linked by arcs corresponding to wires. Rent's rule [5] is an empirical relationship that relates the number of wires crossing an arbitrary boundary containing a group of elements:

$$P = P_0 N^b \tag{1}$$

where P is the number of wires that need to cross the border, N is the number of elements (each of which has P_0 pins) contained within the border, and b is an assumed constant. Analysis of several large designs have indicated that typical values lie in the range $b \in [0.5, 0.75]$ [5].

Now consider an interconnect chip with N elements. If the interconnect is flat (non-hierarchical), applying Equation 1 once gives the following number of outputs:

$$P_{\text{out}} = P_L N^b$$

The interconnection matrix can then be realized by a crossbar with $P_L N$ inputs and P_{out} outputs. The total number of switches is then

$$N_S = P_L^2 N^{b+1} \Rightarrow \beta = N^b \tag{2}$$

which, for large N, represents a sizeable improvement over the full-crossbar interconnect.

Another approach to connecting the N elements is incremental in nature: suppose k elements are connected simultaneously with a crossbar switch, then k of these meta-elements are connected, and so on. This procedure gives rise to the k-ary tree in Figure 2, of height $L = \log_k N$.

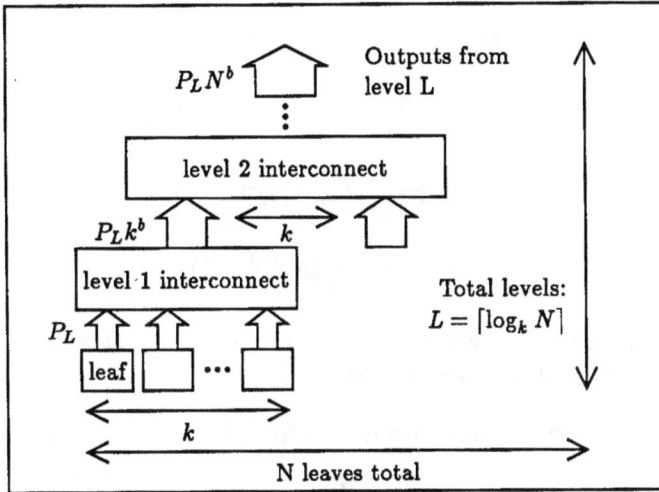

Figure 2   **Hierarchical interconnect parameters.**

The number of switches required to implement the ith level of interconnect is then given by:

$$N_{S_i} = \text{(inputs to each matrix at level } i) \cdot$$
$$\text{(outputs from level } i) \cdot \text{(number of } i^{\text{th}} \text{ level matrices)}$$

Level 1: $N_{S_1} = (kP_L)(P_L k^b)\left(\dfrac{N}{k}\right)$

Level 2: $N_{S_2} = \left(k(P_L k^b)\right)\left((P_L k^b)k^b\right)\left(\dfrac{N}{k^2}\right)$

Level i: $N_{S_i} = kP_{i-1}^2 k^b\left(\dfrac{N}{k^i}\right)$

where

$$P_i = \text{outputs of matrix at level } i = \begin{cases} P_L k^b & \text{if } i = 1 \\ k^b P_{i-1} & \text{otherwise} \end{cases}$$

The solution to this recurrence is $P_i = P_L k^{bi}$, which gives the number of switches at level i to be

$$N_{S_i} = P_L^2 k^{b+1} k^{2b(i-1)} \frac{N}{k^i} = P_L^2 N k^{(2b-1)i} k^{1-b}$$

The total number of switches is the sum over all levels in the hierarchy:

$$N_S = \sum_{i=1}^{\log_k N} P_L^2 N k^{1-b} k^{(2b-1)i}$$

This gives

$$N_S = P_L^2 \frac{k^b}{k^{2b-1} - 1} (N^{2b} - N)$$

$$\beta = \frac{k^b}{k^{2b-1} - 1} (N^{2b-1} - 1)$$

Now $2b - 1 \leq b$ for all $b \in [0,1]$, and, asymptotically, this expression gives us fewer switches than Equation 2.

In addition, an interesting case results from $b = 0.5$, which gives

$$\beta = \frac{\sqrt{k}}{\ln k} \ln N$$

It is useful to consider some typical numbers. The number of switches for the hierarchical interconnect compared to the flat Rent's rule interconnect is

$$\frac{\beta_{\text{hier}}}{\beta_{\text{flat}}} = \frac{N^{b-1} - N^{-b}}{k^{b-1} - k^{-b}} = \frac{N_{S_{\text{hier}}}}{N_{S_{\text{flat}}}}$$

This expression is plotted in Figure 3 for $k = 2$.

Further gains are realized by interleaving processing elements with the interconnect hierarchy. In general, for a tree with branching factor k, the addition of one fixed fanout element at each interconnect node yields

$$\beta = \beta_{\text{hier}} \frac{k - 1}{k}$$

where β_{hier} is the quantity computed previously (and N is the number of leaves at the *bottom* of the tree).

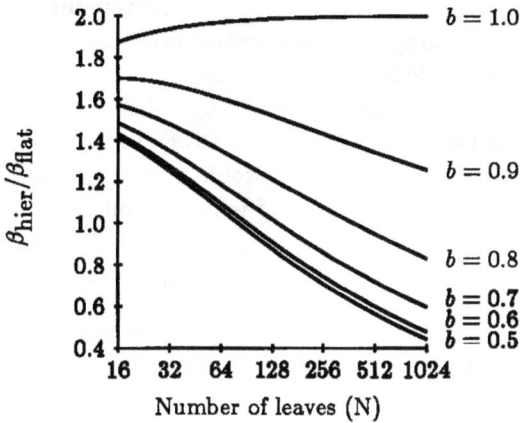

Figure 3 Number of switches in hierarchical interconnect relative to flat Rent's rule interconnect.

3 Hierarchy and Abstraction in System Design

In addition to assisting in the management of the basic issues of wiring complexity and length, the hierarchy in the Proto-chip can be used to take advantage of system-level organization in the network to be embedded. In particular, the top-down design style expounded in structured design methodologies [8] can be exploited; by using for interconnect the same hierarchical specification present in the design's functional description, the mapping from design to silicon is greatly simplified. In short, most circuits that we are interested in prototyping are hierarchical, and therefore the general Proto-chip architecture uses hierarchy as well. Furthermore, such circuits generally are easily scalable to larger systems, as is the Proto-chip architecture itself. Several other issues concerning hierarchy in system synthesis are discussed in the literature [12].

A hierarchical design was chosen for three other reasons. First, the electrical properties of the resulting circuits are improved, due to (i) lower capacitances, because the wires are shorter and there are fewer switches, and (ii) the ability (and space, as will be shown in Section 5.2) to place buffer stages within the interconnect itself. This possibility of integrating processing with interconnect also is appealing for networks that are inherently specified recursively (as are, for example, many neural-network architectures). Second, the partitioning of a circuit into a hierarchy permits (i) non-uniform hierarchies to be tailored *a priori* to specific architectures (e.g. neural-networks, processing surfaces, systolic arrays), (ii) high-density special cells to be incorporated, and (iii) subtle embeddings of subcircuits to be stored in libraries and easily integrated into subsequent designs. Third, hierarchy simplifies the embedding problem by providing a convenient interface with CAD tools, most of which encourage or enforce a hierarchical style. In addition, most interchange formats (CIF, NTK, EDIF, etc.) intrinsically support hierarchy.

4 Embedding a Circuit Graph

The ultimate workability of this prototyping system is determined by the ease with which the desired circuit graph can be embedded onto the interconnect matrix. Basically, the approaches to this problem are:

1. Exploit the hierarchy present in the circuit specification as much as possible. When this approach results in an unroutable system, resort to one of the other approaches to solve the embedding of the problem spot.

2. Exploit structural information available from the design tool. For example, a schematic editor also identifies approximate relative placement of subcircuits; schematics usually are drawn with similar objectives (for example, minimal global and random wiring) as required by the embedding algorithm.

3. Ignore any hierarchy in the specification, and attempt to synthesize a partition of the circuit graph that can be embedded directly.

4. Accept a user-provided interactive routing strategy.

An algorithm based on the third approach has been developed, and is described in detail in the Appendix. The network-embedding compiler accepts a flat circuit graph, and generates a hierarchical decomposition, with the provision for optional human assistance. Work is in progress to incorporate the first two techniques into the compiler.

5 Compiling a Proto-chip

Three issues still need to be resolved: the internal contents and organization of the leaf cells, the physical placement of leaves and interconnect on the silicon (the structure of the chip), and the implementation of a suitable interconnect element.

These issues have been deferred until now because they represent degrees of freedom (number and kind of leaves, matrix branching factor and Rent's rule exponent) that are bound at *fabrication* time, whereas the network to be embedded is dynamically configured at run-time by the network compiler described in the previous section. This sequence opens the possibility of designing a metacompiler, with which a user fabricates a *custom* Proto-chip with parameters and components optimized for the end application. As specialized examples, leaf cells containing photoreceptors and simple analog circuits could be configured into image processing surfaces [14,15], and leaf cells containing bit-serial adders and multipliers allow implementation of configurable data-flow architectures [16].

A simple metacompiler has been designed, using the WOL composition system and design tool [13]. The user provides a leaf cell, which must satisfy simple size constraints (both height and width must be integer multiples of certain constants), and hierarchical composition parameters (fanout at each level, which is more general than the Rent's rule model presented in Section 2). The system generates a Proto-chip ready for fabrication, and produces a programming file of topology parameters needed by the network-embedding compiler.

5.1 Leaf cells

One approach to maximizing the versatility of the Proto-chip while maintaining a uniform hierarchy is to make the leaf cells themselves programmable. To illustrate this technique, the design of a transistor-level leaf intended for prototyping low-level analog circuits and designing new logic families will be described.

The analog leaf (Figure 4) consists of two blocks of partially preconnected n-transistors and p-transistors. Each programming switch connects a horizontal wire, connected to at least one transistor gate,

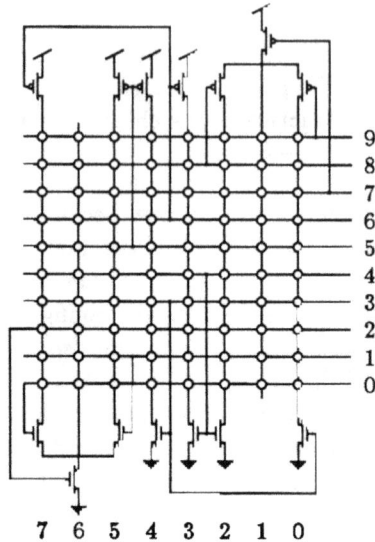

Figure 4   Programmable analog leaf cell. Each circle in the
10×8 matrix is a single programmable switch, which, when
closed, connects the vertical and horizontal wires at that point.

to a vertical wire connected to at least one of a p-drain and n-drain.
Commonly used analog building blocks, such as current mirrors and
differential pairs, are prewired in an attempt to reduce the complexity
and size of the programming matrix. If a full crossbar were used to
interconnect seven three-terminal p-transistors to an equal number of
n-devices, $(3 \times 7)^2 = 441$ switches would be required; with the merged
devices, only 80 switches are used. Yet the compact design is sufficient to construct a rich set of analog circuits, including wide-range
transconductance amplifiers (Figure 5), and horizontal resistors, as
well as more conventional NAND/NOR and similar gates (Figure 6).
In addition, a leaf cell is capable of configuring a fully independent
floating three-terminal transistor.

If strictly digital combinational logic is desired, alternate approaches
yield a higher density of gates per leaf [17]. Our primary research
interest has been the development of analog circuits, but the analog leaf can be usefully applied to the prototyping of new digital logic
forms (for example, delay-insensitive asynchronous circuit elements or
CMOS set/reset logic (CSRL) [9]). In addition, the hierarchical archi-

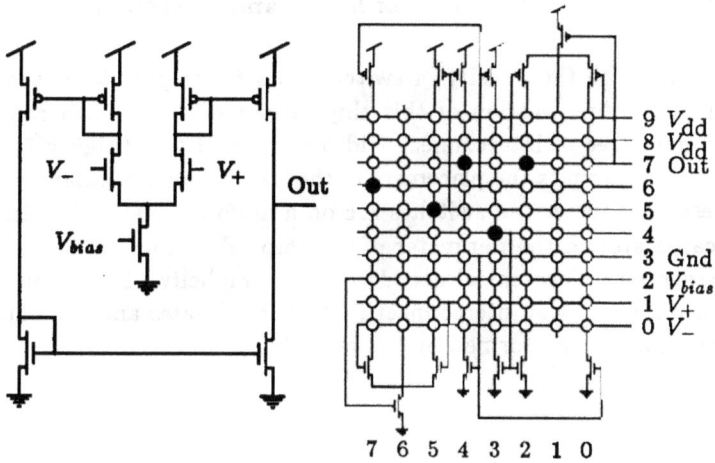

Figure 5   Wide-range transconductance amplifier. Note that the transistors that are not utilized are disabled by applying the appropriate gate bias voltage.

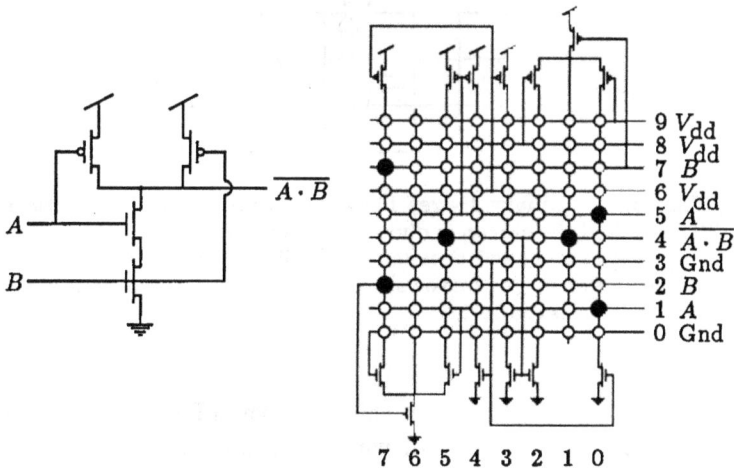

Figure 6   NAND gate implemented with analog leaf.

tecture permits hybrid Proto-chips, in which digital leaves containing gates or PLD-type devices are interleaved with analog elements.

5.2 Physical placement of leaves and interconnect

To maximize the density of switches and to simplify their programming, the first version of this chip fabricated by MOSIS contained a binary tree of interconnect laid out as an H-tree (Figure 7). This structure permits the placement of the leaf cells on a regular grid. Address lines to set the switches are on a uniform mesh, with decoders located around the perimeter of the chip. This approach was chosen instead of a hierarchical decoder for its simplicity. The mapping between hierarchical interconnect matrix coordinates and flat cartesian coordinates is performed by the embedding compiler.

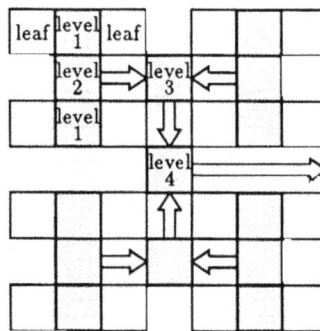

Figure 7   Binary H-tree hierarchical interconnect structure. The area of the silicon is divided between (i) leaf cells, (ii) crossbar interconnect switches (shaded), (iii) routing wires with optional buffer stages (arrows), and (iv) space for processing elements interleaved in hierarchy.

A photomicrograph of the test chip is shown in Figure 17. The analog leaf cell (Figure 4) was used, with a leaf pinout of $P_L = 10$, and a Rent's rule exponent of $b = 0$. This low fanout was chosen because the analog circuits of interest have a considerably smaller number of connections at the system level than do digital circuits. The entire chip was generated recursively by the metacompiler. A 3μ version was fabricated with $L = 4$, $(N = 16)$ on MOSIS run M75A.

5.3 Interconnect switches

By far the most expensive components of the Proto-chip, in terms
of area and performance, are the switches in the interconnect matrix. For the test chip, CMOS transmission gates were used as the
active switch elements, controlled by a static RAM cell based on a
modified CSRL design [9] (Figure 8). The modification permits the
CSRL to remain powered when a read operation is being performed.
Two global signal lines broadcast the Q and $\overline{Q}$ data signals along
the perimeter of the chip. The vertical decoder connects these lines
to a single horizontal row; the horizontal decoder enables the input–
output pass transistors to an entire column. If a read operation is
being performed, transistor M_W remains ON, and the single selected
cell drives its contents onto the global $Q/\overline{Q}$ bus, which is buffered at
pads when it comes off-chip. If a write operation is requested, the
vertical row decoder ANDs its output with the $\overline{R}/W$ request, and
disables transistor M_W in the row addressed. The other power-down
transistor is disabled by the horizontal column decoder. The cross-
coupled inverter is then in an undriven state, and is loaded with $Q/\overline{Q}$
from the bus. The sequence is reversed to reapply power and make
the RAM cell static.

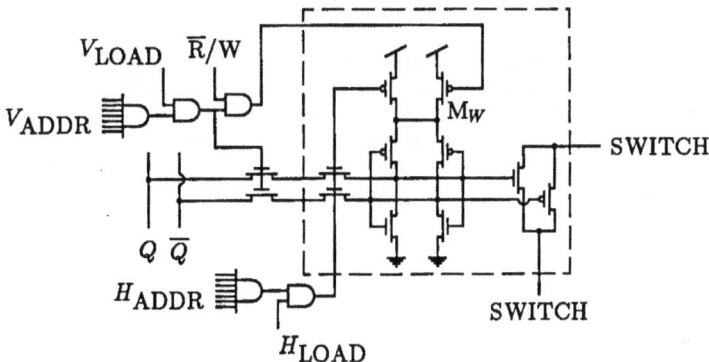

Figure 8   CSRL static RAM and transmission gate switch.
The enclosed circuit is replicated at each interconnect site in
Figure 4; the decoders are shared by the entire row or column.

The size of the prototype switch is $39 \times 49\lambda$ (includes RAM and pass-
gate using $10 \times 2\lambda$ transistors), for a total area of $4300\mu^2$ ($\lambda = 1.5\mu$).

6 Experiments with the Proto-chip

Preliminary testing of the Proto-chip yielded encouraging results. To test the performance of the interconnect, a 3–inverter ring oscillator was constructed and ran at just over 2 MHz, with 10 pF of external capacitive load applied to all three internal electrical nodes. The experimentally measured capacitance of the interconnect was 500 fF/switch, due primarily to the diffusion sidewall capacitance of the transmission-gate drains. The ON resistance of the switches was 2.5 kΩ; the resistance as a function of terminal voltage is shown in Figure 9.

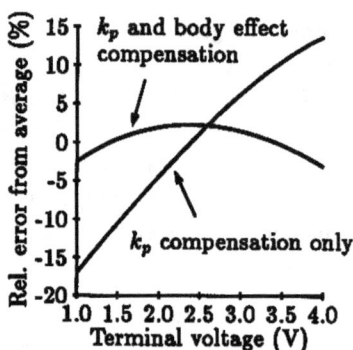

Figure 9   Transmission gate small-signal resistance variation. Zero-order compensation is obtained by scaling both devices inversely with their majority carrier mobilities (and hence k_p). A slightly different scaling can be used to minimize the variation caused by the first-order threshold voltage changes due to the body effect.

Because this version of the chip was intended for development of sub-threshold analog circuits, the resistance of the switches was considered to be unimportant. The capacitance, however, was another issue. In order to try to minimize its effect, the transistors in the leaves were scaled such that their total drain capacitance (diffusion capacitance *plus* leaf interconnect matrix capacitance) to transconductance ratio was the same as for an ordinary minimum-sized device (Figure 10). Compensation for capacitances in higher levels of the matrix could be provided by buffer stages inserted in the signal routing channels (under the arrows in Figure 7).

$$Z = 3, C_D = 1 \qquad Z \approx 16, C_D = 5$$

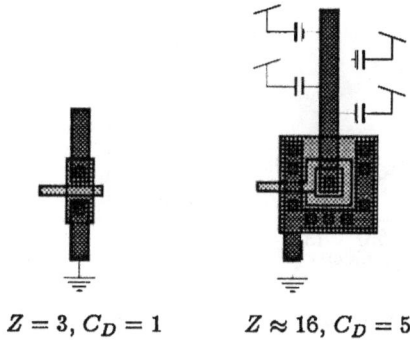

Figure 10 Ring transistors with small drain diffusion areas are used to maintain a constant current drive to capacitive load ratio. The load in the case of the leaf cell comprises both diffusion (sidewall) capacitances of the transmission-gate switches, and wire capacitances.

The resistance of the switches becomes important for high-current digital networks, the transistors of which have relatively high drain conductances. As an artificially arranged example, a leaf cell configured as an inverter with two switches between n- and p-drains (Figure 11) exhibits the transfer function shown in Figure 12; the corresponding switch resistance is shown in Figure 13. For a correctly designed circuit, the transfer curve is more conventional (Figure 14).

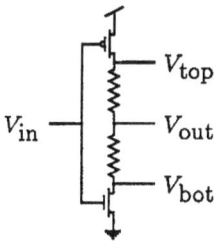

Figure 11 Inverter with transmission gates in series with high-current path.

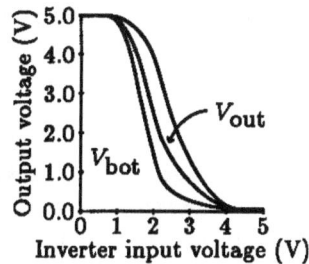

Figure 12 Transfer curve for inverter in Figure 11, showing substantial voltage drop across transmission gates.

Particularly with subtle leaf cells such as the analog leaf, sophisticated design may be required at this low level. Consequently, the network-

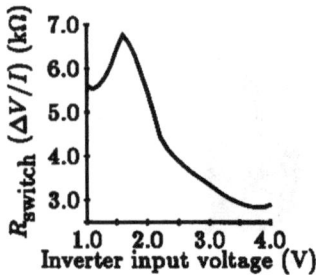

Figure 13   Effective large-signal
resistance of transmission gates
in Figure 11.

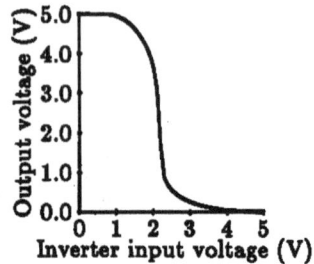

Figure 14   Transfer curve for
an inverter designed with no
transmission gates in series with
high-current path.

embedding compiler searches a library of commonly used, predefined,
and pretested leaves. The library currently contains simple and wide-
range transconductance amplifiers, current mirrors, horizontal resistors, single transistors, two-input NAND and NOR gates, inverters,
analog multipliers, CSRL stages, and is constantly being expanded.

7   Conclusions and Future Work

The test Proto-chip has demonstrated the workability of a dynamically configurable array for prototyping purposes. The system described in this paper is easy to use, provides fast turnaround (from
schematic to ready-to-test chip in under 1 minute), good performance
and simulation accuracy, and is able to handle networks of reasonable
scale.

The ability to specify and instrument a circuit interactively and to
perform a simulation is a great convenience to the user. Work is
currently in progress to integrate the network compiler and Proto-
chip into a schematic capture/software simulator [2,7] environment.
As the schematic editor is already part of our VLSI tool suite (for
netlist generation and network comparison), this project is a step
toward a truly unified design environment.

The next design iteration of the Proto-chip will provide a binary H-
tree laid out as a hexagonally tesselated structure, and is intended
for experiments on our retinal-model image processing architectures.
It will include dedicated local routing and will intersperse processing

elements in the interconnect hierarchy. Also, alternate interconnect technologies are being considered, driven by performance and density requirements.

Finally, we are exploring new applications of this methodology. For example, we are using the metacompiler to generate a system to model the analog computations that take place in the dendritic arborization of biological neurons.

Appendix: The Graph Embedding Algorithm

This appendix describes an algorithm to embed a flat circuit network on a $k = 2$ (binary) hierarchical interconnect network. For simplicity of description, all processing elements are considered to be at the leaves of the tree, and all leaves are assumed to be identical. In practice, diversity among leaves is partially accommodated by making them programmable. In addition, the algorithm can be modified easily to accommodate a non-uniform hierarchy.

This graph embedding problem is similar to the standard-cell place-and-route problem, with one significant difference: in the Proto-chip, the routing channel resources are predefined and fixed, as is the structure of the interconnect. The basic approaches, however, remain the same. A top-down partitioning [6,1] attempts to divide a graph into two nearly equal parts, while approximately minimizing the number of wires crossing between them. A bottom-up technique considers candidate groups of elements for *merging*, based on some cost metric of amalgamating these cells into one [11,3]. Both of these approaches can be extended to incorporate the constraints of the hierarchical interconnect to prune the search tree of candidates for subdivision and merging, respectively. A third approach, based on simulated annealing [4] also is possible, but has not yet been explored in application to a hierarchical interconnect.

We selected a bottom-up approach employing dynamic programming for several reasons. First, a concise mathematical description of the algorithm provided a direct implementation, and facilitated an analysis of the problem complexity. More generally, the bottom-up approach can be easily applied to non-uniform hierarchies, and permits the use of cell libraries to take advantage of previously solved embeddings.

It is useful to view a circuit as a bipartite graph with two sets of vertices, corresponding to elements and electric nodes. Edges between these sets indicate the connectivity of the circuit. The nodal connectivity matrix C is defined by $C_{ij} = 1$ if element E_j connects to E_i.

Two elements are candidates for merging if they share at least one common node. The element adjacency matrix A is computed from

$$A = C^T C$$

$$A_{ij} = \text{number of nodes shared by elements } i \text{ and } j$$

A lower bound on the fanout of the element obtained by merging elements i and j is given by

$$F_{\min} = A_{ii} + A_{jj} - 2A_{ij}$$

If $F_{\min}$ is greater than the fanout of the hierarchical interconnect matrix, elements i and j cannot be merged, at least at this level.

An upper bound on the fanout is

$$F_{\max} = A_{ii} + A_{jj} - A_{ij}$$

Clearly, if $F_{\max}$ is less than the matrix fanout, elements i and j can be merged. Any intermediate cases must be judged on the basis of the actual fanout of the meta-element and the interconnect of the matrix.

If the elements can be joined, the new meta-element's fanout is computed, and the new element is added (as a new column) to the C matrix. The process is iterated until a meta-element is formed that contains all the elements that were present at the beginning of execution of the algorithm. To provide a trace of this process, from which a hierarchical embedding can be determined, an ownership matrix M is maintained, where

$$M_{ij} = \begin{cases} 1 & \text{if element } j \text{ contains element } i \text{ as one of its children} \\ 0 & \text{otherwise} \end{cases}$$

The termination condition for the algorithm can be stated in terms of M^*, the transitive closure of M, which may be interpreted as $M^*_{ij} = 1$ if element i appears in the tree rooted by j. This transitive closure can be computed directly from M, because the columns of M are topologically sorted.

The algorithm can be improved by rejecting as candidates for merger any two cells that have children in common. A necessary and sufficient condition for such independence is

$$(M^*\hat{u}_i)^T(M^*\hat{u}_j) = \vec{0} \implies (M^{*^T}M^*)_{ij} = 0$$

The final criterion for candidates for merger is that they be previously untested. To this end, two sets of elements are maintained:

N = {all elements currently defined}

N' = {elements added in the last iteration of the algorithm}

Clearly, $N' \subseteq N$, and the novelty criterion is satisfied by considering only pairs in the set $N \times N'$ for merging.

Heuristic modifications can accelerate this algorithm tremendously. In particular, much time is spent considering nodes of large fanout; dividing elements into equivalence classes based on their connectivity to global signals frees the algorithm to consider primarily the short local interconnections, which correspond to nodes that are swallowed when two elements are merged. Also, in situations where the interconnect has substantial excess routing resources, it often is adequate to eliminate an element from further consideration after it has been amalgamated into one (or more) higher-level elements.

Empirically, this algorithm works well for sparsely connected circuits, or for those that are difficult to embed (in the sense that only a small number of embeddings are possible, and these usually are not obvious to the human operator). It still works, but becomes computationally inefficient, for highly connected networks, because the number of elements created becomes very large (in the absence of the simplifying heuristics above). In general, these inefficiencies are a result of the breadth-first search pattern of the algorithm, and occur in cases when just about any trial embedding will work. These networks usually are easily embedded with simple operator intervention to the algorithm.

A rough understanding of the performance characteristics can be gained by examining the complexity of the embedding problem. Consider a simplified network of the form shown in Figure 15.

For the first iteration of the algorithm, there are k equivalent neighbors of the center element to consider for merging. For the second

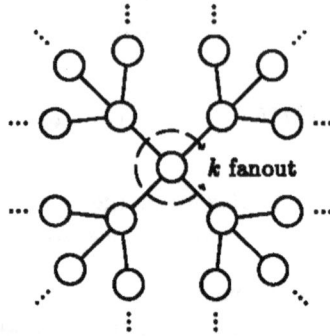

Figure 15 Network model for complexity analysis. Each of N elements has a fanout of k; for this analysis, wires run between pairs of elements only.

iteration, there are $2(k-1)$ neighbors, but each of these is $O(k)$ organized from possible first-iteration mergers. Thus, there are $O(k^2)$ second-iteration merges. By induction, it can be shown that the number of neighbors at merge iteration n is:

$$d_n = 2d_{n-1} - 2 = 2^n \left(\frac{k}{2} - 1\right) + 2$$

each of which is organized $O(k^n)$.

For a graph of height $\log_2 N$, this gives a total number of possible merges, adding the N possible starting points, of:

$$M = N 2^{\log_2 N} \left(\frac{k}{2} - 1\right) k^{\log_2 N} = \left(\frac{k}{2} - 1\right) N^{2 + \log_2 k}$$

Note that this is a gross overestimate, as the number of eligible neighbors in the real (that is, finite) circuit approaches 1 as $n \to \log_2 N$. Even so, the algorithm has a polynomial time complexity, due to the finite fanout of the individual elements.

The case of fanout to multiple elements can be handled by the transformation shown in Figure 16, where multiply connected nodes become pseudoelements. For nodes of fanout $\leq k$, the previous analysis is unaffected. Nodes with fanout $> k$ are best considered global signals, and are best accommodated explicitly in the allocation of routing resources. In other words, it is the consideration of elements that connect to large-fanout nodes (such as power rails and clocks) that makes the simple algorithm inefficient.

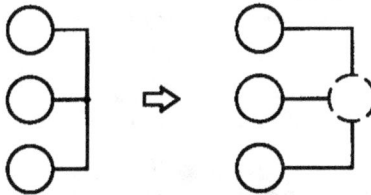

Figure 16   Transformation from shared node to pseudoelement.

References

[1] C. Fiduccia and R. Mattheyses. A linear–time heuristic for improving network partitions. In *19th Design Automation Conference*, 1982.

[2] D. Gillespie. *LOG, The Chipmunk Logic Simulator's User's Guide*. Technical Report, California Institute of Technology, 1984. 5130:TR:84.

[3] W. R. Heller, G. Sorkin, and K. Maling. The planar package planner for system designers. In *19th Design Automation Conference*, pages 253–260, 1982.

[4] S. Kirkpatrick, C. D. Gelatt, Jr., and M. P. Vecchi. Optimization by simulated annealing. *Science*, 220(4598):2554–2558, 13 May 1983.

[5] B. Landman and R. Russo. On a pin vs. block relationship for partitions of logic graphs. *IEEE Transactions on Computers*, 1469–1479, December 1971.

[6] U. Lauther. A min-cut placement algorithm for general cell assemblies based on a graph representation. In *16th Design Automation Conference*, pages 1–10, 1979.

[7] J. P. Lazzaro. *anaLOG: A Functional Simulator for VLSI Neural Systems*. Master's thesis, California Institute of Technology, Computer Science 5229:TR:86, 1986.

[8] C. A. Mead and L. A. Conway. *Introduction to VLSI Systems*. Addison-Wesley, 1980.

[9] C. A. Mead and J. Wawrzynek. A new discipline for CMOS design: an architecture for sound synthesis. In *1985 Chapel Hill Conference on Very Large Scale Integration*, pages 87–104, 1985.

[10] C. A. Mead. *Analog VLSI and Neural Systems*. Addison-Wesley, (in preparation).

[11] R. L. Rivest. The "PI" (placement and interconnect) system. In *19th Design Automation Conference*, pages 475–481, 1982.

[12] S. M. Rubin. *Computer Aids for VLSI Design*. Addison-Wesley, 1987.

[13] M. A. Sivilotti. *A User's Guide to the WOL Design Tools*. Technical Report 5237:TR:86, California Institute of Technology, 1986.

[14] M. A. Sivilotti, M. A. Mahowald, and C. A. Mead. Real-time visual computation using analog CMOS processing arrays. In *1987 Stanford Conference on Very Large Scale Integration*, MIT Press, Cambridge, MA, 1987.

[15] J. E. Tanner and C. A. Mead. An integrated analog optical motion sensor. In S.-Y. Kung *et al*, editors, *VLSI Signal Processing, II*, pages 59–76, IEEE Press, New York, 1986.

[16] J. Wawrzynek. A reconfigurable concurrent VLSI architecture for sound synthesis. In S.-Y. Kung *et al*, editors, *VLSI Signal Processing, II*, pages 385–396, IEEE Press, New York, 1986.

[17] Xilinx Inc. *The Programmable Gate Array Design Handbook*. Xilinx, 1986.

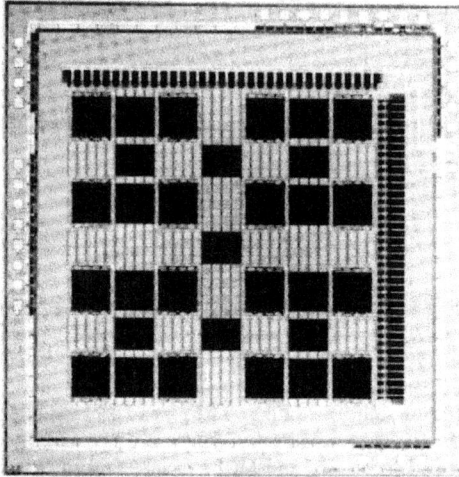

Figure 17 Photomicrograph of test chip.

A Two-Dimensional Visual Tracking Array

Stephen P. DeWeerth and Carver A. Mead

Department of Computer Science
California Institute of Technology
Pasadena, California 91125

The density and concurrency available in VLSI make it an excellent technology for implementing visual image-processing. By incorporating phototransistors and analog processing elements onto a single die, the large signal bandwidths required for real-time computations can be achieved. This paper describes a VLSI chip that computes the "center of intensity" of a two-dimensional visual field. One application for this network is the localization of a bright spot of light against a dark background. Theoretical and experimental results are presented to describe the operation of the system and its suitability as a input device for tracking servo systems.

1 Introduction

The need for highly parallel computational architectures in visual processing and sensing has been well established. VLSI has proven to be an effective implementation technology for such computation.[4] Through the use of computation on both local and global scales, high-level information can be extracted from a visual image. In particular, a class of features exists that can be encoded using a fixed number of wires, independent of the image array size. An example is the uniform velocity detector developed in our laboratory.[5]

This paper describes another feature extraction: the computation of the center of intensity of the entire visual field. This computation effectively determines the position of a bright spot in a visual image provided that the background is sufficiently dim. We have designed and fabricated an integrated circuit that employs analog processing to perform this computation. Circuitry along the periphery of a two–

dimensional array of photoreceptors is used to compute the mean or median of the position of the receptors weighted by their respective light intensities.

2 Theory

In this section, we shall present circuitry to compute a weighted mean using a resistive voltage divider. We will define the concept of a weighted median, and will show how this quantity can be computed with a similar circuit of threshold elements.

2.1 Parallel Resistor Network

A network of linear resistors configured as a voltage divider is shown in Figure 1. The current in the nth resistor is

$$i_n = g_n(V_n - V_{out}) \tag{1}$$

The sum of the currents at node V_{out} must be zero. Hence,

$$\sum_{n=0}^{N} g_n(V_n - V_{out}) = 0 \tag{2}$$

Solving for V_{out} gives

$$V_{out} = \frac{\sum g_n V_n}{\sum g_n} \tag{3}$$

From Equation 3, we see that the output voltage is the mean of the input voltages weighted by the corresponding conductances.

2.2 Parallel Threshold-Element Network

A threshold-element is a hypothetical infinite-gain current source with the following voltage–current relationship:

$$i = \begin{cases} -\gamma & V_1 < V_2 \\ 0 & V_1 = V_2 \\ \gamma & V_1 > V_2 \end{cases} \tag{4}$$

where γ is the magnitude of the current through the device.

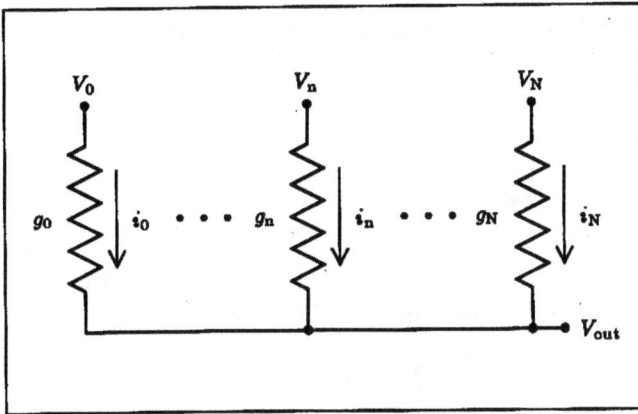

Figure 1 A parallel network of resistors. The sum of the currents through the resistors must be zero due to Kirchhoff's current law at the output node. This effect and the linear voltage–current relationship of the resistors cause the output voltage to equal the mean of the input voltages weighted by the corresponding conductances.

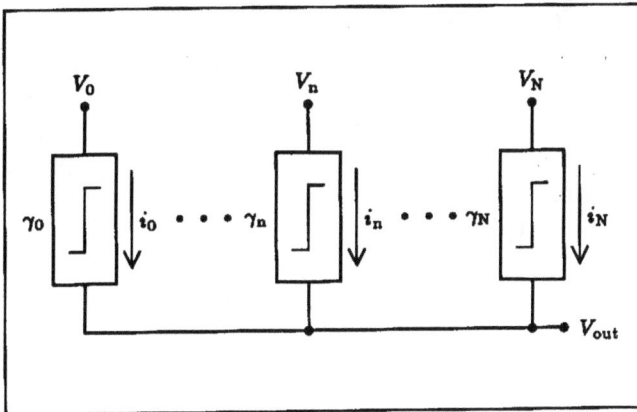

Figure 2 A parallel network of threshold elements. Kirchhoff's current law mandates that the sum of currents through the devices with input voltages less than the output voltage must equal the sum of the currents through the devices with input voltages greater than the output voltage. This relationship causes the network to compute the median of the input voltages weighted by the current magnitudes for the corresponding elements.

We can construct an array of these elements (Figure 2), similar to the resistor array in Figure 1. From the definition in Equation 4, the current through the nth threshold element is

$$i_n = \begin{cases} -\gamma_n & V_n < V_{\text{out}} \\ 0 & V_n = V_{\text{out}} \\ \gamma_n & V_n > V_{\text{out}} \end{cases} \qquad (5)$$

In this array, as in the resistor array, Kirchhoff's current law mandates that the sum of the currents into the V_{out} node must equal zero. This fact, combined with Equation 5, leads to the following equality:

$$\sum_{V_n < V_{\text{out}}} \gamma_n = \sum_{V_n > V_{\text{out}}} \gamma_n \qquad (6)$$

If all the weights (γ_n) are equal, Equation 6 calculates the median of the input voltages. When they are not all equal, this computation defines a "weighted median."

2.3 Encoding Position with Voltages

In Sections 2.1 and 2.2, the input voltages to the parallel dividers were independent of one another. These voltages, however, can be used to encode the position of each element in the array. This encoding can be performed through the use of a linear resistive divider as a voltage divider. In this configuration, N resistors are connected in series. By applying voltages V_0 and V_N to the ends of the resistive line, a linear voltage gradient can be set up along the array, where the voltage at the nth node is given by

$$V_n = \left(\frac{V_N - V_0}{N}\right) n + V_0 \qquad (7)$$

3 Implementation

In this section, we shall present the circuitry needed to implement the tracking arrays. We shall describe the transconductance amplifiers and phototransistors used in this system, and show how these elements can be used to implement both one- and two-dimensional tracking arrays.

3.1 The Photoreceptor

In this design, the vertical parasitic bipolar transistor existing in the standard CMOS fabrication process is used as a phototransistor.[1] Assuming an n-well process, this bipolar transistor is a *pnp* with a grounded collector.

In a phototransistor of this type, at a given wavelength of light,

$$i_{\text{photo}} \propto I \qquad (8)$$

where i_{photo} is the emitter current of the phototransistor and I is the light intensity at the base–emitter junction.

3.2 The Transconductance Amplifier

The transconductance amplifier (Figure 3) is a differential stage producing an output current which is a function of the bias current (set by V_{bias}) and the differential input voltage $(V_1 - V_2)$.[6] The circuit consists of a bias transistor Q_{bias} that sets the maximum output current of the amplifier, a differential pair (Q_1 and Q_2), and a current mirror (Q_3 and Q_4).[1] The bias current is divided by the differential pair according to input voltages V_1 and V_2. The current that flows through the "minus" transistor Q_2 is subtracted directly from the output node. In the other branch, the current flowing through the "plus" transistor Q_1 also must pass through Q_3. The current through Q_4 is the mirror of the current through Q_3. Finally, the current through Q_4 is added to the output node. Thus, the output current is the difference between the currents in the two branches of the differential pair.

A transfer curve for the transconductance amplifier is shown in Figure 4. Saturation behavior is clearly exhibited for large input voltage differentials, indicating that the bias current is flowing through only one of the transistors in the differential pair. This amplifier often is used with very small bias currents, allowing all transistors to be operating in the subthreshold regime.[2] The transfer function when the

---

[1]This configuration is called a current mirror because the current through the input transistor is "mirrored" by the output transistor. The gate voltage of the input transistor is set by the current applied to the device. This voltage also is applied to the output transistor gate, causing the current through the two transistors to be equivalent (ignoring drain–source voltage effects).

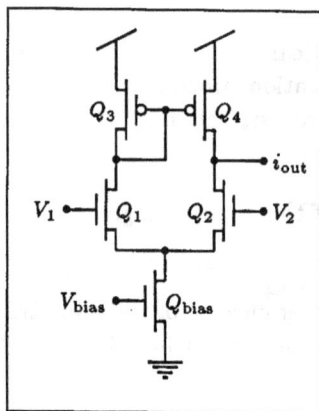

Figure 3 The transconductance amplifier. This circuit is a differential amplifier whose output current is a function of its bias current (the current through Q_{bias}) and differential input voltage. The bias current is set by V_{bias} and divided between Q_1 and Q_2 depending on V_1 and V_2. Q_3 and Q_4 mirror the current through Q_1 giving an output current equal to the difference of the currents through Q_1 and Q_2.

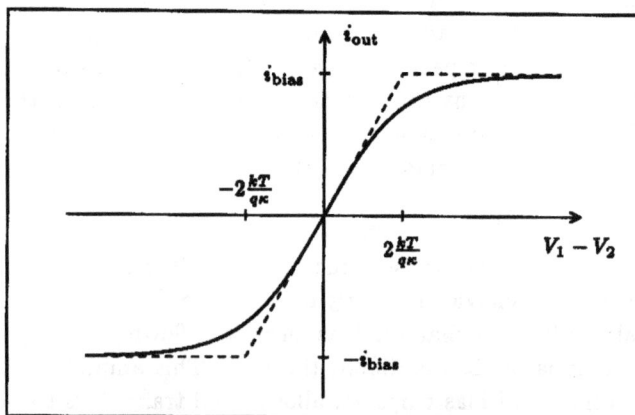

Figure 4 Transconductance amplifier transfer characteristics. The solid line is the actual transfer function for the amplifier. The dotted line is the linear approximation for the amplifier transfer curve.

bias current is subthreshold is

$$i_{\text{out}} = i_{\text{bias}} \tanh \left(\frac{V_1 - V_2}{2\frac{kT}{q\kappa}} \right) \tag{9}$$

For the purpose of calculations in this paper, the tanh function will be approximated by the piecewise linear function described by

$$i_{\text{out}} = \begin{cases} -i_{\text{bias}} & V_1 - V_2 \leq -2\frac{kT}{q\kappa} \\ g(V_1 - V_2) & |V_1 - V_2| < 2\frac{kT}{q\kappa} \\ i_{\text{bias}} & V_1 - V_2 \geq 2\frac{kT}{q\kappa} \end{cases} \tag{10}$$

where $g = i_{\text{bias}} / \left(2\frac{kT}{q\kappa} \right)$ is the transconductance of the amplifier. This approximate transfer function is superimposed in Figure 4.

3.3   A One-Dimensional Amplifier Array

Figure 5 shows a variation on the standard transconductance amplifier in which the bias transistor has been replaced with a phototransistor. Because the phototransistor produces a current (i_{photo}) that is proportional to the light intensity (I), the transfer function of this circuit becomes

$$i_{\text{out}} = i_{\text{photo}} \tanh(V_1 - V_2) \propto I \tanh(V_1 - V_2) \tag{11}$$

with voltages given in units of $2\frac{kT}{q\kappa}$.

Figure 6 shows an array of transconductance amplifiers and phototransistors configured similarly to the arrays of resistors and threshold elements in Figures 1 and 2. The input voltages to the amplifiers are obtained from the resistive line, as described in Section 2.3.

We can set the voltage gradient along the amplifier inputs by changing V_0 and V_N. The output voltage (V_{out}) is a weighted combination of the amplifier input voltages, and is constrained to lie within the input voltage range. Thus, the maximum possible differential voltage at the input of any amplifier is $|V_N - V_0|$. If $|V_N - V_0| \leq 2\frac{kT}{q\kappa}$, all the amplifiers will be operating in their "linear" regimes, and the network will calculate the mean of the input voltages of the amplifiers weighted by the corresponding light intensities:

$$V_{\text{out}} = \frac{\sum i_{\text{photo}_n} V_n}{\sum i_{\text{photo}_n}} = \frac{\sum I_n V_n}{\sum I_n} \tag{12}$$

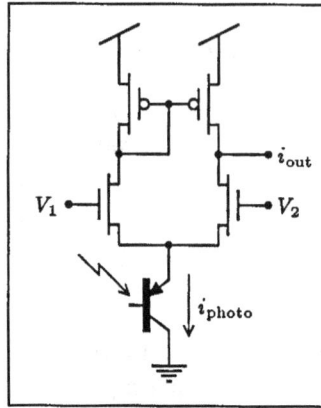

Figure 5 Transconductance amplifier modified to include phototransistor. In this circuit, the bias transistor of the amplifier is replaced with a phototransistor (a vertical parasitic bipolar transistor) resulting in an amplifier bias current that is equivalent to the photocurrent.

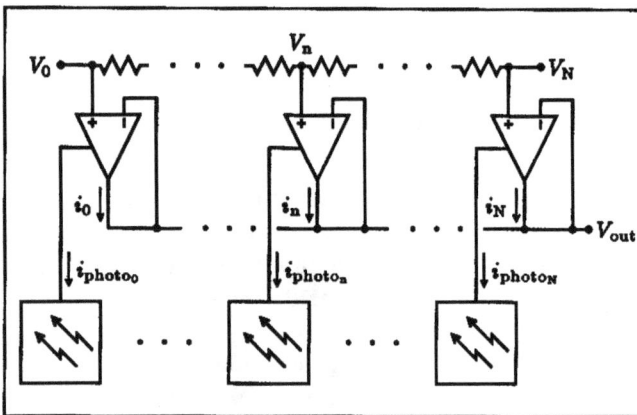

Figure 6 Transconductance amplifier array with position encoding. In this network, amplifiers with photocurrents used as their bias currents are configured as voltage followers with their outputs connected together. A resistive voltage divider encodes the position and these voltages are used as the input voltages to the amplifiers. By changing the differential input voltage $(V_N - V_0)$ the network can perform the range of computations from weighted mean to weighted median.

Because the input voltages encode the positions of the receptors, this calculation is equivalent to computing the center of light intensity. Thus, if the image projected onto the chip is a bright spot of light against a dark background, the array will calculate the position of the spot, and thus, track this stimulus.

If $|V_N - V_0| \gg 2\frac{kT}{q\kappa}$, the gain of each amplifier with respect to distance along the array will be very large. The transconductance amplifiers (except for those with input voltages very near the output voltage) then can be approximated as threshold elements, and the network will approximate the weighted median calculation. In this configuration, the circuit can also track a high-intensity stimulus. Because the amplifier output magnitudes are independent of the positions in the array, however, the receptors further from the bright spot will contribute less to the output than they would in the weighted mean computation. Thus, a higher differential voltage along the resistive divider leads to a more accurate bright-spot localization.

If $|V_N - V_0|$ falls between these two regimes, the calculation will be a combination of the two computation modes. In particular, the section of the array with input voltages near the output voltage (near the center of intensity) will compute a weighted mean, whereas the parts distant from the center of intensity will compute a weighted median.

3.4 A Two-Dimensional Tracking Array

The one-dimensional array of photoreceptors and amplifiers in Section 3.3 can be expanded to two dimensions. Because the two dimensions of the tracking calculations are independent, the value of the intensity at any point along each axis is taken to be the sum of the currents from the receptors in a line perpendicular to that axis. The two-dimensional array can be implemented using dual-output photoreceptors (implemented with current mirrors) and current summing along each row and column (Figure 7).

This configuration has a problem, however; the layout area of a photoreceptor increases by more than a factor of two when the current mirror is added, reducing the resolution and the coverage factor of the array. For this reason, a second scheme of photoreceptor placement using single-output receptors can be used to tile the plane. In this array configuration, each photoreceptor contributes its current

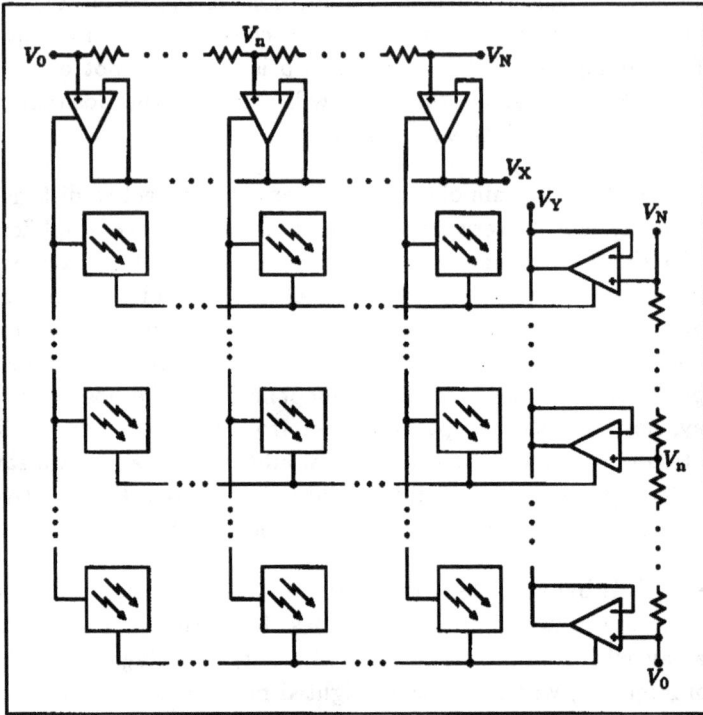

Figure 7 A two-dimensional amplifier array. The photocurrents from the receptors are replicated and summed onto wires running parallel to each axis. These current sums are used as the bias current for the amplifier arrays along the edges of the array. This edge circuitry performs the weighted mean/median computations.

to only one axis. The receptors are spacially alternated so that the currents from adjacent receptors are added to opposing axes. Using this scheme only one-half of the receptors contribute current to each axis, but the effective density of receptors is increased due to the unit size reduction made possible by removing the current mirror.

4 Experimental Set-up and Results

We chose to fabricate the alternating receptor array because of its increased receptor density and resolution. We fabricated the chip with a 200×200-pixel array in $2\,\mu$ CMOS. We implemented the resistive dividers using polysilicon wires running the length of each of the two edges of the chip. The resistance of these lines was approximately $7\,k\Omega$.

We created our experimental setup by removing the light source from a standard photographic enlarger and mounting a tracking chip in its place. We mounted a light-emitting diode (LED) on a two-dimensional precision motion table (accurate to $0.2\,\mu$m) placed below the enlarger, and focused its image through the enlarger lens onto the chip. We could then use the LED as a point stimulus by moving its image across the face of the chip by moving the table. We placed a uniformly reflective background on the motion table surrounding the LED, and created a uniform background intensity by projecting light onto the background from a diffuse light source mounted above the table.

4.1 Position/Output Voltage Relationship

We investigated the output response of the chip by moving the LED parallel to one axis while holding its position constant with respect to the other axis (Figure 8). We set the input voltages (the voltage at the ends of the polysilicon wires) to 2.2 and 3.8 V. As the stimulus passed across the surface of the chip, the output voltage changed as a function of the position. At the ends of the sweep (as the image moved off the edge of the chip), the output voltage leveled off and then began to move rapidly toward the value obtained for a uniformly illuminated background.

4.2 Multiple Curves and Error

In this experiment, we took a family of curves by moving the LED along one axis of the chip at five different positions (separated by 20 pixels) on the perpendicular axis. We performed a least-squares regression on the data points in the −70 to +70 pixel range, and the relative error for each curve was calculated by subtracting the regressed values from the observed values and dividing by the input differential voltage. Also, the maximum deviation (defined as the largest difference between any two curves) among these curves was calculated. Both the relative error curves and the maximum deviation curve are shown in Figure 9.

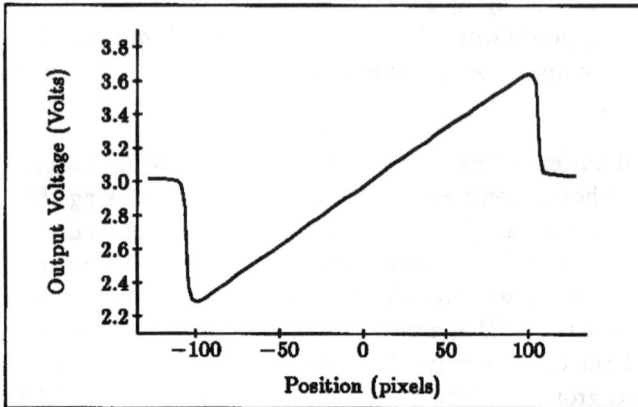

Figure 8 Single axis sweep. This figure shows the output
voltage as a function of the position of the image of an LED
on the chip. The curve represents a sweep of the LED through
one axis while holding its position constant with respect to the
other axis. As the image of the LED moves across the receptor
surface, the output voltage linearly increases. As this stimulus
passes off the edge of the surface, the output voltage returns to
the uniform-intensity output value.

The errors appear to be quite systematic, suggesting that they were
caused to a greater extent by the computational circuitry along the
edges of the array than by mismatched photoreceptors. They could
be a result of a nonhomogeneous resistance along the polysilicon wires
of the voltage divider, or of offsets in the amplifiers due to parametric
variation.

Errors due to parametric variation are very dependent upon the width
of the stimulus. For a given width, the output error is the average of
the errors for the individual components within the boundaries of the
stimulus. Since these errors are statistically distributed around zero,
the average error will decrease as the sample size becomes larger.
Thus, the accuracy of the chip can be improved by increasing the
stimulus width. The width of the stimulus for this experiment was
approximately 6 pixels.

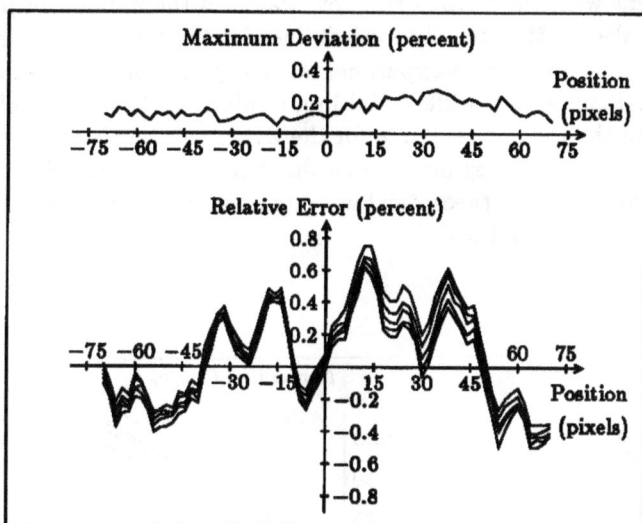

Figure 9 Tracking-array relative error. This figure displays the error for the central region of five curves taken by sweeping the LED along one axis at five different positions on the other axis. The bottom curves show the relative error from the best-fit line for the set of curves. The upper curve shows the maximum deviation between any two curves in the set.

4.3 Varying Differential Input Voltage

We tested the effects of moving from a weighted mean to a weighted median by taking a set of curves with varied differential input voltages applied across the resistive wire. These curves were all taken while holding the stimulus-to-background intensity ratio constant at approximately 1. We varied the differential input voltages from 100 mV to 1.6 V by factors of two while keeping the common mode voltage (and thus the uniform intensity output voltage) constant at 3 V. As in the previous experiments, the width of the stimulus was approximately 6 pixels.

In the appendix, we calculated the input–output voltage relationships of the weighted mean and weighted median functions for a given stimulus width and stimulus-to-background intensity ratio. Figure 10 shows the best-fit lines (using linear regression) to the curves for this

272

experiment with the output voltages scaled to the number of pixels. We have also plotted the theoretical lines for the weighted mean and weighted median for this experimental set-up. As predicted, the experimental curve for a differential input voltage of 100 mV is closest in slope to the theoretical curve for the weighted mean. The slopes of the experimental curves increase as the differential voltage increases, coming closest to the theoretical weighted median curve at a differential input voltage of 1.6 V.

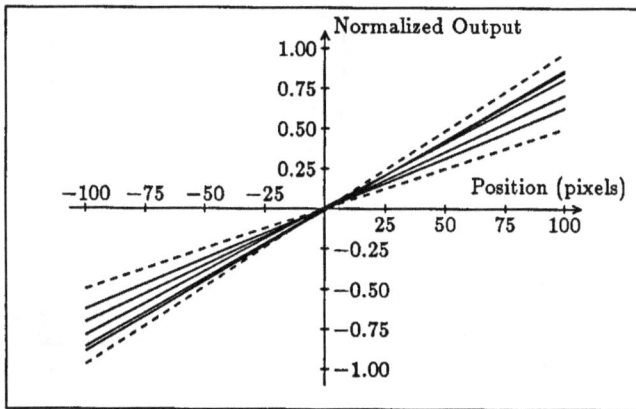

Figure 10 Varying the differential input voltage. The solid lines are the best-fit lines for a family of curves taken by varying the differential input voltage to the chip. The dotted lines are the theoretical predictions for the weighted mean and median computations. As predicted, the slopes of the experimental curves vary from the slope of the mean to the slope of the median as the amplifiers move from resistive elements to high-gain current sources.

5 Conclusions

In this paper we have presented a system that integrates both photo-sensors and processing elements on a single die to calculate the center of intensity of a visual image, or to track a bright spot on a dark background. This circuitry was implemented with a standard digital CMOS process.

For multiple sweeps along an axis, we have shown that the maximum error from a best-fit line is less than 0.8 percent, and the maximum difference among these sweeps (between any two curves) is less than 0.3 percent. In addition, the repeatability has been measured to be better than 0.05 percent, and the monotonicity (differential linearity) has been measured to be better than 0.1 percent. These facts suggest ways to increase the accuracy of the chip. If we treat the chip as a linear tracking system, we are limited by the deviation from the best-fit line. However, by creating a calibration curve for each axis, output voltages can be corrected to the accuracy of the maximum deviation among the sweeps along that axis. Finally, by creating a two-dimensional calibration map of the chip, output voltages can be corrected to within a resolution limited by the monotonicity and an accuracy equal to the repeatability. Through this scheme, it should be possible to greatly improve the accuracy of the system.

We also showed that the functionality of the system, which ranges from a weighted mean to a weighted median, can be changed by varying the differential input voltage that encodes the positions of the photoreceptors.

We are currently working on similar arrays to perform single edge detection using spatial derivatives coupled with the circuitry to compute the weighted mean/median described in this paper. We are also incorporating the tracking chip into sensory–motor feedback servo systems that will actively track objects in the visual field.

Appendix — Tracking Theory

Networks that compute weighted means and medians can be applied to calculate the position of any high-intensity stimulus on a low-intensity background. This tracking is accomplished by encoding the intensity of the stimulus and background as the weights of the computation, and the position of the elements as the values being weighted.

In order to simplify the computation of these fuctions, we assume that the stimulus and background intensities are constant and uniform over their areas. We define the variables α and β to be the total stimulus and background intensities, respectively, and assume that in the region of the stimulus, the measured intensity is the sum of the two. We scale the input and output voltages to a range of N

(the number of pixels) centered at zero so that the index step for the summations is 1. We define the stimulus to be of width w pixels centered at the input voltage. Therefore, the value of the background intensity at every pixel is $\frac{\beta}{N}$, the value of the stimulus intensity at each pixel in its range is $\frac{\alpha}{w}$, and the sum of all the intensities is $\alpha + \beta$.

We can calculate the tracking function for the weighted mean by separating the summations into two parts (one for stimulus and one for background) as follows:

$$
\begin{aligned}
V_{\text{out}} &= \frac{1}{\sum I_n}\left(\sum I_n V_n\right) \\
&= \frac{1}{\alpha + \beta}\left(\sum_{V=V_{\text{in}}-\frac{w-1}{2}}^{V_{\text{in}}+\frac{w-1}{2}} \frac{\alpha}{w}V + \sum_{V=-\frac{N-1}{2}}^{\frac{N-1}{2}} \frac{\beta}{N}V\right) \\
&= \frac{1}{\alpha + \beta}\left(\frac{\alpha}{w}\sum_{V=V_{\text{in}}-\frac{w-1}{2}}^{V_{\text{in}}+\frac{w-1}{2}} V + \frac{\beta}{N}\sum_{V=-\frac{N-1}{2}}^{\frac{N-1}{2}} V\right) \\
&= \frac{1}{\alpha + \beta}\left(\frac{\alpha}{w}(V_{\text{in}}w) + \frac{\beta}{N}(0)\right) \\
&= \frac{\alpha}{\alpha + \beta}V_{\text{in}}
\end{aligned}
$$

We compute the tracking function for the weighted median by again dividing the summations into two parts. We must further make the assumption that when the center of the stimulus is between pixels the network will linearly interpolate. (This is acceptable because the stimulus will then fall partially on a pixel at either end.) This leads to the following two equations for the sum of the intensities on either side of the output voltage:

$$
\sum_{V_n < V_{\text{out}}} I_n = \frac{\beta}{N}\left[V_{\text{out}} - \left(-\frac{N-1}{2}\right)\right] + \frac{\alpha}{w}\left[V_{\text{out}} - \left(V_{\text{in}} - \frac{w-1}{2}\right)\right]
$$

$$
\sum_{V_n > V_{\text{out}}} I_n = \frac{\beta}{N}\left[\left(\frac{N-1}{2}\right) - V_{\text{out}}\right] + \frac{\alpha}{w}\left[\left(V_{\text{in}} + \frac{w-1}{2}\right) - V_{\text{out}}\right]
$$

Kirchhoff's current law mandates that these two sums must be equal. By combining and rearranging the equations, we get V_{out} in terms of V_{in} as follows:

$$
V_{\text{out}} = \frac{N\alpha}{N\alpha + w\beta}V_{\text{in}}
$$

Both of the above models are only accurate when the entire stimulus falls upon the computational surface. In addition, due to the summations used for the stimulus terms in the median computation, that computation is only valid as long as the output voltage falls within the input voltage range covered by the stimulus.

As the stimulus-to-background intensity ratio increases, the effectiveness of both computations improves (i.e., the ratio of V_{out} to V_{in} increases). As the stimulus gets narrower, the effectiveness of the median computation improves. Finally, since, by definition, the stimulus is smaller than the background, the weighted median computation is more effective at tracking than is the weighted mean.

Acknowledgements

This research was supported by grants from the System Development Foundation and the Office of Naval Research, and a hardware grant from the Hewlett-Packard Corporation. The authors are grateful to Michael Emerling, Massimo Sivilotti, and Mary Ann Maher for their careful reviews of this document. We are also indebted to Lars Nielson, Richard Lyon, and John Tanner for their many helpful discussions.

References

[1] C. A. Mead. A sensitive electronic photoreceptor. In *1985 Chapel Hill Conference on Very Large Scale Integration*, pages 463–471, Computer Science Press, Rockville, Maryland, 1985.

[2] C. A. Mead. *Analog VLSI and Neural Systems*. Addison-Wesley (in preparation).

[4] M. A. Sivilotti, M. A. Mahowald, and C. A. Mead. Real-time visual computation using analog CMOS processing arrays. In *1987 Stanford Conference on Very Large Scale Integration*, pages 295–312, MIT Press, Cambridge, MA, 1987.

[5] J. E. Tanner. *Integrated Optical Motion Detection*. PhD thesis, California Institute of Technology, 1986. 5223:TR:86.

[6] E. A. Vittoz. Micropower techniques. In Y. Tsividis and P. Antognetti, editors, *Design of MOS VLSI Circuits for Telecommunications*, pages 104–144, Prentice-Hall, Englewood Cliffs, NJ, 1985.

Toward a Mathematical Theory of Single-Layer Wire Routing

F. Miller Maley

Department of Computer Science
Princeton University
Princeton, New Jersey 08544

This paper presents two theorems concerning the routing of wires among obstacles in the plane when rough routings of the wires are given. These theorems underlie previously published algorithms for single-layer wire routing [5] and layout compaction with automatic jog insertion [7]. One theorem characterizes the routability of a layout in terms of straight "cuts" between obstacles in the layout. A layout is routable if and only if for each pair of obstacles, the amount of wiring forced by the topology to pass between them is not too great. The second theorem states that a routable layout has an "ideal" routing that simultaneously minimizes the length of every wire, and it defines ideal routings in terms of straight "half-cuts" from obstacles to wires. Each wire in an ideal routing is a minimum-length wire such that for every obstacle, the amount of wiring passing between the wire and the obstacle is not too great. The statements and proofs of these theorems rely on elementary homotopy theory, specifically the lifting of wires and cuts from the routing region to its simply connected covering space.

1. Gridless Single-Layer Routing

Problems of laying out interconnections are ubiquitous in the design of VLSI circuits, and most are very difficult [15]. Even when the interconnecting wires run in a single plane, the routing problem is generally NP-complete [12]. Recently, however, theorists have identified significant classes of efficiently solvable wire-routing problems. The domain addressed in this paper, which I call *single-layer wire routing*, includes most of the so-called river routing problems [1, 2, 3, 14, 16] as special cases. It involves finding paths in the plane that stay separated from obstacles and from each other by certain distances in some given metric. The topology of the layout is specified in advance; one is given *rough routings* for the wires with respect to the obstacles. To obtain the final, *detailed routings*, one may deform each rough routing in any continuous fashion, provided that it crosses over no obstacles and its endpoints remain on the obstacles (*terminals*) it connects. See Figures 1 and 2. Previous definitions of this problem [2, 5, 11] differ; they restrict wires and obstacles to a grid.

The algorithmic power in this domain comes from theorems that provide necessary and sufficient conditions for routability. From constructive proofs of these theorems one can derive efficient routing algorithms. Moreover, the routability conditions themselves support algorithms for testing routability [5] and for one-dimensional compaction to improve module placement [7]. Despite their applications, however, no fundamental theorems of single-layer routing have yet seen published proof. This paper presents an approach to single-layer wire routing that yields two strong theorems, one characterizing routability and one characterizing minimum-length routings. These theorems allow the routing and compaction algorithms of [5] and [7] to be understood, generalized to gridless models, and proven correct.

1.1. Two theorems, informally

The fundamental property of single-layer routing is that local routability conditions determine global routability. A layout has a legal routing if and only if no "channel" between two obstacles has too many wires passing through it. We formalize the idea of channel as a straight *cut*: a linear path between two obstacles. Each cut has (1) a *capacity*, the amount of space available along it for crossings with wires, and (2) a *flow*, the space required by the wire crossings that are topologically forced, and cannot be eliminated by deforming the wires. When a cut's flow exceeds its capacity, the layout becomes unroutable—except in the case of a *degenerate* cut whose endpoints fall on a single obstacle or on the two terminals of a wire. Routability, in my model, can be characterized as follows.

Theorem A. *A layout is routable if and only if no nondegenerate straight cut of the layout has flow exceeding its capacity.*

Another important facet of single-layer routing is suggested by Figure 2. In the optimal routing of a routable layout, the wires are stretched tightly around one another like thick elastic bands. Each wire is as short as possible given the presence of the other wires; wire length is minimized for all wires simultaneously. We can characterize the optimal routing using straight *half-cuts*, which are linear paths from obstacles to wires. Like cuts, half-cuts have capacity and flow, and there are certain *trivial* half-cuts between wires and their terminals that must be ignored. A wire is called *evasive* if it has no

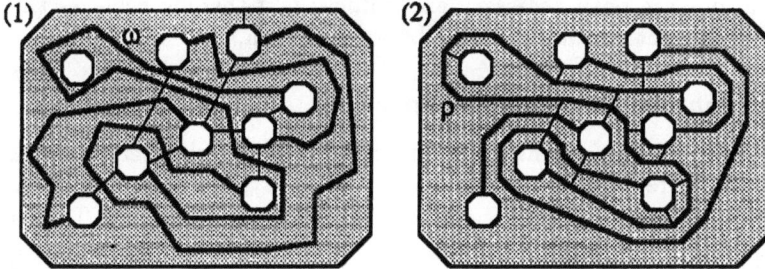

Figures 1 and 2. A routable design (1) and its optimal routing (2) in an octagonal metric. Dark polygons are obstacles, grey paths are wires, and the shaded area is the routing region. Light lines are cuts and half-cuts whose flow and capacity are equal. If any of the cuts shown in (1) were shorter, the design would be unroutable. If any of the half-cuts shown in (2) were shorter, the route ρ of ω would be *infeasible* (not part of any proper routing).

nontrivial straight half-cut whose flow exceeds its capacity.

Theorem B. *Each wire in a routable layout can be routed by means of a minimum-length evasive wire. The result is a proper routing that optimizes the length of every wire.*

Theorems A and B are strongly intuitive, and may seem obvious. Yet they are far from easy to prove, or even to state accurately!

1.2. Applications

As formalized in Section 4, Theorems A and B have several interesting algorithmic consequences. Theorem A is the basis for several routability testing algorithms [8]; in particular, it implies that the routability tester of [5] generalizes to gridless situations. Theorem B applies likewise to justify and extend the routing algorithm of [5], which previously had no correctness proof. Using these methods, one can test routability and perform single-layer routing in worst-case time $O(n^2 \log n)$ on input of size n. Recent work on routing under the euclidean metric [4] also relies on results like A and B.

Most importantly, the routability conditions established by Theorem A and the router based on Theorem B support and generalize the compaction algorithm presented in [7]. That algorithm, recently improved [9] to run in worst-case time $O(n^3 \log n)$, compacts one-dimensionally using routability conditions as the constraints on module positions; it restores the wires via single-layer routing only after the module positions have been determined. Since the routability conditions are both necessary and sufficient, it thereby achieves the optimal layout width attainable by jog insertion.

1.3. Contributions of this paper

What make Theorems A and B difficult are the self-interactions of wires. A wire can be forced to pass through a channel several times, and the different parts of the wire all constrain one another. Thus a wire behaves in some ways like several wires and in some ways like a single wire. This sort of ambiguity makes writing correct proofs and algorithms very difficult. I contend that to obtain rigorous proofs for Theorems A and B, one must use mathematics appropriate to the problem statement, such as *homotopy theory*.

This paper demonstrates methods of analyzing wires and cuts by *lifting* them from the routing region to its simply connected covering space. Using the covering space, we formalize the notions of the amount of wiring "forced" to pass across a cut or half-cut (its flow), the regions that are "forbidden" to a wire, and the "necessary" crossings of a cut by wires, all of which play major roles in single-layer routing problems. Finally, we rigorously formulate Theorems A and B in a natural and useful model, and outline their proofs. The definitive reference for this material is [8].

2. Basic Definitions

We begin with some definitions concerning paths. A path α is a continuous function with domain $I = [0,1]$, and should not be confused with its image $Im\ \alpha$. When we speak of a path intersecting a set or another path, however, we refer implicitly to the image of that path. The **endpoints** of a path α are the points $\alpha(0)$ and $\alpha(1)$. If these endpoints coincide, the path is a **loop**. A piecewise linear, injective path is called **simple**. If α is a path and $s,t \in I$, then the **subpath** $\alpha_{s:t}$ is defined by

$$\alpha_{s:t}(x) = \alpha\big(s + x(t - s)\big).$$

If two paths $\alpha, \beta : I \to X$ satisfy $\alpha(1) = \beta(0)$, their **concatenation** $\alpha \star \beta$ is the path $\gamma : I \to X$ defined by $\gamma_{0:\frac{1}{2}} = \alpha$ and $\gamma_{\frac{1}{2}:1} = \beta$. The euclidean arc length of a path α is denoted $|\alpha|$.

Two paths are **path-homotopic** if one can be continuously deformed into the other while keeping its endpoints fixed. Formally, the path $\alpha : I \to X$ is path-homotopic to $\beta : I \to X$, written $\alpha \simeq_P \beta$, if there is a continuous function $F : I \times I \to X$ such that $F(s,0) = \alpha(s)$

and $F(s, 1) = \beta(s)$ for all $s \in I$, and the functions $t \mapsto F(0, t)$ and $t \mapsto F(1, t)$ are constant. The relation of being path-homotopic, called **path homotopy**, is an equivalence relation, and the equivalence class of a path α is denoted $[\alpha]_P$. A space X is **simply connected** if X is path-connected (every pair of points can be joined by a path) and every loop in X is path-homotopic to a constant path. In a simply connected space, two paths α and β are path-homotopic if and only if they have the same endpoints: $\alpha(0) = \beta(0)$ and $\alpha(1) = \beta(1)$.

Our primary objects of study—wires, cuts, half-cuts, and their liftings—are paths in **2-manifolds with boundary**. A 2-manifold with boundary is any topological space M that looks locally like part of the plane: each point of M has a neighborhood homeomorphic either to R^2 or to the half-plane $H^2 = \{(x, y) \in R^2 : y \geq 0\}$. The points of the latter type form the **boundary** of the manifold M, denoted $Bd\, M$. I call the components of $Bd\, M$ the **fringes** of M. A path in a manifold M is a **link** if it touches the boundary of M at its endpoints alone $(\alpha^{-1}(Bd\, M) = \{0, 1\})$, and a **half-link** if it begins on the boundary of M but never returns $(\alpha^{-1}(Bd\, M) = \{0\})$. The **terminals** of a link or half-link in M are the fringes of M containing its endpoints.

There is a natural notion of deformation that applies to links in any manifold. It allows the endpoints of the links to move, but only along their respective fringes. Two links α and β in a manifold M are **link-homotopic**, written $\alpha \simeq_L \beta$, if β can be obtained by a continuous motion of α that keeps its endpoints on $Bd\, M$. Formally, there must be a continuous function $F : I \times I \to M$ such that $F(s, 0) = \alpha(s)$ and $F(s, 1) = \beta(s)$ for all $s \in I$, and $F(0, t) \in Bd\, M$ and $F(1, t) \in Bd\, M$ for all $t \in I$. **Link homotopy** is an equivalence relation on the links in M; we write the equivalence class of a link α as $[\alpha]_L$.

3. The Design Model

This section defines carefully the wire-routing problem introduced in Section 1. I choose it for its amenability to mathematical study, not for any reasons of practical utility. In this model the abstraction of a layout is a structure called a *design*. Despite the name, a design represents only the geometry of a circuit layer, and none of its functional or electrical characteristics except wire-to-terminal connections. In this section I introduce designs, explain when one design is a routing of another, and define which designs are *proper* circuit representations (i.e., design-rule correct). To keep things simple, all the objects we discuss will be piecewise linear.

3.1. Designs and routing

We route wires within a 2-manifold called a **sheet**. A sheet includes one or more disjoint **inner** polygons, no one of which encloses another, an **outer** polygon that encloses the inner ones, and all the space that lies inside the outer polygon but outside all the inner polygons. The outer and inner polygons form the boundary of the sheet, and hence are its **fringes**. Wires and cuts, for us, are links in a sheet.

Both the input and output of our routing problem take the form of a *design*. Designs are essentially collections of wires, which can either be considered rough routings (input) or detailed routings (output). A **wire** in a sheet S is a simple link in S with two convex inner fringes as terminals. A **design** on a sheet S is a finite set of wires in S whose images are disjoint and whose terminals are all distinct.

To route a design, one continuously deforms its wires so as to obtain a new design. In other words, one replaces each wire by a link-homotopic wire. Any link v that is link-homotopic to a wire ω is called a **route** for ω. If v is simple, then v is a wire in its own right, and we call it an **embedding** of ω. If Ω and Υ are designs on the same sheet, we say Υ is a routing or **embedding** of Ω if there exists a bijection $f : \Omega \to \Upsilon$ such that $\omega \simeq_L f(\omega)$ for every wire $\omega \in \Omega$.

3.2. Proper designs

The main problem concerning designs is that of finding a *proper* embedding for a design: one that represents a legal circuit layer. Since designs are purely geometric, whether a design is proper depends only upon the regions that its **details**—its wires and fringes—occupy or influence. These regions are called *extents*. Every design assigns a positive **width** to each wire and fringe, no wire being wider than either of its terminals. We consider a route for a wire to have the same width as the original wire. The **extent** of a detail of width d is the set of points in R^2 lying within $d/2$ units of it. Keeping to the piecewise linear category, we measure distances not with the euclidean metric, but rather with a *piecewise linear norm*.[1] This **wiring norm**, denoted $\| \cdot \|$, is a parameter of the entire model.

Two details are assumed to interact if and only if their extents overlap. Sometimes interactions among details are good, as when a wire connects to its terminals. To sort out the desirable overlaps, we group the details of a design into *articles*. Let Ω be a design on the sheet S. An **article** of Ω is either a fringe of S that is not a terminal of Ω, or the union of the terminals and image of some wire in Ω. The extent of an article is the union of the extents of its details. Articles must not approach too closely: a design is proper only if its articles have disjoint extents.

A subtler rule prevents wires and obstacles from approaching *themselves* too closely, lest they form loops or slivers in the layout. A set $X \subseteq R^2$ **divides** a sheet S if two fringes of S fall in different components of $R^2 - X$. An article of a design on a sheet S is **divisive** if its extent divides S. No article of a proper design may be divisive.

To summarize: A design is **proper** if its articles are nondivisive and have disjoint extents. A design is **routable** if it admits a proper embedding, and the wires in the proper embedding are called **feasible** embeddings of the wires in the original design.

4. Statement of Theorems

The design model supports a rich theory of single-layer wiring, only the highlights of which are presented here. Our main theorems effectively characterize the routable designs and the optimal feasible embeddings of their wires. They do so primarily in terms of the flows and capacities of straight cuts and half-cuts, thus reducing global properties of a design to local ones. The purpose of this section is to state those theorems, excluding only the definition of flow, which is a significant result in itself. Flow is defined in Section 6. For now, remember that the flow across a cut or half-cut represents the total amount of wiring forced by the topology to cross it.

4.1. Routability theorem

A design is routable unless it requires too much wiring to cross some **straight cut**: a linear link in its sheet. As a first approximation, a cut has too much wiring if its flow exceeds its capacity. Let a design Ω on a sheet S be given, and let χ be any link in S. Denote the width of a detail D of the design Ω by $width(D)$. If the terminals of χ are X and Y, then the **capacity** of χ in Ω is

$$cap(\chi, \Omega) = \|\chi\| - width(X)/2 - width(Y)/2,$$

where $\|\chi\|$ is the arc length of χ measured in the wiring norm. The capacity represents the amount of wiring in Ω that can fit across χ. We call χ **unsafe** if its flow exceeds its capacity, and otherwise **safe**.

Usually the presence of an unsafe straight cut makes a design unroutable. There are exceptions, however, for cuts that connect points in the same fringe or article. A link χ in a sheet S is **trivial** if χ is path-homotopic to a path in $Bd\ S$. Trivial cuts have no bearing on the routability of any design. More generally, a link χ is **degenerate** in the context of a design Ω if χ is path-homotopic to a path in a single article of Ω. Degenerate cuts can be unsafe without making their design unroutable, but they are the only such cuts.

Design Routability Theorem. *A design is routable if and only if all its nondegenerate straight cuts are safe.*

This theorem is our characterization of routability. To state it more concisely, we call a design **safe** if all its nondegenerate straight cuts are safe. Safety and routability are thus equivalent for designs. That routability implies safety is relatively easy to prove, but the converse is hard.

The Design Routability Theorem is the basis of efficient algorithms for testing routability [5] and for layout compaction [7]. To meet these needs it must be strengthened to show that routability depends only upon the safety or degeneracy of certain *critical* cuts, whose number is at most quadratic in the complexity of the sheet. Given the Design Routability Theorem and the machinery of Sections 6.3 and 7.1, results like this are relatively easy to establish [8].

4.2. Routing theorem

The best way to route a safe design is to choose for each wire a minimal route that leaves enough space for the other wires. To formalize this statement, we look at *straight half-cuts*. Let ω be a route for a wire in a design Ω on a sheet S. A **straight half-cut for ω at t** is a linear half-link σ in S such that $\sigma(1) = \omega(t)$. If X denotes the terminal of σ, the **capacity** of σ is defined by

$$cap(\sigma, \Omega) = \|\sigma\| - width(X)/2 - width(\omega)/2.$$

It represents the amount of routing space available between X and ω. To a first approximation, the route ω leaves enough space for wires to cross σ if and only if σ is **safe**, meaning that the flow across σ does not exceed its capacity.

As in the case of cuts, there are some half-cuts we ignore. To define them we look at certain links associated with a half-cut. If σ is a half-cut for ω at t, then it divides ω into two half-links $\omega_{0:t}$ and $\omega_{1:t}$, which connect with σ to form links. In our notation, those **associated links** are $\sigma \star \omega_{t:0}$ and $\sigma \star \omega_{t:1}$. A half-cut is **trivial** if it has a trivial associated link, and **degenerate** if it has a degenerate associated link. Neither trivial nor degenerate half-cuts affect the placement of wires, but only the trivial half-cuts need to be shunned. An **evasive** route, one that stays far enough away from all fringes, is a route whose nontrivial straight half-cuts are safe.

Evasive routes can interfere with one another, and need not be embeddings. To make them into wires, we minimize their length and control their parameterization. A path α is **canonical** if $|\alpha_{0:t}| = t\,|\alpha|$ for all $t \in I$. An **ideal** route is a canonical, evasive route of minimum length. When routing a safe design, one can do no better than to use an ideal route for each wire.

Design Routing Theorem. *The ideal routes of the wires in a safe design form a proper design, and they have minimal euclidean arc length among all feasible embeddings of those wires.*

This theorem tells us that no tradeoffs need be made in wire length when routing a design, and leads to efficient routing algorithms. The definition of ideal route is not especially constructive as it stands, but the proof of the Design Routability Theorem generates a wealth of information about ideal routes that suggests methods of routing.

The rest of the paper develops the theory of the design model, focusing on the concept of flow, and outlines proofs of the two design theorems. Here we have the first rigorous derivation of routability conditions for single-layer routing with homotopy constraints, and the first precise characterization of the optimal routings on a layer.

5. Covering spaces

The central tool in the study of the design model is the lifting of wires and cuts from the routing region, a sheet, to its simply connected covering space, called a **blanket**. This section defines blankets and lifting, and it examines the topology of blankets with regard to liftings of links. We show how link-homotopic links can be lifted and projected so that routes may be analyzed in terms of their liftings. We also note that a simple link in a blanket splits the blanket into two pieces. This observation underlies the definition of flow.

5.1. Covering spaces and lifting

A covering space of a space X is a space that is locally indistinguishable from X, but whose parts may be connected together differently. Formally, a surjective map $p : M \to X$ is a **covering map** if every point $x \in X$ has a neighborhood U such that each component of $p^{-1}(U)$ is mapped homeomorphically by p onto U. The space M is called a **covering space** of X. One can often think of M as lying "above" X, and the covering map as projecting M downward onto X.

One can study maps into a space, and especially paths, by transporting them into a covering space. If $p : M \rightarrow X$ is a covering map, a **lifting** of a map $\alpha : A \rightarrow X$ to M is any map $\widetilde{\alpha} : A \rightarrow M$ such that $p \circ \widetilde{\alpha} = \alpha$. For example, if α is a path in X, then $\widetilde{\alpha}$ is a path in M that "sits over" α. Not all maps have liftings, but paths can always be lifted. If $\alpha(t)$ is any point along a path $\alpha : I \rightarrow X$, and $m \in M$ is any point such that $p(m) = \alpha(t)$, then α has a lifting $\widetilde{\alpha}$ such that $\widetilde{\alpha}(t) = m$. In fact, this lifting is unique.

Proposition 1. *Two liftings of a map are equal if they agree at one point.*

Standard results of homotopy theory [10] imply that every connected manifold with boundary—for example, a sheet—has a simply connected covering space which is, for all topological purposes, unique. We may therefore speak of "the" blanket of a sheet. In addition, blankets are highly symmetrical. If $p : M \rightarrow S$ is the covering map taking a blanket onto a sheet, then for each two points $x, y \in M$ such that $p(x) = p(y)$, there is a homeomorphism that carries x onto y and preserves p. More generally, if $f, g : A \rightarrow M$ are liftings of the same map, there is a homeomorphism $h : M \rightarrow M$ such that $h \circ g = f$ and $p \circ h = p$. This homeomorphism makes the liftings f and g topologically indistinguishable. Consequently, when lifting maps from a sheet to its blanket, the choice of the first lifting is immaterial.

5.2. The topology of blankets

To work with blankets, one must know a fair amount about their geometry and topology. Fortunately, for most purposes one can treat a blanket as a simply connected subset of the plane, bounded by its fringes. Notions such as angles, linearity, etc. can be defined in a blanket by lifting and/or projecting via the covering map. This section considers some crucial blanket properties that have no obvious counterparts in planar topology.

We look first at link homotopy. Two links α and β in a blanket are link-homotopic only if $\alpha(0)$ and $\beta(0)$ share a fringe and $\alpha(1)$ and $\beta(1)$ share a fringe. A more concise statement is the following.

Lemma 2. *Two links in a blanket are link-homotopic if and only if they have the same terminals.*

The boundary of a blanket covers the boundary of its sheet; every fringe of the blanket projects onto a single fringe of the sheet. Hence liftings and projections of links are links, and similarly for half-links. If link-homotopic links in a blanket are projected to the sheet, they remain link-homotopic. Conversely, if two links in a sheet are link-homotopic, they can always be lifted to preserve that relation.

Lemma 3. *Let α and β be link-homotopic links in a sheet S, and let M be a blanket of S. There is a bijective correspondence between the lifts of α to M and the lifts of β to M such that corresponding lifts are link-homotopic.*

Next we look at the separation properties of blankets. In general, paths in a blanket separate the blanket into **scraps**, which are simply connected, open submanifolds of the blanket. (In particular, the blanket itself is a scrap.) The following proposition is fundamental.

Proposition 4. *Removing a simple link from a scrap leaves exactly two components, both scraps.*

The scraps that result when a simple link is removed from a blanket are called the **sides** of the link. One side can be identified as the **left** and one as the **right**.

As a consequence of Proposition 4, a simple link in a blanket partitions the fringes of the blanket into three categories: the terminals of the link, the fringes lying entirely in its left side, and the fringes lying entirely in its right side. No loop in a blanket encloses any fringe. Consequently, two simple links with the same terminals divide the fringes in the same way.

Proposition 5. *Link-homotopic simple links in a blanket partition the fringes identically.*

6. Crossings and Flow

This section defines flow and several related concepts, and derives some elementary properties of flow. The flow across a cut counts the *necessary* crossings of the cut by wires: those that cannot be removed by routing the wires. We formalize the notion of necessary crossing and show how to determine which necessary crossings are essentially *dissimilar*, or separate. The flow accounts for the wiring space that these necessary crossings must occupy. We define flow for all cuts and half-cuts, and this class is quite large: A link (or half-link) in a sheet is a **cut** (or **half-cut**) if its liftings to the blanket are simple.

We also compare flow to a measure called *congestion*. The congestion of a cut counts forced crossings differently: it looks at wire routes that minimize the number of crossings they make with the cut. We show that congestion and flow are equal for simple cuts, thus providing evidence that flow is the proper measure of wiring density.

6.1. How one link cuts another

When river routing in simply connected channels, one knows when a wire is forced to cross a cut. If the endpoints of the wire fall on opposite sides of the cut, then it has to cross. Otherwise, geometric constraints might force the wire to cross, but the topological constraints do not. In a similar way one can determine whether one link in a blanket is forced to cross another.

Definition 6. A simple link α in a blanket M **cuts** another link β in M if the endpoints of α and β lie on four distinct fringes of M and the endpoints of β lie on different sides of α.

We now examine why Definition 6 captures the notion of a forced crossing. If α cuts β, then the terminals of β lie in different components of $M - Im\,\alpha$, which means that α and β intersect. By Lemma 2, every link in $[\beta]_L$ has the same terminals as β, so α intersects every link in $[\beta]_L$. On the other hand, if α does not cut β, then one side of α includes parts of both terminals of β. Some link γ within that side connects those terminals, and by Lemma 2 we have $\gamma \simeq_L \beta$. Thus α cuts β if and only if α intersects every link in $[\beta]_L$.

The cutting relation has other nice properties. Provided that α is simple, whether α cuts β depends only upon the homotopy classes $[\alpha]_L$ and $[\beta]_L$, for link-homotopic links have the same terminals (Lemma 2) and partition fringes identically (Proposition 5). Finally, if both α and β are simple, then α cuts β if and only if β cuts α.

6.2. Necessary crossings

Next we develop the idea of forced crossings for links in sheets. In a blanket, one link can be forced to cross another link only once, while in a sheet, multiple crossings can be necessary. We therefore consider necessity to be a property of crossings, not links. If α and β are paths in the same space, a **crossing** of α by β is a pair $(s,t) \in I \times I$ such that $\alpha(s) = \beta(t)$. The pair (s,t) is ordered; (t,s) would be a crossing of β by α. To determine whether a crossing between two links in a sheet is necessary, we lift those links to the blanket a way that reflects that crossing, and check whether one lifting cuts the other.

Definition 7. Let (s,t) be a crossing of a cut χ by a link ω, both in the sheet S. Let M be a blanket of S with covering map $p : M \to S$, and choose any lifting $\widetilde{\chi}$ of χ to M. Because $p(\widetilde{\chi}(s)) = \omega(t)$, the link ω has a unique lifting $\widetilde{\omega}$ such that $\widetilde{\chi}(s) = \widetilde{\omega}(t)$. We say that $\widetilde{\chi}$ and $\widetilde{\omega}$ **reflect** the crossing (s,t). The crossing (s,t) of χ by ω is **necessary** if $\widetilde{\chi}$ cuts $\widetilde{\omega}$. Two crossings of χ by ω are **similar** if the corresponding liftings of ω are identical.

Necessary crossings can never be removed by wire routing, though similar crossings can be merged. For suppose that (s,t) is a necessary crossing of a cut χ by a wire ω; let $\widetilde{\chi}$ and $\widetilde{\omega}$ be liftings that reflect (s,t). If v is any route for ω, it has a lifting $\widetilde{v} \in [\widetilde{\omega}]_L$ by Lemma 3. And since $\widetilde{\chi}$ cuts $\widetilde{\omega}$, it also cuts $\widetilde{v}$, and hence crosses $\widetilde{v}$. The links χ and v make the same crossing. In this sense the crossing (s,t) cannot be eliminated. The lifting $\widetilde{v}$ may, however, make fewer crossings with $\widetilde{\chi}$ than $\widetilde{\omega}$ does. Thus if (s,t) is similar to some other crossings of χ by ω, those crossings can be reduced to one by routing ω.

Dissimilar necessary crossings cannot be merged. Suppose that among the crossings of χ by ω are n necessary crossings, no two of which are similar. Then $\widetilde{\chi}$ cuts n liftings $\widetilde{\omega}_1, \ldots, \widetilde{\omega}_n$ of ω. For any route $v \in [\omega]_L$ there are, by Lemma 3, distinct liftings $\widetilde{v}_1, \ldots, \widetilde{v}_n$ such that $\widetilde{v}_i \simeq_L \widetilde{\omega}_i$ for each i. Each is cut by $\widetilde{\chi}$ and hence crosses $\widetilde{\chi}$. Using uniqueness of liftings (Proposition 1), you can check that no two of these crossings are equal. Hence χ makes at least n crossings with v.

6.3. Definition of flow

Our observations about necessary and similar crossings motivate the definition of flow. Let χ be a cut and ω a link in the same sheet. The **winding** of χ with ω, denoted $wind(\chi, \omega)$, is the number of similarity classes of necessary crossings of χ by ω. (Similarity of crossings is an equivalence relation.) Equivalently, for any lifting $\widetilde{\chi}$ of χ to the blanket, $wind(\chi, \omega)$ is the number of liftings of ω that $\widetilde{\chi}$ cuts. Every route for ω makes at least $wind(\chi, \omega)$ crossings with χ. The **flow** across χ in a design Ω is a sum of the windings of χ with the wires in Ω, weighted by wire widths. In symbols,

$$flow(\chi, \Omega) = \sum_{\omega \in \Omega} width(\omega)\, wind(\chi, \omega).$$

Each necessary crossing (actually, each similarity class thereof) requires space equal to the width of the wire involved.

We now derive two important properties of flow. First, a cut has the same flow in all embeddings of a design. To prove this result, it suffices to show that the windings of a cut with link-homotopic wires are equal. If two wires are link-homotopic, their liftings are in bijective correspondence, with corresponding liftings being link-homotopic (Lemma 3). Hence any cut lifting cuts the same number of liftings of each wire. The second property is even more useful, though equally easy to prove.

Proposition 8. *Link-homotopic cuts have equal flow.*

By Lemma 3, link-homotopic cuts have link-homotopic liftings, which therefore cut the same wire liftings. Hence link-homotopic cuts have the same winding with any wire, and thus the same flow.

Proposition 8 allows us to define the flow across a half-cut. Let v be a route for a wire ω in a design Ω. If σ is a half-cut for v at t, the **flow** across σ in the design Ω is defined to be the common flow in Ω of the cuts that are link-homotopic to $\sigma \star v_{t:1}$. (One can always find such a cut: lift $\sigma \star v_{t:1}$ to the blanket, find a simple link having the same terminals as that lifting, and project it to the sheet.) This definition of flow makes intuitive sense, even though it involves nonsimple cuts. Because v is a route of ω, no wire in Ω makes a necessary crossing with v. Hence necessary crossings of $\sigma \star v_{t:1}$ somehow represent necessary crossings of σ.

6.4. Flow versus congestion

In the remainder of this section I lend some support to the definition of flow by showing how it subsumes the earlier and more elementary definition of the *congestion* of a cut. Let us denote the number of crossings between two paths α and β by $count(\alpha, \beta)$. Of course, the set of crossings can be infinite or even uncountable, but in the cases of interest it will be finite. The **entanglement** of a cut χ with a wire ω is defined to be the minimum number of crossings of χ by any route for ω. In symbols,

$$tangle(\chi, \omega) = \min\{\, count(\chi, v) : v \simeq_L \omega \,\}.$$

Entanglement is akin to winding. The **congestion** of χ in the design Ω is the total entanglement of χ by wires in Ω, where each crossing is weighted according to the width of its wire. Formally, we have

$$cong(\chi, \Omega) = \sum_{\omega \in \Omega} width(\omega)\, tangle(\chi, \omega).$$

In any embedding of Ω, the total amount of wiring passing across χ is at least $cong(\chi, \Omega)$. Flow and congestion are closely related.

Proposition 9. *The congestion of a cut is no less than its flow, and the two measures are equal if the cut is simple.*

Our argument that dissimilar necessary crossings cannot be removed or merged shows that flow is a lower bound on congestion. To prove that the congestion of a simple cut does not exceed its flow, one constructs for each wire a route that crosses the cut only as many times as their winding. This construction takes place in the blanket.

Flow has two advantages over congestion. First, its definition makes it easy to relate the flows of different cuts, simply by examining how the liftings of the cuts separate the fringes of the blanket. Proposition 8 is just one of many examples. Congestion is much harder to work with because it involves minimizing over all routes of a wire, and each cut may have its own minimizing routes. Second, flow retains most of its nice properties (like Proposition 8) in the realm of nonsimple cuts, while the congestion of a nonsimple cut varies in more complicated ways. Consequently, the definition of flow extends naturally to half-cuts, while the definition of congestion does not.

7. Proof Sketches

Space constraints limit us to a brief look at the key ideas and lemmas in the proofs of the Design Routability and Routing Theorems. The main line of argument, which we now follow, leads to the claim that all safe designs are routable using ideal routes.

7.1. Safety of cuts and half-cuts

We can prove the safety of a cut or half-cut by comparing its flow and capacity to those of cuts known to be safe. The proof of the next result gives the flavor of these derivations.

Proposition 10. *Let χ be a cut in a sheet S, and let Ω be a design on S. Suppose that (s, t) is a necessary crossing of χ by a route ρ for a wire ω in Ω. If the half-cuts $\chi_{0:s}$ and $\chi_{1:s}$ for ρ at t are unsafe in Ω, then so is the cut χ.*

Proposition 10 follows from the two relations

$$cap(\chi, \Omega) = cap(\chi_{0:s}, \Omega) + cap(\chi_{1:s}, \Omega) + width(\omega), \qquad (1)$$
$$flow(\chi, \Omega) \geq flow(\chi_{0:s}, \Omega) + flow(\chi_{1:s}, \Omega) + width(\omega). \qquad (2)$$

(Subtract the second from the first: if both $\chi_{0:s}$ and $\chi_{1:s}$ are unsafe, one finds that $cap(\chi, \Omega) - flow(\chi, \Omega)$ is negative.) Equation (1) is immediate from the definition of capacity and the fact that $\|\chi\| =$

$\|\chi_{0:s}\| + \|\chi_{1:s}\|$. To derive (2), lift χ and ρ to reflect the crossing (s,t). One can prove that whenever a lifting of a wire in Ω cuts either $\widetilde{\chi}_{0:s} \star \widetilde{\rho}_{t:1}$ or $\widetilde{\chi}_{1:s} \star \widetilde{\rho}_{t:1}$, it does not cut the other, but does cut $\widetilde{\chi}$. Consequently every wire lifting that contributes to the flow across $\chi_{0:s}$ or $\chi_{1:s}$ contributes an equal amount to the flow across χ. Furthermore, the lifting $\widetilde{\omega} \in [\widetilde{\rho}]_L$ of ω guaranteed by Lemma 3 cuts $\widetilde{\chi}$ but not $\widetilde{\chi}_{0:s} \star \widetilde{\rho}_{t:1}$ or $\widetilde{\chi}_{1:s} \star \widetilde{\rho}_{t:1}$, as you can verify. This lifting accounts for the term $width(\omega)$ in inequality (2).

Next we relate simple cuts to straight cuts. Beginning with an arbitrary simple cut, shrink the cut like a rubber band, keeping its endpoints fixed, until it becomes a path-homotopic **chain** of linear links interspersed with paths along fringes. When the cut splits into two cuts by contacting a terminal of a wire ω, a necessary crossing by ω may be lost. In this way the total flow across the cuts can decrease. But the total capacity decreases also, by the width of that terminal. Because no wire is wider than its terminals, every decrease in flow is compensated by an equal or greater decrease in capacity. Hence if the original cut was unsafe, there must be an unsafe cut within the final chain. Further arguments show that nondegeneracy can also be preserved. Thus one obtains the following results.

Proposition 11. *A design that contains a nondegenerate, unsafe, simple cut also contains a nondegenerate, unsafe, straight cut.*

Corollary 12. *Nondegenerate simple cuts in safe designs are safe.*

The argument behind Proposition 11 appears to be central to single-layer routability; if the model is tweaked in a way that breaks this argument, the Routability Theorem generally fails as well. A version of this argument appears in [2] as a proof sketch. With careful use of blankets one can make it absolutely rigorous.

7.2. Construction of ideal routes

Now we come upon the second major function of blankets, namely, to help us determine where wires should be routed. In single-layer channel routing [14, 16] one defines *barriers* that the wires must avoid if they are to leave sufficient room for other wires. This construction fails in a multiply connected routing region like a sheet, but it can be carried out within its blanket. In effect, we construct evasive and ideal routes by routing within the covering space.

Definition 13. Let v route a wire ω in a design Ω on a sheet S, and let $\widetilde{v}$ be a lifting of v to the blanket M of S. Let $p : M \to S$ be the covering map, and say $\widetilde{v}$ runs from fringe X to fringe Y. Suppose

$F \notin \{X, Y\}$ is a fringe of M. All simple links $\tilde{\alpha}$ from F to Y have the same value of $flow(p \circ \tilde{\alpha}, \Omega)$, call it f. (For a proof, combine Lemma 2 and Proposition 8.) The **barrier** for $\tilde{v}$ around F is the set of endpoints $\tilde{\sigma}(1)$ of straight half-links $\tilde{\sigma}$ with $\tilde{\sigma}(0) \in F$ and

$$\|p \circ \tilde{\sigma}\| - width(p(F))/2 - width(\omega)/2 < f. \tag{3}$$

We call the barrier for $\tilde{v}$ around F a **left** or **right** barrier according to which side of $\tilde{v}$ (that is, which scrap of $M - Im\,\tilde{v}$) contains F. The barriers for $\tilde{v}$ depend only upon ω and the terminals of $\tilde{v}$; hence all links in $[\tilde{v}]_L$ have the same barriers.

Definition 13 reveals the geometry of barriers, but is otherwise rather opaque. The motivation for Definition 13 may be discerned by supposing that the projection $\sigma = p \circ \tilde{\sigma}$ is a half-cut for v at a point t; say $\tilde{\sigma}(1) = \tilde{v}(t)$. Then the left-hand side of (3) is just $cap(\sigma, \Omega)$. In addition, the links $\tilde{\alpha}$ in Definition 13 are liftings of cuts in $[\sigma \star v_{t:1}]$, and so $f = flow(\sigma, \Omega)$ by definition. Hence inequality (3) simply says that the half-cut σ is unsafe. The requirement that F not be a terminal of $\tilde{v}$ means that σ is *nontrivial*. In sum, $\tilde{v}$ avoids its barriers if and only if v has no nontrivial, unsafe, straight half-cuts.

Lemma 14. A route is evasive if and only if one of its liftings avoids its barriers.

To create an evasive route for a wire, therefore, it suffices to lift the wire to the blanket, determine the barriers, and deform the lifting so that it intersects no barrier. The barriers themselves do not change. Proposition 15 ensures the success of this construction.

Proposition 15. In a safe design, no left barrier for a route lifting intersects a right barrier for the same route lifting.

A sketch of the proof reveals the need for our results on safety. Using the notation of Definition 13, we prove the contrapositive. Suppose that a left barrier for $\tilde{v}$ overlaps a right barrier for $\tilde{v}$. We may assume that $\tilde{v}$ passes through a point x in their intersection. Then there exist half-links $\tilde{\sigma}$ and $\tilde{\tau}$ that begin on opposite sides of $\tilde{v}$, end at x, and otherwise conform with Definition 13. Both $\sigma = p \circ \tilde{\sigma}$ and $\tau = p \circ \tilde{\tau}$ are unsafe half-cuts for v. The concatenation $\tilde{\chi} = \tilde{\sigma} \star \tilde{\tau}_{1:0}$ cuts $\tilde{v}$, and hence the simple cut $\chi = \sigma \star \tau_{1:0}$ makes a necessary crossing with v at $p(x)$. Proposition 10 now shows that χ is unsafe. By Corollary 12, therefore, the underlying design cannot be safe.

Proposition 16. Every wire in a safe design has an evasive route.

Finally, with the help of Ascoli's Theorem [13], one deduces the existence of evasive routes that are canonical and minimal in length.

Proposition 17. *Every wire in a safe design has an ideal route.*

The ideal route of a wire need not be unique, though it usually is.

7.3. Ideal routes form a proper design

Most of the remainder of the proofs of the design theorems consists in deriving properties of ideal routes. Since ideal routes are defined so abstractly, there is a lot of work to be done. Just to get started, for instance, one must prove that they are piecewise linear. I will skip these technical matters and describe instead an interesting characteristic of ideal routes called *tautness*. For a route to be taut means, in essence, that it only turns where a half-cut is preventing it from approaching a fringe. Figure 2 should prime your intuition.

Definition 18. Let Ω be a design on the sheet S, and let ω route a wire in Ω. A **strut** for ω at $t \in (0, 1)$ is a nondegenerate, straight half-cut σ for ω at t such that $flow(\sigma, \Omega) = cap(\sigma, \Omega)$ and if $\tilde{\sigma}$ and $\tilde{\omega}$ are liftings reflecting the crossing $(1, t)$, then $\tilde{\omega}$ turns toward $\tilde{\sigma}(0)$ at t. The link ω is **taut** if there is a strut for ω at every point $t \in I$ at which ω turns.

The concept of "turning toward" used in Definition 18 is this:[2] For $\tilde{\omega}$ to turn toward $\tilde{\sigma}(0)$ at t, it must make an angle at t, and $\tilde{\sigma}(0)$ must lie on the side of $\tilde{\omega}$ that contains points internal to this angle and arbitrarily close to $\tilde{\omega}(t)$. The more obvious condition, that $\sigma(0)$ be internal to the angle formed by ω at t, is not strong enough.

Our next result is proven by the same sorts of techniques as Proposition 15, but adapted for half-cuts instead of cuts.

Proposition 19. *Ideal routes are taut.*

The struts of ideal routes play a key role in the remaining arguments and in the wire-routing algorithm of [5] and [8].

We resolve the interactions among ideal routes by adapting the argument of [16] to a multiply connected routing region. For the most part, this means working in the blanket rather than in the sheet. Whenever one compares two ideal routes, they could turn out to coincide; the covering space is essential for sorting out the interactions of different parts of that route—or equivalently, between different liftings of that route. The end result is the following.

Proposition 20. *Replacing each wire in a safe design by one of its ideal routes yields a proper design.*

The proof of Proposition 20 is fairly modular. First, one shows that no two ideal routes of wires in a safe design cross over or touch one another. If they did, one could find within a strut for one an unsafe, nondegenerate, straight half-cut for the other, contrary to its evasiveness. Applying this result to two copies of the same route, one finds that the route has no self-intersections; ideal routes are wires. Second, the barriers ensure that no ideal route approaches any fringe too closely. Third, one proves that no two ideal routes approach one another too closely. For if they did, one could build an unsafe, nondegenerate, *simple* half-cut for one using of a strut for the other, and, by the method of Proposition 11, reduce it to an unsafe, nondegenerate, *straight* half-cut. Finally, one must confront the issue of self-avoidance, which does not normally arise in channel routing. Fortunately, one can convert it to a question about a route approaching *itself* too closely, and apply the previous result.

Proposition 20 is the heart of the two design theorems. To complete those theorems, we must prove that ideal routes are optimal in two senses: no similar construction can be carried out for the wires in an unsafe design, and no shorter embeddings are feasible for the wires in a safe design. For these results, see [8].

8. Conclusions

Lifting to a covering space appears to be the right tool for analyzing single-layer wire routing with homotopy constraints. It yields elegant definitions, such as that of a necessary crossing, that offer insight into the routing problem, and it solves all the problems that arise in generalizing from simply connected to multiply connected routing regions. This technique has two drawbacks: it generates rather lengthy proofs, and does not apply easily to models in which the routing region is not a compact manifold. Further research and greater mathematical sophistication should alleviate these problems.

Acknowledgements

This paper is based on a thesis submitted in August, 1987 in partial fulfillment of the requirements for the degree of Doctor of Philosophy in the Department of Electrical Engineering and Computer Science at the Massachusetts Institute of Technology. The work described herein was supported in part by a Graduate Fellowship from the Office of Naval Research and in part by the Defense Advanced Research Projects Agency under Contract N00014-80-C-0622. The author is currently supported by a Mathematical Sciences Postdoctoral Research Fellowship from the National Science Foundation (Grant No. DMS-8705835).

Notes

[1] See, for instance, [13] for a definition of *norm*. A norm on a vector space (such as R^2) gives rise to a distance metric that respects its linear structure; the metric is invariant under translation and is linear on each line. Piecewise linearity implies that the set of points of norm 1 is a polygon.

[2] For the proof of the Design Routing Theorem, one needs a stronger definition of tautness that considers cuts constraining the route's endpoints as well as half-cuts constraining its middle. In [8] "turning" is defined at the endpoints of a link, and the definition of tautness should have been correspondingly strengthened, though due to an error it was not.

References

[1] B. S. Baker and R. Y. Pinter, "An algorithm for the optimal placement and routing of a circuit within a ring of pads," *24th FOCS* (November 1983), pp. 360–370.

[2] R. Cole and A. Siegel, "River routing every which way, but loose," *25th FOCS* (October 1984), pp. 65–73.

[3] D. Dolev, K. Karplus, A. Siegel, A. Strong, and J. D. Ullman, "Optimal wiring between rectangles," *13th STOC* (May 1981), pp. 312–317.

[4] S. Gao, M. Jerrum, M. Kaufmann, K. Mehlhorn, W. Rülling, and C. Storb, "Homotopic one-layer routing", private communication (July 1987).

[5] C. E. Leiserson and F. M. Maley, "Algorithms for routing and testing routability of planar VLSI layouts," *17th STOC* (May 1985), pp. 69–78.

[6] C. E. Leiserson and R. Y. Pinter, "Optimal placement for river routing," *SIAM Journal on Computing*, Vol. 12, No. 3 (August 1983), pp. 447–462.

[7] F. M. Maley, "Compaction with automatic jog introduction," *1985 Chapel Hill Conference on VLSI* (May 1985), pp. 261–283.

[8] F. M. Maley, *Single-Layer Wire Routing*, Ph.D. thesis, MIT EECS Department (August 1987), available as MIT/LCS/TR-403.

[9] K. Mehlhorn, private communication (August 1987).

[10] J. R. Munkres, *Topology: A First Course*, Prentice-Hall, Englewood Cliffs, New Jersey (1975).

[11] R. Y. Pinter, *The Impact of Layer Assignment Methods on Layout Algorithms for Integrated Circuits*, Ph.D. Thesis, MIT EECS Department (August 1982), available as MIT/LCS/TR-291.

[12] D. Richards, "Complexity of single-layer routing," *IEEE Transactions on Computers*, Vol. C-33, No. 3 (March 1984), pp. 286–288.

[13] H. L. Royden, *Real Analysis*, 2nd ed., Macmillan (1968).

[14] A. Siegel and D. Dolev, "The separation for general single-layer wiring barriers," *Carnegie-Mellon Conference on VLSI* (October 1981), pp. 143–152.

[15] T. Szymanski, "Dogleg channel routing is NP-complete," *IEEE Transactions on CAD*, Vol. CAD-4, No. 1 (January 1985), pp. 31–41.

[16] M. Tompa, "An optimal solution to a wire-routing problem," *Journal of Computer and System Sciences*, Vol. 23, No. 2 (October 1981), pp. 127–150.

A Faster Compaction Algorithm with Automatic Jog Insertion *

Kurt Mehlhorn and Stefan Näher

FB10, Informatik
Universität des Saarlandes
6600 Saarbrücken
Federal Republic of Germany

Abstract: In this paper we refine work of Maley [13] on one-dimensional compaction with automatic jog insertion. More precisely, we give an algorithm with running time $O((n^2 + k)\log n)$, where $k = O(n^3)$ is a quantity which measures the difference between the input and the output sketch, and so improve upon Maley's $O(n^4)$ algorithm. The compaction algorithm takes as input a layout sketch; the wires in a layout sketch are flexible and only indicate the topology of the layout. The compacter minimizes the horizontal width of the layout whilst maintaining its routability. The exact geometry of the wires is filled in by a router after compaction.

0. Introduction

A *compactor* takes as input a VLSI-layout and produces as output an equivalent layout of smaller area. An effective compaction system frees the designer from the details of the design rules and hence increases his productivity and on the other hand produces high quality layouts. For these reasons, compaction algorithms have gained widespread attention in the VLSI-literature [4,7,9,10,13,22] and are the basis for several computer-aided circuit design systems [3,4,11,19,21].

In this paper we restrict ourselves to one-dimensional compaction where the area is reduced by moving the features of the layout in a single direction only. For convenience, we assume this direction to be horizontal. For an approach to two-dimensional compaction we refer the reader to [7].

For the purpose of compaction a layout is separated into modules, which are rigid in shape and size, and wires, which are flexible. Most compaction algorithms use this flexibility in a very limited way. For example, the compactor in HILL [11,19] treats vertical wire segments as rigid objects during horizontal compaction and only changes the length of horizontal wire segments. Some compaction algorithms allow vertical wires to bend during compaction by the insertion of jogs. This is either done interactively by the designer ([4,21]) or automatically [20,21,22]. The latter procedures are however ad hoc and not guaranteed to be effective.

recent advances in homotopic routing to use in compaction. A *homotopic router* takes a layout sketch, consisting of the exact placement of the modules and

---
* Research supported by DFG under contract SFB 124, TP B2

the topology of the wires (cf. Figure 2), as its input and produces as output a (detailed) routing of the sketch in some wiring model (cf. Figures 3,4 and 5). For several different wiring models (one-layer routing in grids [8,14], gridless one-layer routing [5,14], knock-knee mode routing [6]) it is known that efficient homotopic routers exists. More precisely, it was shown that in these cases a simple cut condition, henceforth called routability condition, is sufficient for the routability of the sketch.

Maley [13] proposed to view, at least for the purposes of compaction, the wires in a VLSI layout, only as indicators of the layout topology, and to compact the layout maintaining the routability condition. The wires are constructed in their final form by a homotopic router. In this way, he puts automatic jog insertion on a sound theoretical basis. Maley shows that the compaction problem can be solved in time $O(n^4)$, where n is the size of the sketch. In this paper, we improve the running time to $O((n^2 + k) \log n)$ where k is a quantity which measures how much the input and output sketch differ. We expect $k = O(n^2)$ in practice and always have $k = O(n^3)$. Our algorithm is not only faster than Maley's but also easier to understand. We want to emphasize however, that our correctness proof rests completely on the foundations laid by Maley.

The intuition behind our algorithm is quite simple. We put the input sketch between two rigid vertical bars and then move the right bar to the left. Initially, all modules except the right bar stay at their initial position. At some point, a tight cut (density = capacity) between a module and the right bar will arise. Therefore, this module starts moving together with the bar. At some later point, some other cut becomes tight, either between a module and the right bar or between a module and the module which moves already with the bar. So a second module starts to move with the bar. We continue in this fashion until the left bar starts to move. At this point, we have computed a configuration of minimal x-width.

The simplicity of our approach makes it very flexible. In particular, we can also handle maximum distance constraints, we may compact in arbitrary directions and not just in x-direction and we can support plowing. A *plow* is a line segment which intersects no feature but apart from that can be in arbitrary position. A plow operation consists of moving the plow in a certain direction for a certain distance, cf. Figure 1. The running time of our plowing algorithm lies between $O(n \log n)$ and $O(n^3 \log n)$ depending on how much it changes the sketch. Consecutive plowing operations may use different directions. Our plowing algorithm is more general than the one presented in [20] because it works on the symbolic instead of the geometric level and hence can change wire geometries more cleverly, secondly, the plow can be a line segment of arbitrary orientation and is not constrained to be iso-oriented, and thirdly, the plow can move in an arbitrary direction and not just parallel to the coordinate axes.

This paper is structured as folows. In section I, we give the relevant definitions and state the problem and the results precisely. In section II, we prove the correctness of our algorithm and describe an efficient implementation of it. The main novel idea in the implementation is an efficient data structure for the maintenance of the capacities and densities of all cuts. In section III we describe several extensions (in particular, plowing) and in section IV we offer a short conclusion.

I. Definitions and Results

A *sketch* is a triple (F, W, P) consisting of a finite set F of *features*, which are points (= point feature) and open straight line segments (= line feature), a finite set W of *wires*, which are simple paths in the plane, and a partition P of the features F. Each block of the partition is called a *module*. Figure 2 shows an example of a sketch. When the partition P is understood we will refer to a pair (F, W) as a sketch. The features and wires of a sketch must satisfy the following conditions:

(1) Distinct features do not intersect and the endpoints of each line feature are point features.

(2) No wire may cross itself.

(3) Each wire touches exactly two features, which are point features lying at the endpoints of the wire. They are called the *terminals* of the wire.

(4) There are two line features, called the left and right bar, which are infinite vertical lines and lie to the left and right of all other features.

A *point in a sketch* is a point lying on a feature. Modules form the rigid part of a layout and wires represent the flexible interconnections.

Sketches comprise the information of placement and global routing. A (detailed) *routing* of a sketch (F, W, P), $W = \{p_1, \ldots, p_m\}$, is a sketch (F, W', P), $W' = \{q_1, \ldots, q_m\}$, such that q_i is *homotopic* to p_i, i.e., p_i and q_i have the same endpoints and p_i can be transformed continuously into q_i without moving its endpoints and without allowing its interior to touch a feature in F, and such that the q_i's satisfy the constraints of the particular wiring model used. We consider two wiring models in this section, the grid model and the free model, and comment on the knock-knee model in the conclusion.

In the *grid model* wires are vertex-disjoint paths in the rectangular grid of unit-spacing. Figure 3 (4) shows a routing of the (compacted) sketch of Figure 2 in the grid model. In the *free model* wires are arbitrary paths in the plane satisfying the following minimum separation constraints (for a point set T, $U(T) = \{x;\ dist(x, t) < 1/2$ for some $t \in T\}$ denotes the open $1/2$-neighborhood of T):

1) $U(q_i) \cap U(q_j) = \emptyset$ for $i \neq j$

2) $U(q_i) \cap U(f) = \emptyset$ where f is any feature which is not a terminal of q_i

3) $U(q_i)$ is simply connected.

Figure 5 shows a routing of the compacted sketch of Figure 2 in the free model.

A *cut* is any open line segment connecting two points of the sketch, say $p = (x_p, y_p)$ and $q = (x_q, y_q)$, and not intersecting any feature. We denote the cut with endpoints p and q by $\overline{pq}$. The *density* of cut $\overline{pq}$ is the number of crossings of $\overline{pq}$ by wires which are enforced by the topology of the sketch, cf. Figure 6. Crossings of $\overline{pq}$ which can be removed by deforming the wires do not contribute to the density. The *capacity* of a cut in the free model is the Euclidean length (of the corresponding closed segment) minus one and the capacity in the grid model is given by $\max\{|x_p - x_q|, |y_p - y_q|, 1\} - 1$. A cut is called *safe* if its density does not exceed its capacity and it is called *tight* or *saturated* if its density is equal to its capacity. The following theorem was proved by Cole/Siegel and Leiserson/Maley for the grid model and by Gao et al. and Maley for the free model.

Theorem 1. *A sketch has a routing iff all cuts of the sketch are safe.*

Actually, the results are slightly stronger. Let us call a cut $\overline{pq}$ *critical*, if either p and q are point features or at most one of them lies on a line feature and the line segment $\overline{pq}$ is perpendicular to that line feature. Then a sketch is routable iff all critical cuts are safe.

We are now ready to define the (one-dimensional) compaction problem. The goal of compaction is to displace the modules in x-direction such that the resulting sketch is routable and has minimal x-width (the x-width of a sketch is the horizontal distance between the left and right bar). Let $S = (F, W, P)$ be a *routable sketch*. We denote the displacement of feature $f \in F$ by $d(f)$ and call the vector $d \in \mathbb{R}^F$ of displacements a *configuration*. Of course, not all displacements make sense. Firstly, features in the same module must be displaced by the same amount and therefore we must have $d(f) = d(g)$ for any two features in the same module. Secondly, features should not cross over during compaction and we therefore must have $x_p + d(f) < x_q + d(g)$ for any two points $p = (x_p, y_p)$ and $q = (x_q, y_q)$ where $x_p < x_q$ and $y_p = y_q$ and p lies on a feature f and q lies on a feature g. Let d be a configuration satisfying the two constraints above. We can now define the sketch $S(d)$ in a natural way. A point p on feature f with coordinates (x_p, y_p) in the initial sketch has coordinates $(x_p + d(f), y_p)$ in $S(d)$ and the wires in $S(d)$ have the "same" homotopies as in S; cf. [13] for a more precise definition. Figures 4 and 5 show compacted versions of the sketch of Figure 2. The *configuration space* $C(S) \subseteq \mathbb{R}^F$ of a sketch S consists of all configurations d such that d satisfies the two constraints above and $S(d)$ is routable. Note that the zero-vector 0 belongs to $C(S)$ since the sketch S is assumed to be routable. The *essential configuration* space $C_0(S)$ of a sketch S consists of that connected component of $C(S)$ which contains the zero-vector, i.e., a configuration d belongs to $C_0(S)$ if the sketch $S(d)$ can be obtained by continuously deforming the sketch S going only through routable intermediate sketches. The compaction problem can now be formalized as follows.

(One-Dimensional) *Compaction Problem* [13]
Input: A routable sketch $S = (F, W, P)$
Output: A configuration $d \in C_0(S)$ such that $S(d)$ has minimal x-width

Theorem 2 [13] .
Let $S = (F, W, P)$ be a routable sketch. Then the essential configuration space $C_0(S)$ of the sketch S is a convex polyhedron.

Maley gave an explicit description of the polyhedron $C_0(S)$ and made it the basis of his solution for the compaction problem. In order to state his result we need some more notation. Let us assume that the wires in a sketch are given by polygonal paths. For a wire $w \in W$, let b_w be the number of line segments in the polygonal path for w, let $b = \sum_{w \in W} b_w$, and let $m = |F|$ be the number of features.

Theorem 3 [13] .
The compaction problem can be solved in time $O(m^4 + m^2 b \log mb) = O(n^4)$ where $n = m + b$ is the size of the input sketch.

In this paper we improve upon theorem 3 and show

Theorem 4. *The compaction problem can be solved in time* $O(mb + (m^2 + k)\log m) = O(n^3 \log n)$, *where* $n = m + b$, *and* k *is the number of times that a module moves across a cut during the compaction process;* $k = O(m^3)$ *always.*

We prove theorem 4 in section II. The quantity k measures in a certain sense how much the compaction algorithm changes the sketch. k can be as large as $O(m^3)$, and as small as 0 (if the input sketch has already minimal x-width). In practical compaction problems, where the input sketch is frequently nearly optimal, we expect k to be much smaller than $O(m^3)$. Under the assumption that a cut is crossed by only $O(1)$ features on average we have $k = O(m^2)$ and hence a running time of $O(n^2 \log n)$.

II. The new compaction algorithm

Let $S = S(0)$ be a routable sketch. Recall that there are two vertical bars enclosing the sketch. The idea underlying our algorithm is very simple. We move the right bar to the left. Whenever a tight cut between an already moving feature and a still motionless feature arises the motionless feature starts to move along with the right bar. We continue in this fashion until the left bar starts to move. At this point we have minimized the x-width of the sketch. We will now describe the algorithm in more detail.

At each position P ($= x$-coordinate) of the right bar we have a partition of the features into two sets L and R. Initially, R consists only of the right bar. In general, R consists of a certain set of modules, i.e., if a feature f belongs to R then the entire module containing f is a subset of R. Moreover, there exists a rooted tree on the modules in R such that the right bar is the root of the tree and such that for every non-root M there is a point p on M and a point q on the parent of M such that $\overline{pq}$ is a tight cut in the current configuration and p has smaller x-coordinate than q. We will frequently use "p lies to the left of q" instead of "p has smaller x-coordinate than q".

The *current configuration* is defined implicitly by the algorithm below. The initial configuration is the initial sketch S. Suppose now that the current configuration $d(P)$ is defined for some position P of the right bar and let $\Delta > 0$ be such that the partition (L, R) does not change in the interval $(P - \Delta, P)$. Then the configuration $d(P - \Delta)$ is obtained from $d(P)$ by moving all features in R to the left by the amount Δ. For ease of notation we write S_P instead of $S(d(P))$. It remains to describe when and how the partition (L, R) changes.

Lemma 1. *Let P be the current position of the right bar, and let S_P be routable and such that there is no tight critical cut $\overline{pq}$ with $p \in f \in L, q \in g \in R$ and p to the left of q in S_P. Let $\Delta > 0$ be minimal such that a non-vertical critical cut becomes tight in $S_{P-\Delta}$.*

a) $S_{P-\delta}$ is routable for $0 \leq \delta \leq \Delta$.

b) Let $\overline{pq}$ be a cut which becomes tight in $S_{P-\Delta}$. Then $p \in f \in L, q \in g \in R$ and p lies to the left of q in $S_{P-\Delta}$.

Proof: (omitted) ∎

Let P and Δ be as in the premise of lemma 1. Let R' be defined by program 1.

---

(1) $\mathbb{R}' \leftarrow R; L' \leftarrow L$;
(2)   **while** $\exists$ tight critical cut $\overline{pq}$ with $p \in f \in L', q \in g \in R'$
        and p to the left of q in $S_{P-\Delta}$
(3)   **do** $R' \leftarrow \{f'; f$ and f' belong to the same module$\}$
(4) $L' \leftarrow F - R'$
(5)   **od**

_____ **Prog. 1** _____

Then $S_{P-\Delta}$ is routable by part a) of lemma 1 and there is no tight critical cut $\overline{pq}$ with $p \in f \in L', q \in g \in R'$ and p to the left of q in $S_{P-\Delta}$, i.e., the premise of lemma 1 is again satisfied.

It remains to define a rooted tree on the modules in R' with the desired properties. Let $\overline{pq}$ be a tight critical cut with $p \in f \in L', q \in g \in R'$ and p to the left of q in $S_{P-\Delta}$. We make the module containing g the parent of the module containing f. Also, if the cut $\overline{pq}$ became tight in $S_{P-\Delta}$ then there might be a point $r \in h \in R$ such that p lies on $\overline{rq}$ in $S_{P-\Delta}$, $\overline{rq}$ is a tight cut and the module containing g is the parent of the module containing h. In this case, the cut $\overline{rp}$ is also tight and we make the module containing f the new parent of the module containing h; cf. Figure 8.

Theorem 5. _The algorithm above solves the compaction problem._

Proof: (omitted)   ■

The correctness of our algorithm is now established. We turn to its implementation next. First we introduce the concepts of cut changing (C) and tightening (T) events and show how to compute them efficiently. Then we reformulate the compaction algorithm in terms of C-events and T-events.

Let f and g be features with one of them, say f, being a point feature. If g is also a point feature then $\overline{fg}$ denotes the line segment connecting f and g, if g is a line feature then $\overline{fg}$ denotes the line segment through f perpendicular to g. A line segment $\overline{fg}$ is a cut if it intersects no other feature.

A C-event occurs when a cut appears or disappears. Note that the next C-event is the next position of the right bar where two existing cuts $\overline{fg}$ and $\overline{gh}$ become collinear (see Figure 9).

The _T-event_ of a cut $\overline{fg}$ is the next position of the right bar where $\overline{fg}$ becomes tight. This position is computed under the assumption that the density of $\overline{fg}$ and the partition (L, R) does not change until the event occurs. Note that the T-event only exists for cuts connecting a feature in L with a feature in R. For cuts connecting two features in the same set the capacity does not change and hence no T-event exists (or equivalently, takes place at $-\infty$).

We keep all C-events and T-events in a priority queue, called the _global event queue_ Q. The next event is the minimal element of Q. Let P and Δ satisfy the premise of Lemma 1. Then it is clear that the next event occurs no later than at position $P - \Delta$, because if no C-event occurs before $P - \Delta$ then a T-event will occur at $P - \Delta$.

Knowing the density of a cut $\overline{fg}$ and the current status of its two endpoints (element of L or R) the T-event of $\overline{fg}$ can easily be computed in time $O(1)$. To compute the C-events we have to work harder.

We maintain for every point feature f a special data structure which we call the *C-structure* of f. It is defined as follows. Consider all cuts incident to f. If we extend these cuts to straight lines, they partition the plane into sectors (or wedges). (cf. Figure 10)

These sectors will change their size during the compaction, since moving point features in R will cause the corresponding lines to move too. Let $S(f, g, h)$ be the sector between the lines through g and h (cf. Figure 10). Every time a sector collapses, i.e., points f, g and h become collinear, a C-event occurs.

Now compute for every sector $S(f, g, h)$ the position of the right bar $P(f, g, h)$ when this sector collapses. We call $P(f, g, h)$ the *collapsing time* of $S(f, g, h)$. The collapsing time is computed under the assumption that the status (= element of L or R) of the three points f, g, h will not change before the event.

$P(f, g, h)$ is either $-\infty$ or the solution of a simple system of 3 linear equations which can be solved in time $O(1)$. For example, in the situation of figure 10 with $f \in L, h \in R$ and $g \in L$ we have $P(f, g, h) = \delta$ with

$$\begin{vmatrix} 1 & 1 & 1 \\ x_f & x_g & x_h + \delta \\ y_f & y_g & y_h \end{vmatrix} = 0$$

where x_q, y_q are the initial coordinates of point feature $q \in \{f, g, h\}$. All other situations lead to similar equations. We store each sector $S(f, g, h)$ according to its collapsing time $P(f, g, h)$ in a local priority queue at f. Then we have

Lemma 2. a) *The C-structure of a point feature f needs space $O(m)$ and can be constructed in time $O(m \log m)$.*

b) *The minimal event of all local priority queues is the next C-event.*

Every time a C-event occurs, i.e., a sector $S(f, g, h)$ collapses, the C-structure of f has to be updated as follows. Let $S(f, x, h)$ be the sector left of $S(f, g, h)$ and let $S(f, g, y)$ be the sector right of $S(f, g, h)$ (cf. Figure 10).

- remove $S(f, x, h)$, $S(f, g, h)$ and $S(f, g, y)$ from the local priority queue
- compute $P(f, x, h)$ and $P(f, h, y)$
- insert $S(f, x, h)$ and $S(f, h, y)$ into the local queue according to their collapsing time

If a cut $\overline{fg}$ disappears it has to be deleted from the C-structures of f and g (if point features). Furthermore the T-event of $\overline{fg}$ must be removed from the global priority queue Q.

If a cut $\overline{fg}$ appears it has to be inserted into the C-structure of f and g and the density and the T-event of $\overline{fg}$ must be computed.

All of this can be done in time $O(\log n)$ (We will show later how the density can be computed in time $O(\log n)$).

If a T-event for some potential cut $\overline{fg}$ occurs f is removed from L and put into R. This may change the collapsing times of all sectors in the C-structure of f. Therefore we rebuild the entire C-structure of f in this case which has cost $O(m \log m)$.

Furthermore we have to update the collapsing time of the two sectors incident to f for all C-structures of point features $h \neq f$. This again has cost $O(m \log m)$ since we perform $O(m)$ update operations on local priority queues.

Every time a new cut $\overline{fg}$ appears its density has to be computed. Let $\overline{fh}$ and $\overline{hg}$ be the two cuts that became collinear (cf. figure 11). Then we have

$$density(\overline{fg}) = density(\overline{fh}) + density(\overline{hg}) + \#\text{nets with terminal } h$$
$$- 2 \cdot \#\text{turning nets}$$

To compute the number of turning nets we maintain the sketch during the compaction as follows. Consider the nets in the sketch as tight rubber bands. Then each net forms a polygonal path whose edges are cuts and boundary edges of modules and whose vertices are corners of modules. Now assign to every cut or edge $\overline{fg}$ the bundel of nets that use the line segment $\overline{fg}$ in the rubber band model.

For every corner h of a module we store all cuts and edges incident to h in a list $L(h)$ in clockwise ordering and associate with each entry $\overline{fh}$ the size of its bundel $b(\overline{fh})$. When two cuts $\overline{fh}$ and $\overline{hg}$ become collinear and create the new cut $\overline{fg}$ the number of turning nets can be computed as follows (cf. figure 12).

Let a be the number of nets using cuts before $\overline{fh}$ in the list $L(h)$ and b be the number of nets using cuts after $\overline{hg}$ in $L(h)$, i.e.

$$a = \sum\{b(\overline{xh}) \mid \overline{xh} \in L(h) \text{ and } \overline{xh} < \overline{fh}\}$$
and $b = \sum\{b(\overline{hy}) \mid \overline{hy} \in L(h) \text{ and } \overline{hy} > \overline{hg}\}$.

Then the number of turning nets is $\min(a, b)$, since wires do not cross. Note that a and b can be computed in time $O(\log n)$ if the lists $L(h)$ are implemented as balanced search trees. If a C-event occurs the above defined data structure for maintaining the sketch must be updated. If a cut $\overline{fg}$ disappears the corner h of a module comes to lie on it (cf. figure 13) and the following actions have to be performed:

$$b(\overline{fh}) \leftarrow b(\overline{fh}) + b(\overline{fg})$$
$$b(\overline{hg}) \leftarrow b(\overline{hg}) + b(\overline{fg})$$
$$\text{delete } \overline{fg} \text{ from } L(f)$$
$$\text{delete } \overline{fg} \text{ from } L(g)$$

All of this takes time $O(\log n)$.

The case where a new cut $\overline{fg}$ is created is a bit more complicated. Assume that cuts $\overline{fh}$ and $\overline{hg}$ become collinear. Before this event happens the situation is as shown in figure 14.

In this figure, a_1 is the number of nets using cuts before $\overline{fh}$, $a_2 = b(\overline{fh})$ and a_3 is the number of nets using cuts between $\overline{fh}$ and the horizontal line through h.

b_1 is the number of nets using cuts after $\overline{hg}$, $b_2 = b(\overline{hg})$ and b_3 is the number of nets using cuts between the horizontal line through h and $\overline{hg}$.

All the numbers can be computed in time $O(\log n)$ if $L(h)$ is implemented as a balanced tree. For example $a_1 = \sum\{b(\overline{xh}) \mid \overline{xh} \in L(h) \text{ and } \overline{xh} > \overline{fh}\}$.

Our goal is to compute $b(\overline{fg})$ and the new values for $b(\overline{fh})$ and $b(\overline{hg})$ (we denote these values by $b'(\overline{fg}), b'(\overline{fh}), b'(\overline{hg})$). It is easy to see that

$$b'(\overline{fh}) = b(\overline{fh}) - b'(\overline{fg}) \text{ and } b'(\overline{hg}) = b(\overline{hg}) - b'(\overline{fg}).$$

The bundel of the new cut $\overline{fg}$ is the intersection of the bundels of $\overline{fh}$ and $\overline{hg}$. $b'(\overline{fg})$ is the cardinality of this intersection. The following case analysis for the computation of $b'(\overline{fg})$ relies heavily on the assumption the nets do not cross. A generalization to knock-knee mode is discussed at the end of this paper.

case 1: $a_2 = 0$ or $b_2 = 0$
i.e., at least one of the bundels of $\overline{fh}$ or $\overline{hg}$ is empt. Then $b'(\overline{fg}) = 0$.

case 2: $a_2 \neq 0$ and $b_2 \neq 0$
Assume w.l.o.g. that $a_3 \geq b_3$
(the other case is symmetric, just exchange the a's and b's)

2.1: $a_3 > b_2 + b_3$
then there is no net using both $\overline{fh}$ and $\overline{hg}$ and hence $b'(\overline{fg}) = 0$

2.2: $a_3 \leq b_2 + b_3$

2.2.1: $a_1 \geq b_1$
Then all nets in the bundel of $\overline{fh}$ are also contained in the bundel of $\overline{hg}$ and we have $b'(\overline{fg}) = a_2$

2.2.2: $a_1 < b_1$
In this case there are $b_1 - a_1$ nets in the bundel of $\overline{fh}$ that are not contained in the bundel of $\overline{hg}$ and we have $b'(\overline{fg}) = a_2 - b_1 + a_1$.

This completes the case analysis. We have shown that the additional cost for maintaining the sketch is $O(\log n)$ per C-event. Thus the total cost of a C-event is $O(\log n)$.

Note that both data structures presented above (for computing C-events and for maintaining the sketch) have space requirement linear in the number of all currently existing cuts which is $O(n^2)$ in the worst case but which can be exepcted to be much smaller in practice.

We are now ready to formulate the entire compaction algorithm.

Preprocessing: (* takes time $O(bn + n^2 \log n)*$)

(1)   compute the visibility graph $G = (V, E)$ of the input sketch
(2)   **for** all cuts $\overline{fg} \in E$
(3)   **do** compute the density
(4)   the capacity
(5)   **od**
(6)   **for** all cuts and edges $\overline{fg}$
(7)   **do** compute the size of its bundel $b(\overline{fg})$
(8)   insert $\overline{fg}$ into $L(f)$
(9)   insert $\overline{fg}$ into $L(g)$

(10)   **od**
(11)   **for** all point features f
(12)   **do** construct the C-strructure of f **od**

Initialization: (∗ takes time $O(n \log n)$∗)

(1) $R \leftarrow \{\text{"right bar"}\}$
(2) $L \leftarrow$ all features not in R
(3)   initialize the global event queue with all C-events and T-events $\neq -\infty$
       (∗ Note that then are only $O(m)$ events in the beginning ∗)

Compaction: (∗ takes time $O((n^2 + k) \log m)$∗)

consider the minimal event in the global event queue Q; it is either a C- or a T-event.

C-event: point features f, g, h become collinear

– update the C-structure of f, g, h as described before
  (∗ takes time $O(\log n)$∗)
– update the sketch as described before
  (∗ takes time $O(\log n)$∗)
  case 1: the cut $\overline{fg}$ disappears
         – delete the T-event of $\overline{fg}$ from Q
  case 2: a new cut $\overline{fg}$ is created
         – compute its density and T-event
           and insert it into Q
  (∗ both cases need time $O(\log n)$∗)

T-event: let $\overline{fg}, f \in L, g \in R$ be the cut that became tight.

– $R \leftarrow R \cup \{f\}$; $L \leftarrow L \backslash \{f\}$
– update the C-structure of f and of all point features visible from f
  (∗ takes time $O(n \log n)$∗)
– update the global event queue Q (delete old C-events,
  insert new C-events) (∗ takes time $O(n \log n)$∗)
– compute the T-events for all cuts incident to f, insert them into Q
(∗ the total cost of a T-event is $O(n \log n)$∗)

Theorem 6. *The above algorithm solves the compaction problem in time $O(bn + (n^2 + k) \log n)$ and space $O(V)$, where k is the number of C-events and V is the maximal size of the visibility graph of the sketch.*

Proof: The space bound follows from the fact that both the data structure for computing the events and that for maintaining the sketch need space proportional to the size of the visibility graph of the current sketch. We have already seen that the algorithm spends time $O(\log n)$ in the C-event case and $O(n \log n)$ time in the T-event case. Now the time bound follows from the observation that there are at most $O(n)$ T-events (every time a T-event occurs a feature moves from L to R). ∎

The number of C-events may be as large as $O(m^3)$. However, if the initial sketch is already optimal then $k = 0$ and if the compaction process brings about only local changes of the layout then $k = O(m^2)$ since cuts are crossed by only $O(1)$ features in this case. On the other hand, a large value of k can only occur if the compaction changes the layout globally. In this case, the increased running time is well spent.

III. Extensions

In this section we present extensions and modifications of our compaction algorithm which can be implemented without great effort.

1. Maximum distance constraints

In practice there are often constraints which postulate that the distace of two layout components must not exceed a certain value. They are either user-defined or result from the technology. With a small modification our algorithm can also handle such maximum distance constraints. We only have to define the T-event of a maximum distance constraint, say between features f and g, as the next position of the right bar where the distance reaches its maximal value. If the T-event of a maximal distance constraint between f and g occurs then $f \in R, g \in L$ and f is to the left of g. We add g to R at this point. All other parts of the algorithm remain unchanged.

2. Compaction in arbitrary directions

None of the data structure presented uses the fact that compaction is done in x-direction. So we may allow arbitrary directions; we only have to change the part of the algorithm where the collapsing times of sectors of C-structures are computed, because now not only the x-coordinates of point features, but also the y-coordinates change over time. But the collapsing time of a sector is still the solution of a simpe system of linear equations and can be computed in time $O(1)$.

3. Plowing

Plowing is a powerful concept of the Magic Layout Editor [17]. It can be used interactively to rearrange the geometry of a cell, compact a sparse layout or to create new space in a dense layout. The user places a plow (horizontal or vertical line segment) into the layout and gives the direction and distance the plow is to move. According to these specifications the plow is then moved through the layout and the material behind it is stretched (if necessary) and the material in front of it is compressed. For details see [20].

Our compaction method can support plowing as well. After the user has defined the plow and the direction and distance it is to move, the set R is initialized with the plow and the global event queue is initialized with all C-events and T-events $\neq -\infty$. Since there are at most $O(n)$ such events (local visibility), the initialization step takes time $O(n \log n)$.

Now the compaction algorithm is executed as described before. Its running time is bounded by $O(k \log n)$ where k is the number of C-events occuring during the plowing operation.

The data structures for computing the events and for maintaining the sketch have to be computed only once for an entire sequence of plowing operations since they do not depend on the compaction direction and the initial values of L and R (as discussed in part 2 of this section); thus our plowing procedure can be used as an efficient interactive tool. The cost of an entire sequence of consecutive plowing operations only depends on how much the sketch is changed by this sequence.

The plowing operation supported by our compaction algorithm is more general than the one presented in [20]. Firstly, it works on the symbolic level instead of the geometric level and hence can change wire geometries more cleverly, secondly, the plow can be a line segment of arbitary orientation and is not constrained to be iso-oriented, and thirdly, the plow can move in an arbitrary direction and not just parallel to the coordinate axes.

4. Compaction of channels

If we use our algorithm for minimizing the length of a channel of given width its performance is better than in the general case. Assume that a horizontal channel configuration is to be compacted in x-direction, then there are no C-events since no cut disappears and none is created. Thus we have $k = 0$ in theorem 6 and the running time is $O(n^2 \log n)$.

Of course, the algorithm of Leiserson/Pinter solves this problem in linear time.

IV. Conclusions

We presented a compaction algorithm which achieves optimal one-dimensional compaction with jog-insertion. Its running time is $O(n^2 \log n)$ in the best case and deteriorates to $O(n^3 \log n)$ if the compaction process changes the layout dramatically. The algorithm is conceptually very simple; only its implementation is nontrivial. The algorithm follows the push (minimum distance constraints) and pull (maximum distance constraints) paradigm. Another recent application of this paradigm can be found in [15].

We discussed the compaction problem for one-layer routing. The algorithm can be applied to multi-layer routings by applying it to each layout plane separately. Of course, the vias and the layer assignment must be part of the input sketch in this case.

It is very likely that our algorithm also applies to routing in knock-knee mode (wires are edge-disjoint paths on a grid). Kaufmann/Mehlhorn have shown that a sketch can be routed if all cuts are safe and the free capacity (= capacity - density) of every cut is even. Because of the evenness condition the routability condition is only sufficient but not necessary in this case. Suppose now that we are given a sketch which satisfies the routability condition. Then the free capacity of every nonsaturated cut is a multiple of two and hence one should be able to move the right bar by a multiple of two. If the resulting sketch were still even, a fact which we have not shown yet, then we could iterate and our algorithm would find the sketch of smallest x-width also satisfying the routability condition (of course, this may not be the routable sketch of smallest x-width). The algorithms of Kaufmann/Mehlhorn and Brady/Brown can then turn the sketch into a 4-layer detailed routing. In this way, layer assignment would be postponed until after compaction.

V. References

[1] Brady, M., and Brown, P., "VLSI Routing: Four Layers Suffice", Advances in Computing Research, Ed. F. Preparata, 1984, pp. 245–258

[2] Cole, R., and Siegel, A., "River routing every which way, but loose", Proceedings of the 25th Annual Symposium on Foundations of Computer Science, October 1984, pp. 65–73

[3] Dunlop, A.E., "SLIP: symbolic layout of integrated circuits with compaction", Computer Aided Design, Vol. 10, No. 6, 1978, pp. 387–391

[4] Hsueh, M.Y., "Symbolic Layout and Compaction of Integrated Circuits", Ph.D. thesis, EECS Division, University of California, Berkeley, 1979

[5] Gao, S., Jerrum, M., Kaufmann, M., Mehlhorn, K., Rülling, W., Storb, C., "On homotopic river routing", BFC 87, Bonn, 1987

[6] Kaufmann, M. and Mehlhorn, K., "Local Routing of Two-Terminal Nets", 4th STACS 87, LNCS 247, pp. 40–52

[7] Kedem, G., and Watanabe, H., "Optimization techniques for IC layout and compaction", Technical Report 117, Computer Science Department, University of Rochester, 1982

[8] Leiserson, C.E., and Maley, F.M., "Algorithms for routing and testing routability of planar VLSI layouts", Proceedings of the 17th Annual ACM Symposium on Theory of Computing, 1985, pp. 69–78

[9] Leiserson, C.E., and Pinter, R.Y., "Optimal placement for river routing", SIAM Journal on Computing, Vol. 12, No. 3, 1983, pp. 447–462

[10] Lengauer, T., "Efficient algorithms for the constraint generation for integrated circuit layout compaction", Proceedings of the 9th Workshop on Graphtheoretic Concepts in Computer Science, 1983

[11] Lengauer, T., and Mehlhorn, K., "The HILL system: a design environment for the hierarchical specification, compaction, and simulation for integrated circuit layouts", Proceedings, Conference on Advanced Research in VLSI, 1984

[12] Lengauer, T., "On the solution of inequality systems relevant to IC layout", Journal of Algorithms, Vol. 5, No. 3, 1984, pp. 408–421

[13] Maley, F.M., "Compaction with Automatic Jog Insertion", 1985 Chapel Hill Conference on VLSI

[14] Maley, F.M., personal communication, 1987

[15] Mehlhorn, K.,and Rülling, W., "Compaction on the Torus", Techn. Report, FB10, Informatik, Universität des Saarlandes, Saarbrücken, Dec. 87

[16] Mehlhorn, K., "Data Structures and Algorithms", Vol. 3, Springer Publ. Company, 1984

[17] Ousterhout, J.K., Hanachi, G., Mayo, R.N., Scot, W.S., Taylor, G.S., "The Magic VLSI Layout System", Proceedings of the 21st Design Automation Conference, June 1984

[18] Pinter, R.Y., "The Impact of Layer Assignment Methods on Layout Algorithms for Integrated Circuits", Ph.D. thesis, MIT Department of Electrical Engineering and Computer Science, 1982

[19] Rülling, W.,"Einführung in die Chip-Entwurfssprache HILL", Techn. Bericht 04/1987, SFB124, Univ. des Saarlandes, 1987

310

[20] Scott, W.S., and Ousterhout, J.K., "Plowing: interactive stretching and compaction in Magic", Proceedings of the 21th Design Automation Conference,
1984, pp. 166–172

[21] Williams, J.D., "STICKS — a graphical compiler for high level LSI design",
National Computer Conference, 1978, pp. 289–295

[22] Xiong, X-M. "Optimized One-Dimensional Compaction of Building-Block Layout", Berkeley Memorandum UCB/ERL M87/45, 1987

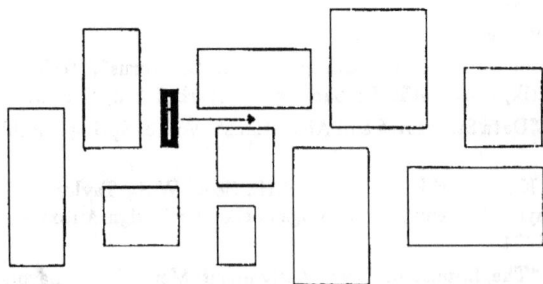

Figure 1: Example of a plowing operation.

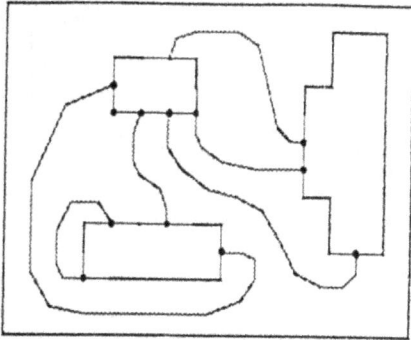

Figure 2: A typical sketch.

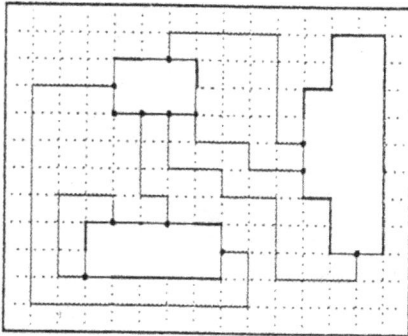

Figure 3: A routing of the sketch in figure 2 using wires of minimal length. The width is 15.

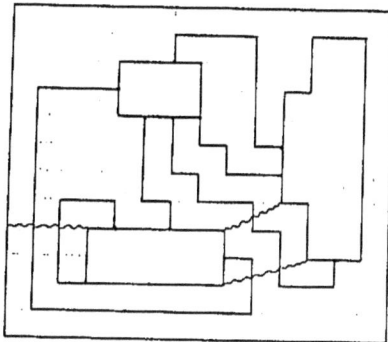

Figure 4: A routed and compacted version of the sketch of figure 2 (grid model); tight cuts are shown as wiggled lines. The width is 14.

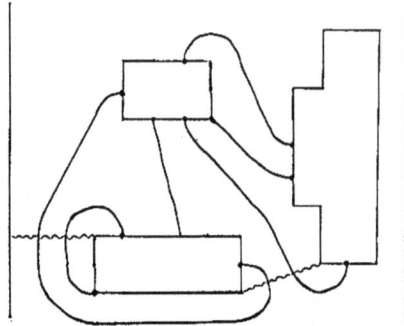

Figure 5: A routed and compacted version of the sketch of figure **2** (free model). The width is $11 + 2\sqrt{2}$.

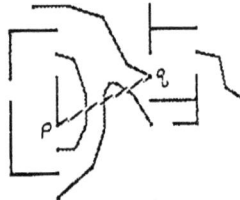

Figure 6: A portion of a sketch with cut $\overline{pq}$. The flow across $\overline{pq}$ is 1.

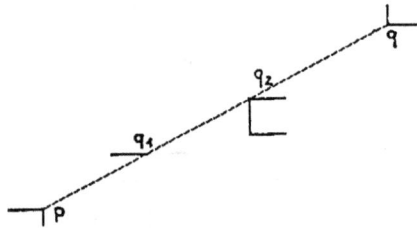

Figure 7: The features $q_1, q_2, q_3, \ldots$ on the line segment $\overline{pq}$.

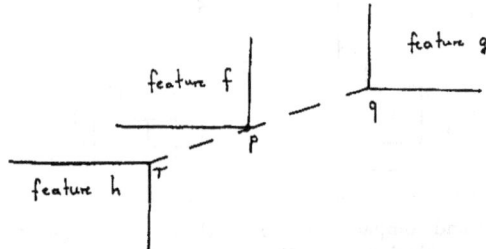

Figure 8: $r(p, q)$ lies on feature $h(f, g)$. The features h and g belong to R and the cut $\overline{rq}$ is tight. When $p \in f \in L$ comes to lie on $\overline{rq}$ in $S_{P-\Delta}$ both cuts $\overline{rp}$ and $\overline{pq}$ are tight.

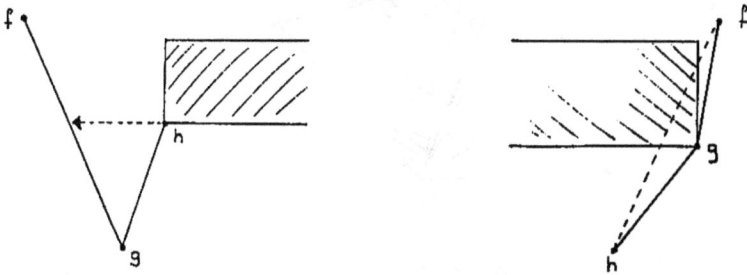

Figure 9

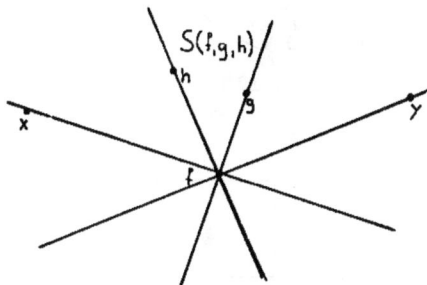

Figure 10

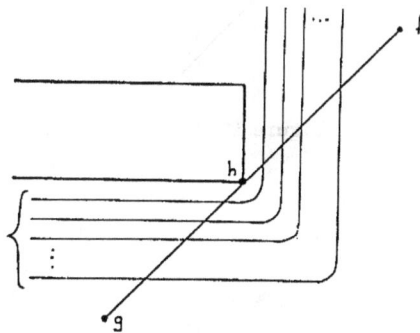

Figure 11

314

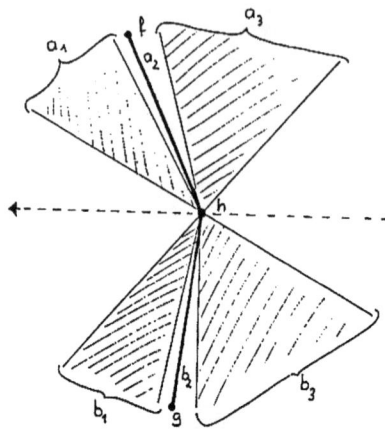

Figure 12

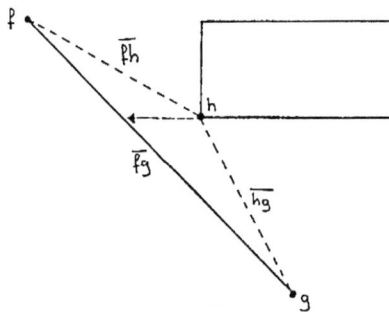

Figure 13

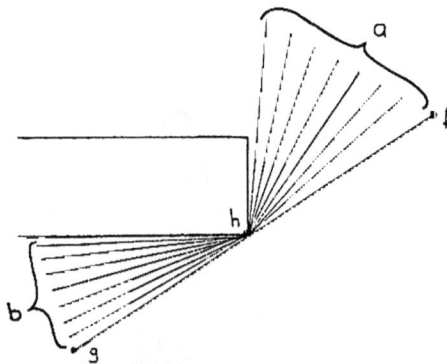

Figure 14

Layout Permutation Problems and Well-Partially-Ordered Sets

Michael R. Fellows

Michael A. Langston

Department of Computer Science
University of Idaho
Moscow, ID 83843

Department of Computer Science
Washington State University
Pullman, WA 99164-1210

We prove the existence of decision algorithms with low-degree polynomial running times for a number of well-studied VLSI layout problems. Some were not previously known to be in P at all; others were only known to be in P by way of exhaustive dynamic programming formulations with unboundedly high-degree polynomial running times. Our methods include the application of the recent Robertson-Seymour theorems on the well-partial-ordering of graphs under both the minor and immersion orders. We also prove polynomial-time self-reducibility for each of the problems we address.

1. Introduction

In the design and manufacturing of VLSI systems, practical problems are often characterized by fixed-parameter instances. The parameter may represent, for example, the number of tracks permitted on a chip, the number of processing elements to be employed, the number of channels required to connect circuit elements or the load on communications links. In fixing the value of such parameters, we focus on the physically realizable nature of the system rather than the purely abstract aspects of the model.

In this paper, we employ Robertson-Seymour posets and self-reducibility strategies to prove small-degree polynomial-time decision and construction complexity for a variety of fixed-parameter layout, placement and routing problems, dramatically lowering known time-complexity upper bounds. Our main results are summarized in the table that follows, where k denotes the relevant fixed parameter.

General Problem Area	Problem	Best Previous Upper Bound	Our Result
circuit layout	GATE MATRIX LAYOUT [LL]	open for $k \geq 3$ $O(n)$ for $k \leq 2$ [DKL]	$O(n^2)$
linear arrangement	MIN CUT LINEAR ARRANGEMENT [GJ] MODIFIED MIN CUT [Su] VERTEX SEPARATION NUMBER [Le] TOPOLOGICAL BANDWIDTH [Ch, MPS] HYPERGRAPH MODIFIED CUTWIDTH [Le, MS] HYPERGRAPH VERTEX SEPARATION NUMBER [Le]	$O(n^{k-1})$ [MaS] $O(n^k)$ [Su] $O(n^{k^2+2k+4})$ [EST] $O(n^k)$ [MPS] $O(n^{k^2+3k+c})$ [Su] $O(n^{k^2+2k+4})$ [Su]	$O(n^4)$ $O(n^2)$ $O(n^2)$ $O(n^2)$ for max degree 3 $O(n^2)$ $O(n^2)$
network searching and clearing	SEARCH NUMBER [MHGJP] NODE SEARCH NUMBER [KP]	$O(n^{2k^2+4k+8})$ [EST] $O(n)$ for $k \leq 3$ [MHGJP] $O(n^{2k^2+4k+c})$ [Su]	$O(n^2)$ $O(n^2)$
embedding and routing	2-D GRID LOAD FACTOR [Ro] BINARY TREE LOAD FACTOR [Ro]	open open	in P in P

In the next section, we briefly outline the advances from graph theory and graph algorithms that make these results possible. Sections 3, 4 and 5 describe our results on three representative types of problems, illustrating a range of techniques based on well-partial-orders. A few related issues are briefly discussed in the final section.

2. Background

All graphs we consider are finite and undirected, but may have loops and multiple edges. A graph H is less than or equal to a graph G in the *minor* order, written $H \leq_m G$, if and only if a graph isomorphic to H can be obtained from G by a series of these two operations: taking a subgraph and "contracting" edges. For example, the construction that follows shows that $W_4 \leq_m Q_3$ (although $W_4 \not\leq_t Q_3$ where $\leq_t$ denotes less than or equal to in the topological order).

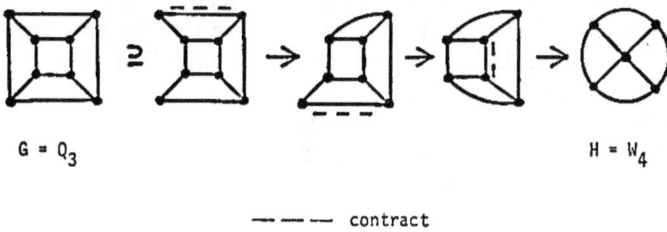

G = Q_3 H = W_4

— — — — contract

Note that the relation $\leq_m$ defines a partial ordering on graphs. A family F of graphs is said to be *closed* under the minor ordering if the facts that G is in F and that $H \leq_m G$ together imply that H must be in F. The *obstruction set* for a family F of graphs is defined to be the set of graphs in the complement of F that are minimal in the minor ordering. Therefore, if F is closed under the minor ordering, it has the following characterization: G is in F if and only if there exists no H in the obstruction set for F such that $H \leq_m G$.

Theorem 1. [RS5] (formerly known as Wagner's Conjecture [Wa]) Any set of finite graphs contains only a finite number of minor-minimal elements.

Theorem 2. [RS4] For every fixed graph H, the problem that takes as input a graph G and determines whether $H \leq_m G$ is solvable in polynomial time.

Remarkably, Theorems 1 and 2 guarantee only the *existence* of a polynomial-time decision algorithm for any minor-closed family F of graphs. The proof of Theorem 1 is inherently *nonconstructive* in a precise, mathematically-strong sense [FRS]. That is, when we are able to apply Theorem 1 to (a finite description of) F, we are assured of a finite obstruction set for F without being given (by the arguments that establish the theorem) a means for identifying the elements of the set, the cardinality of the set, or even the order of the largest graph in the set.

Another interesting feature of Theorems 1 and 2 is the low degree of the polynomials bounding the decision algorithms' running times. Letting n denote the number of vertices in G, the general bound is $O(n^3)$. If F excludes even one planar graph, then the bound decreases to $O(n^2)$. Curiously, as of this writing, these polynomials possess enormous constants of proportionality, rendering them impractical for problems of any nontrivial size [Jo]. Therefore, Theorems 1 and 2 can be viewed largely as results that constitute a tool for determining problem complexity, and thus serve to direct attention to a search for alternate, practical algorithms. Whether the algorithms directly promised by these theorems can be made effective in practice (assuming the obstruction set for the family of graphs of interest is known) is an important open question.

A graph H is less than or equal to a graph G in the *immersion* order, written $H \leq_i G$, if and only if a graph isomorphic to H can be obtained from G by a series of these two operations: taking a subgraph and "lifting" pairs of adjacent edges [Ma]. For example, the construction that follows shows that $C_4 \leq_i K_1 + 2K_2$ (although $C_4 \not\leq_m K_1 + 2K_2$ and $C_4 \not\leq_t K_1 + 2K_2$).

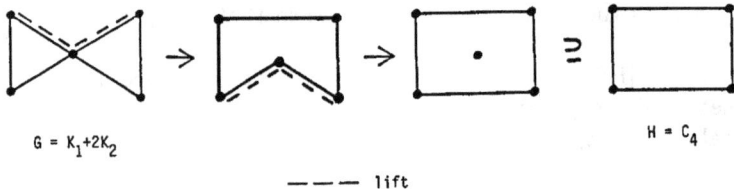

The relation $\leq_i$, like $\leq_m$, defines a partial ordering on graphs with the associated notions of closure and obstruction sets.

Theorem 3. [RS3] (formerly known as Nash-Williams' Conjecture [Na]) Any set of finite graphs contains only a finite number of immersion-minimal elements.

Theorem 4. For every fixed graph H, the problem that takes as input a graph G and determines whether $H \leq_i G$ is solvable in polynomial time.

Proof Sketch. Letting k denote the number of edges in H, we replace $G = \langle V, E \rangle$ with $G' = \langle V', E' \rangle$, where $|V'| = k|V| + |E|$ and $|E'| = 2k|E|$. Each vertex in V is replaced in V' with k vertices. Each edge e in E is replaced in G' with a new vertex and $2k$ edges connecting it to all of the vertices that replace e's endpoints. We can now apply the disjoint-connecting paths algorithm of [RS4], since it follows that $G \geq_i H$ if and only if G' contains a set of vertex-disjoint paths, one for each edge in H. $\square$

Theorems 3 and 4, like Theorems 1 and 2, guarantee only the existence of a polynomial-time decision algorithm for any immersion-closed family F of graphs. The method we use in proving Theorem 4 yields a general time bound of $O(n^{h+6})$, where h denotes the order of the largest graph in F's obstruction set. Thus the present line of attack for the immersion order is analogous to an earlier strategy for the minor order [FL1, Jo], enabling us to establish polynomial-time decidability without knowing the degree of the polynomial bounding the decision algorithm's running time. For the results in this paper, however, all bounds will be $O(n^{h+3})$, since the problem graphs of interest permit only a linear number of edges. Moreover, even better bounds are likely when more is known about each problem's obstruction set [FL3].

Observe that these results for both orders are nonconstructive at another level. Even if an appropriate obstruction set were given for the decision problem, there is absolutely no guarantee that constructing a solution to an optimization version of the problem can be accomplished within any given time-complexity class [FL2]. These developments, therefore, call in to question the previous intuitive wisdom that only the complexity of decision problems need be addressed (see, for example, [KUW]), and motivate a serious consideration of the practical aspects of employing a decision algorithm to construct a solution, a process termed *self-reducibility*, that has until now been primarily a subject of theoretical curiosity (see, for example, [MP, Sc]). Unlike the general methods provided by Theorems 1, 2, 3 and 4, self-reducibility strategies remain very problem-specific.

3. Relevant Prior Results on GATE MATRIX LAYOUT

Consider the fundamental layout problem known as GATE MA-TRIX LAYOUT (GML). It has been the focus of much attention, arising in several VLSI layout styles, including gate matrix, PLAs under multiple folding, Weinberger arrays and others. Although constructing an actual layout requires a number of algorithmic steps, the combinatorial problem of interest can be stated as follows: we are given an $n \times m$ Boolean matrix M and an integer k, and are asked whether we can permute the columns of M so that, if in each row we change to * every 0 lying between the row's leftmost and rightmost 1, then no column contains more than k 1s and *s.

From an optimization standpoint, most progress until very recently has been negative: the decision version of the general problem is NP-complete [KF]; unless $P = NP$, no absolute approximation algorithm is possible [DKL]; and very strong circumstantial evidence suggests that even relative approximation algorithms cannot exist [RL]. Moreover, instances of GML exist with only 2 satisfactory permutations and $m!-2$ unsatisfactory ones [DKL], preventing brute-force attacks (such as the well-known methods that work for problems like clique, vertex cover, etc.) that focus on a predetermined candidate solution set. Nevertheless, it has been shown that for any fixed value of k, an arbitrary instance of GML can be mapped to an equivalent instance with only two 1s per column, then modeled as a graph whose family of "yes" instances is closed under the minor order and excludes a planar graph [FL1].

Theorem 5. [FL1] For any fixed k, GML can be decided in $O(n^2)$ time.

From a practical perspective, however, the goal is to construct a satisfactory permutation of the columns of M, if one exists. (Given such a permutation, it is easy to complete the layout itself with a simple greedy rule [HS, OMKKF].) Thus arises the issue of self-reducibility. Happily, it turns out that a decision algorithm can be employed much like an oracle to modify M until the "consecutive 1s property" is achieved, at which time the PQ-Tree algorithm of [BL, Go] can be applied to identify an appropriate column permutation.

Theorem 6. [BFL2] For any fixed k, a satisfactory solution can be constructed, if any exist, in $O(n^4)$ time.

Progress has also been made on closing in on the obstruction sets (and hence the decision and construction algorithms) for small values of k.

Theorem 7. [BFKL, FKL] There are exactly 2 obstructions for GML when $k = 2$, and 110 when $k = 3$.

4. MIN CUT LINEAR ARRANGEMENT and Load Factors

An *embedding* of a graph G into a graph H is a one-to-one map $f\colon V(G) \to V(H)$ together with an assignment, to each edge uv of G, of a path from $f(u)$ to $f(v)$ in H. Graph embeddings have been studied for a wide variety of applications [Ro]. For problems of VLSI layout, H generally represents physical constraints on the layout of G. The *minimum load factor* of G relative to H is the minimum, over all embeddings of G in H, of the maximum over all edges e of H of the number of paths of the embedding that contain e.

For example, in the case where H is the one-dimensional grid, the minimum load factor of G with respect to H is called the *cutwidth* of G. In the NP-complete MIN CUT LINEAR ARRANGEMENT problem [GJ], we are given a graph G and an integer k, and are asked whether the cutwidth of G is no more than k. Unlike GATE MATRIX LAYOUT, however, neither the family of "yes" instances nor the family of "no" instances is minor-closed when k is fixed. Fortunately, we can take advantage of the immersion order.

Lemma. For all fixed k and for all fixed H, the family of graphs for which the minimum load factor relative to H is less than or equal to k is immersion-closed.

Proof. Let an embedding of G in H with load factor no more than k be given. If G' is a subgraph of G, then it is clear that the restricted embedding has load factor no more than the original. If G' is obtained from G by lifting two edges uv and vw incident at a vertex $v \varepsilon V(G)$, then an embedding for G' may be described by assigning to the resulting edge uw the composition of the paths from u to v and v to w in H. This does not increase the load factor. $\square$

Theorem 8. For any fixed k, the problems MIN CUT LINEAR AR-RANGEMENT, TWO-DIMENSIONAL GRID LOAD FACTOR and BINARY TREE LOAD FACTOR are decidable in polynomial time.

Theorem 9. For each of the problems just mentioned and any fixed k, an embedding with load factor no more than k can be constructed, if any exist, in polynomial time.

It is possible to say even more about MIN CUT LINEAR AR-RANGEMENT since it has been shown in [MoS2] that any instance can be reduced in quadratic time to an equivalent instance of maximum degree 3 on $O(n^2)$ vertices. The family of "yes" instances for graphs of maximum degree 3 *is* minor-closed [FL2].

Theorem 10. For any fixed k, MIN CUT LINEAR ARRANGEMENT can be decided in $O(n^4)$ time.

Theorem 11. For any fixed k, a layout with cutwidth less than or equal to k can be found, if any exist, in $O(n^6)$ time.

5. MODIFIED MIN CUT LINEAR ARRANGEMENT

A *layout* L of a graph G is a one-to-one function $\ell\colon V(G) \to \{1,\ldots,|V(G)|\}$. The *modified cutwidth* of G is the minimum, over all layouts ℓ of G, of the maximum over $v \varepsilon V(G)$ of the number of edges uw of G with $\ell(u) < \ell(v) < \ell(w)$. Modified cutwidth is one of several closely-related parameters [MaS, MiS] that measure resource requirements such as channels or chip area [Le] needed for a linear arrangement of macro- or micro-circuitry.

The MODIFIED MIN CUT LINEAR ARRANGEMENT (MM-CLA) problem is to determine, for a graph G and integer k, whether the modified cutwidth is no more than k. It has been shown to be NP-complete, even for planar graphs of maximum degree 3 [MoS2].

Unfortunately, MMCLA is neither minor-closed nor immersion-closed when k is fixed (this is true for both "yes" and "no" families). However, MMCLA can be reduced to GML so that Theorem 5 is applicable. The combinatorial reduction we employ is equivalent to mapping an input graph to a modification of its *line graph* so that the family of images of "yes" instances is minor-closed.

Theorem 12. For any fixed k, MMCLA can be decided in $O(n^2)$ time.

Theorem 13. For any fixed k, a layout with modified cutwidth less than or equal to k can be found, if any exist, in $O(n^4)$ time.

6. Other Results

The range of combinatorial VLSI problems to which an approach based on well-partially-ordered sets applies is remarkable indeed! Self-reducibility has thus far also been possible with problem-specific methods (a caution: there are problems from different domains that appear to thwart all attempts at self-reducibility [FL2]).

HYPERGRAPH VERTEX SEPARATION NUMBER reduces to
VERTEX SEPARATION NUMBER for graphs, which is closed under the minor order, as are SEARCH NUMBER and NODE SEARCH
NUMBER. This last fact has been independently noted by Papadimitriou [Pa].

We observe that bandwidth problems seem to be more difficult.
The *edge bandwidth* of a graph G is the least k such that there is a
one-to-one map $\ell\colon E(G) \to \{1,\ldots,|E(G)|\}$ so that for any two edges
e, f with a common endpoint, $|\ell(e) - \ell(f)| \le k$. Both of the decision
problems EDGE BANDWIDTH and ordinary BANDWIDTH have
so far resisted this general line of attack. TOPOLOGICAL BAND-
WIDTH restricted to graphs of maximum degree 3, however, yields
to closure under the minor order [FL2].

Acknowledgements

We wish to thank those who have encouraged us in this study,
including James Abello, Manuel Blum, Donna Brown, Jeff Lagarias,
Steve Mahaney and, of course, Neil Robertson and Paul Seymour.
This research has been supported in part by the Washington State
Technology Center, by the Sandia University Research Program, and
by the National Science Foundation under grants ECS–8403859 and
MIP–8603879.

References

[BFKL]   R. L. Bryant, M. R. Fellows, N. G. Kinnersley and M. A.
Langston, "On Finding Obstruction Sets and Polynomial-
Time Algorithms for Gate Matrix Layout," *Proc. 25th
Allerton Conf. on Communication, Control, and Com-
puting*, 1987.

[BFL1]   D. J. Brown, M. R. Fellows and M. A. Langston, "Nonconstructive Polynomial-Time Decidability and Self-Reducibility," *Princeton Forum on Algorithms and Complexity*, 1987.

[BFL2]   ———, "Polynomial-Time Self-Reducibility: Theoretical Motivations and Practical Results," Computer Sci-

324

ence Technical Report CS–87–171, Washington State University, 1987.

[BL]   K. S. Booth and G. S. Lueker, "Testing for the Consecutive Ones Property, Interval Graphs, and Graph Planarity Using PQ-Tree Algorithms," *J. of Computer and System Sciences* 13 (1976), 335–379.

[Ch]   F. R. K. Chung, "On the Cutwidth and Topological Bandwidth of a Tree," *SIAM J. on Algebraic and Discrete Methods* 6 (1985), 418–444.

[CMST]   M-J Chung, F. Makedon, I. H. Sudborough and J. Turner, "Polynomial Time Algorithms for the Min Cut Problem on Degree Restricted Trees," *SIAM J. on Computing* 14 (1985), 158–177.

[DKL]   N. Deo, M. S. Krishnamoorthy and M. A. Langston, "Exact and Approximate Solutions for the Gate Matrix Layout Problem," *IEEE Trans. on Computer-Aided Design* 6 (1987), 79–84.

[EST]   J. Ellis, I. Sudborough and J. Turner, "Graph Separation and Search Number," to appear.

[FKL]   M. R. Fellows, N. G. Kinnersley and M. A. Langston, paper in preparation.

[FL1]   M. R. Fellows and M. A. Langston, "Nonconstructive Advances in Polynomial-Time Complexity," *Info. Proc. Letters* 26 (1987), 157–162.

[FL2]   _____, "Nonconstructive Tools for Proving Polynomial-Time Decidability," *J. of the ACM*, to appear.

[FL3]   _____, "Fast Self-Reduction Algorithms for Combinatorial Problems of VLSI Design," to appear.

[FRS]   H. Friedman, N. Robertson and P. D. Seymour, "The Metamathematics of the Graph Minor Theorem," in <u>Applications of Logic to Combinatorics</u>, American Math. Soc., Providence, RI, to appear.

[GJ]   M. R. Garey and D. S. Johnson, <u>Computers and Intractability: A Guide to the Theory of NP-Completeness</u>, Freeman, San Francisco, CA, 1979.

[Go]   M. C. Golumbic, <u>Algorithmic Graph Theory and Perfect Graphs</u>, Academic Press, New York, NY, 1980.

[GS]   E. M. Gurari and I. H. Sudborough, "Improved Dynamic Programming Algorithms for Bandwidth Minimization and the Min Cut Linear Arrangement Problem," *J. of Algorithms* 5 (1984), 531–546.

[HS]   A. Hashimoto and J. Stevens, "Wire Routing by Optimizing Channel Assignment with Large Apertures," *Proc. 8th Design Automation Workshop* (1971), 155–169.

[Jo]   D. S. Johnson, "The Many Faces of Polynomial Time," in The NP-Completeness Column: An Ongoing Guide, *J. Algorithms* 8 (1987), 285–303.

[KF]   T. Kashiwabara and T. Fujisawa, "NP-completeness of the Problem of Finding a Minimum-Clique-Number Interval Graph Containing a Given Graph as a Subgraph," *Proc. IEEE Symp. on Circuits and Systems* (1979), 657–660.

[KP]   M. Kirousis and C. H. Papadimitriou, "Searching and Pebbling," Technical Report, National Technical University, Athens, Greece, 1983.

[KUW]   R. M. Karp, E. Upfal and A. Wigderson, "Are Search and Decision Problems Computationally Equivalent," *Proc. 17th ACM Symp. on Theory of Computing* (1985), 464–475.

[Le]   T. Lengauer, "Black-White Pebbles and Graph Separation," *Acta Inf.* 16 (1981), 465–475.

[LL]   A. D. Lopez and H-F. S. Law, "A Dense Gate Matrix Layout Method for MOS VLSI," *IEEE Trans. Elec. Devices* 27 (1980), 1671-1675.   ·

[Ma]   W. Mader, "A Reduction Method for Edge-Connectivity in Graphs," *Annals of Disc. Math* 3 (1978), 145–164.

[MaS]   F. S. Makedon and I. H. Sudborough, "On Minimizing Width in Linear Layouts," to appear.

[MHGJP]   N. Megiddo, S. L. Hakimi, M. R. Garey, D. S. Johnson and C. H. Papadimitriou, "On the Complexity of Searching a Graph," IBM Research Report RJ 4987, 1986.

[MiS]   Z. Miller and I. H. Sudborough, "Polynomial Algorithms for Recognizing Small Cutwidth in Hypergraphs," to appear.

[MoS1]   B. Monien and I. H. Sudborough, "Bandwidth Constrained NP-Complete Problems," *Theoretical Computer Science* 41 (1985), 141–167.

[MoS2]   ———, "Min Cut is NP-Complete for Edge Weighted Trees," to appear.

[MP]   A. Meyer and M. Paterson, "With What Frequency Are Apparently Intractable Problems Difficult," Technical Report, MIT, 1979.

[MPS]   F. S. Makedon, C. H. Papadimitriou and I. H. Sudborough, "Topological Bandwidth," *SIAM J. Alg. Disc. Meth* 6 (1985), 418–444.

[Na]   C. Nash-Williams, "On Well-Quasi-Ordering Infinite Trees," *Proc. Cambridge Phil. Soc.* 61 (1965), 697–720.

[OMKKF]   T. Ohtsuki, H. Mori, E. S. Kuh, T. Kashiwabara and T. Fujisawa, "One Dimensional Logic Gate Assignment and Interval Graphs," *IEEE Trans. on Circuits and Systems* 26 (1979), 675–684.

[Pa]   C. H. Papadimitriou, private communication.

[RL]   S. S. Ravi and E. L. Lloyd, "On Approximation Algorithms for PLA Folding," Computer Science Technical Report, Univ. of Pittsburgh, 1984.

[Ro]   A. Rosenberg, "Issues in the Study of Graph Embed-
            dings," *Lecture Notes in Computer Science* 100 (1981),
            150–176.

[RS1]   N. Robertson and P. D. Seymour, "Disjoint Paths—a
            Survey," *SIAM J. Alg. Disc. Meth.* 6 (1985), 300–305.

[RS2]   _____, "Graph Minors—a Survey," in Surveys in Com-
            binatorics (I. Anderson, ed.), Cambridge Univ. Press,
            1985.

[RS3]   _____, "Graph Minors IV. Tree-Width and Well-Quasi-
            Ordering," to appear.

[RS4]   _____, "Graph Minors XIII. The Disjoint Paths Prob-
            lem," to appear.

[RS5]   _____, "Graph Minors XVI. Wagner's Conjecture," to
            appear.

[Sc]   C. P. Schnorr, "Optimal Algorithms for Self-Reducible
            Problems," *Proc. ICALP* (1976), 322–337.

[Su]   I. H. Sudborough, private communication.

[Wa]   K. Wagner, "Uber Einer Eingeshaft der Ebener Com-
            plexe," Math. Ann. 14 (1937), 570–590.

Invited Talk

Large Scale Integrated Circuits for Electronic Imaging

Timothy J. Tredwell

Eastman Kodak Company
Rochester, N. Y.

Over the past 100 years, imaging has been virtually synonymous with silver halide photography. In the past decade, however, electronics has impacted imaging in a variety of applications. This impact occurred first in government markets, where the ability to transmit and digitally process images was more important than cost and image quality. This impact is now being felt in the commercial and consumer arena, with the advent of electronic news gathering, electronic scanners and output printers for document copying, and video cameras and camcorders. Most recently, the first electronic still photography systems have become available. In this talk we will review both present and future systems for electronic still photography and the impact of VLSI on these systems. The requirements for still photographic systems will be discussed, the components of present and future systems will be reviewed, and the image quality of these systems will be assessed.

Traditional silver halide film presents a very difficult challenge to electronic imaging in terms of image quality. Using the 50% point of the MTF curve as a criteria for resolution, the equivalent pixel size and number of pixels in various photographic films can be estimated. The equivalent pixel size for Kodak VR-200 film is 10 microns, and the resolution is over 20 million pixels. In small-format photographic systems the number of elements is lower. Kodak Disc negative film has an equivalent resolution of 2.0 million elements. The threshold for photographic-quality systems is thus in the 2 - 6 megapixel range, although there are still many commercial applications for lower-resolution systems where image quality is not critical.

Still electronic imaging systems consist of a variety of components. These include the image sensor, image processing and compression circuits, image storage, image transmission and hard and soft-copy display. There are significant impacts of VLSI on each. Present still electronic systems are based on components developed for video applications. As an example of such a system, the VLSI components of the Kodak still video system will be discussed. The image sensor is a television resolution sensor with 300,000 pixels. Image processing is performed in the analog domain using analog amplifiers and delay lines. Image storage is achieved through analog magnetic recording onto a 2-inch magnetic disc. Image display is on television and hard copy is achieved using 500-line color thermal printers.

Future electronic still systems will utilize image sensors with more than 1-million pixels, digital image processing and image compression, digital image storage utilizing magnetic, magneto-optical or pure optical recording, and display using high definition monitors and color thermal, electro-photographic or laser-printed hard copy. Although these systems do not exist as complete systems, many of the individual pieces of technology are now being demonstrated. Examples of these components developed at Kodak and elsewhere will be described.

Present electronic photography systems utilize image sensors with resolution designed for television systems. A 300,000 pixel color image sensor developed for use in the Kodak still video system will be described. Each pixel consists of a photodiode for light sensing and a charge-coupled device (CCD) readout stage. Because of the very low capacitance of the sense-node in CCDs (less than 10 fF in a CCD as compared to 1 pF in dynamic memories), noise levels of less than 20 electrons are obtained at video rates. Color sensitivity is achieved by fabricating organic color filter arrays on the wafer in a process similar to photolithography. This process produces a red, blue or green filter above each photodiode. One of the key challenges from a microelectronics point of view is reducing the dark current in the sensor to very low levels in order to obtain low-light imaging capability and zero density of defective cells.

The next generation of electronic photography systems will require image sensors with one to four million elements. A 1.4 megapixel sensor developed at Kodak will be discussed as an example. Utilizing three-dimensional device modeling, it has been possible to scale the CCD process to achieve a 6.8 micron pixel size and to achieve noise levels below 8 rms electrons at video rates. The signal, noise and image quality of these sensors will be described.

VLSI is extensively used in the image processing for still electronic photography. The present generation of still cameras utilizes analog processing for such functions as sampling, construction of luma and chroma signals, defect correction and filtering. The image processing for future generations will utilize real-time digital processing for such functions as dark reference, conversion between linear and logarithmic space, two-dimensional interpolation for luminance and color-difference, defective pixel estimation, color correction, noise-removal and edge enhancement. Custom digital processing circuit blocks addressing these functions will be discussed.

Digital processing is also extensively used for image compression. There is significant activity worldwide in the development of chips for video image compression for motion. Still images, however, have somewhat different requirements. Simulations of various degrees of compression on images will be shown and VLSI implementation discussed. Finally, for digital recording of images it is necessary to provide codes for error correction. This is particularly true for compressed images, in which a single bit error can migrate to a significant portion of the image. Examples of VLSI for digital error correction will be discussed.

There will be significant demand for hard copy of electronic images in commercial and consumer applications of electronic still photography. The most widely used hard copy process for electronic output is thermal printing, in which a linear array of resistors is precisely heated to cause dye transfer to a receiving layer in thermal print paper. Novel LSI chips have been developed for thermal printing both to address the array and to calculate the precise heating required based on the past thermal history of each element.

In summary, the present generation of electronic still photographic systems is limited to a quality which falls far short of traditional photographic products owing to the low resolution of the sensors and the use of analog processing and recording. Future systems utilizing megapixel sensors in combination with digital processing and recording, will produce images which will approach small-format photographic systems in quality. Although the cost of these systems will be beyond the consumer product range for many years to come, such systems will find a number of important applications in commercial and industrial environments.

Verifying a Static RAM Design by Logic Simulation

Randal E. Bryant

Computer Science Department
Carnegie Mellon University
Pittsburgh, Pennsylvania 15213

A logic simulator can prove the correctness of a digital circuit if it can be shown that only circuits implementing the system specification will produce a particular response to a sequence of simulation commands. Three-valued modeling, where the third state X indicates a signal with unknown digital value, can greatly reduce the number of patterns that need to be simulated for complete verification.

As an extreme case, an N-bit random-access memory can be verified by simulating just $O(N \log N)$ patterns. The technique has been applied to a CMOS static RAM design using the COSMOS switch-level simulator. This approach to verification is fast, requires minimal attention on the part of the user to the circuit details, and can utilize more sophisticated circuit models than other approaches to formal verification.

1. Introduction

Although logic simulators are widely used to test circuit designs informally, they have not been recognized as tools for formally proving the correctness of circuits. Conventional wisdom holds that verifying a circuit by simulation is at best impractical and at worst impossible. The large number of possible input and initial state combinations would seem to require an overwhelming amount of simulation to test exhaustively. Furthermore, as Moore has shown [9], a sequential system cannot be fully characterized by observing its response to a sequence of stimuli. This would seem to indicate that, unless supplemented by detailed knowledge of the circuit structure, no amount of simulation can prove the correctness of a sequential system.

Most researchers have turned to automated theorem provers [1,6] to demonstrate that a circuit meets the specification of its desired behavior. With the current state of the art, this process is only partially automated. The user must provide complete specifications of every component of the circuit and guide the program on proof strategies. Furthermore, these programs cannot operate with the detailed, transistor-level models required to verify complex MOS circuits. As an exception to this generalization, Weise [10] has developed a verifier that utilizes a very detailed electrical model. When composing circuits hierarchically, however, his program will at times resort to an exhaustive case analysis. This yields unsatisfactory performance for certain classes of circuits.

Other researchers have applied model checking programs [2] to construct a data structure representing the finite state behavior of the circuit, allowing the user to then prove assertions about the circuit behavior. This approach works especially well for small, asynchronous controller circuits but is impractical for circuits, such as memories, having large numbers of possible states.

The conventional wisdom about logic simulation overlooks the capabilities provided by three-valued logic modeling, in which the state set $\{0, 1\}$ is augmented by a third value X indicating an unknown digital value. Most modern logic simulators provide this form of modeling, if for nothing more than to provide an initial value for the state variables at the start of simulation. Assuming the simulator obeys a relatively mild monotonicity property, a three-valued simulator can verify the circuit behavior for many possible input and initial state combinations simultaneously. That is, if the simulation of a pattern containing X's yields 0 or 1 on some node, the same result would occur if these X's were replaced by any combination of 0's and 1's. This technique is effective for cases where the behavior of the circuit for some operation is not supposed to depend on the values of some of the inputs or state variables. Three-valued modeling can also overcome the machine identification problem of Moore, assuming the user can command the simulator to set all state variables to X.

For performance reasons, most simulators err on the side of pessimism in modeling the effects of X values. That is, they will produce an X at some point even though it can be shown that the circuit would produce a 0 or 1 in all cases covered by the X values on the input and initial state variables. This can cause a verifier based on three-valued simulation to give a *false negative* response, labeling a correct design as defective. Fortunately, these erroneous responses always have the form of producing an X at some point where a 0 or 1 was expected.

On the other hand, a verifier based on three-valued simulation can never produce a *false positive* response, labeling a defective design as correct.

That is, if a circuit passes our verification tests, then no other simulation sequence will uncover additional errors. Although this claim is only demonstrated informally in this paper, it is backed up by a more formal theory and proof [5]. Of course, this style of verification proves the correctness of the actual circuit only if the simulator faithfully models the circuit behavior. Any approach to formal verification must assume that its abstract model of circuit behavior is valid.

Random access memories are particularly amenable to verification by logic simulation. Although an N-bit memory has 2^N possible states, an operation on one memory location should not affect or be affected by the value at any other memory location. Thus many aspects of circuit operation can be verified by simulating the circuit with all, or all but one, bits set to X, covering a large number of circuit conditions with a single simulation operation.

This paper develops these ideas in more detail, using as a case study the verification of a CMOS static RAM circuit design by the switch-level simulator COSMOS [4]. The circuit design was constructed solely as a benchmark for verification. However, it contains the same circuit structures found in actual CMOS static RAM's [7].

This circuit provides a convincing demonstration of the advantages of verification by simulation. No other automatic verifiers are currently capable of verifying this design for nontrivial memory sizes. Most verifiers based on theorem provers do not provide a sufficiently detailed model of transistor operation to capture the behavior of the circuit. Weise's verifier would attempt an exhaustive case analysis of the circuitry forming the entire memory array due to the connections formed by the pass transistors in the column selector. Verifiers based on model checking would attempt to construct a finite automaton containing all 2^N possible memory states.

A high-level specification of the desired circuit behavior can be expressed quite easily. A straightforward translation of the specification into a set of simulation patterns, however, yields false negative responses. Overcoming these problems requires taking into account the details of the row and column addressing structure. Although this places additional burden on the user, it does not compromise the rigor of the verification in any way, and the amount of detail is reasonably small.

The resulting verification requires simulating only $O(N \log N)$ patterns. Even a minimal test of a memory design requires simulating $\Omega(N)$ patterns to make sure that each location can be written and read properly. The added $\log N$ factor seems a modest price to pay for a rigorous verification. Furthermore, we have been able to tune the performance of our simulator to match the characteristics of the simulation patterns arising from formal verification. This tuning makes formal verification require no more simulation time than a minimal design test.

2. Verification Methodology

Specifications are expressed in a notation similar to Floyd-Hoare assertions [8]. Each assertion is an equation of the form

$$Initial \left\{ Action \right\} Result$$

where *Initial* specifies a precondition on the initial circuit state, *Action* specifies a circuit operation, and *Result* specifies a postcondition on the circuit output and state. All conditions are expressed as propositional formulas of the form $L_1 \wedge L_2 \wedge \cdots \wedge L_k$, where each L_i is a *literal* of the form $var = 1$ or $var = 0$, for some circuit input, output, or state variable *var*. For this paper, an *Action* will specify a condition on the inputs for a single cycle of circuit operation. An assertion states that for any initial circuit state satisfying *Initial*, and any circuit operation satisfying *Action*, the resulting circuit state and output should satisfy *Result*.

With a three-valued simulator, a circuit can be shown to satisfy an assertion by simulating a single pattern. Starting with every input and state variable set to X, the input and state variables appearing in the formulas *Initial* and *Action* are set to their specified values. Then the circuit is simulated for one cycle, and the values on the output and state variables appearing in *Result* are compared to their specified values. The monotonicity requirement imposed on the simulator guarantees that a circuit satisfies the assertion if it passes this test.

3. System Specification

The circuit to be verified is an $N \times 1$ bit static RAM. Memories with larger word sizes can be verified similarly, by verifying each bit of each word individually while setting all other bits to X. Figure 1 illustrates the general plan of the circuit. Assuming $N = 2^n$, the circuit has address inputs $A_{n-1}, \ldots, A_0$, a data input *Din*, and a control input *write* that is set to 1 for a write and to 0 for a read operation. The circuit has a single output *Dout*. Each memory cell i contains a feedback path with a pair of inverters

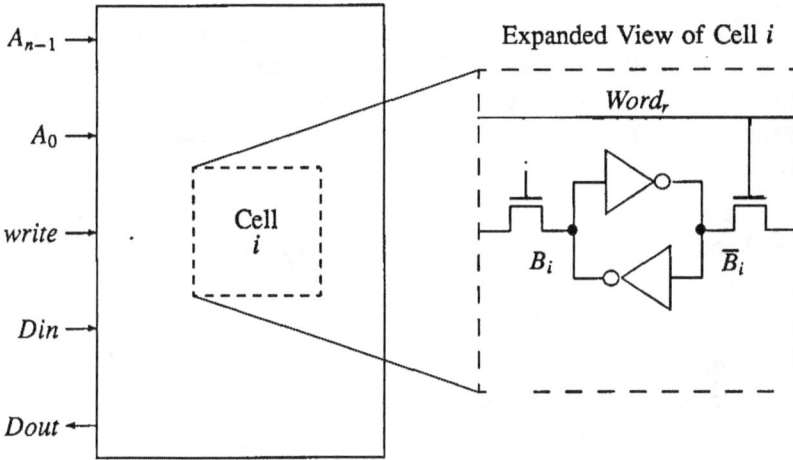

Figure 1: *Static RAM Circuit*

connecting nodes B_i and $\overline{B}_i$, along with a pair of access transistors [7]. As a shorthand, the formula $Store(i, v)$ expresses the fact that value $v \in \{0, 1\}$ is stored in memory cell i:

$$Store(i, v) \quad \equiv \quad B_i = v \wedge \overline{B}_i = \neg v.$$

Unlike many other sequential systems, the desired behavior of a memory circuit can be specified quite easily. First, a write operation should cause the addressed memory cell to be updated. For all $v \in \{0, 1\}$ and all $0 \leq i < N$:

$$True \left\{ Din = v \wedge A = i \wedge write = 1 \right\} Store(i, v), \tag{1}$$

where the notation $A = i$ is a shorthand indicating that for $0 \leq k < n$, each input line A_k equals i_k, the corresponding bit in the binary representation of i. These assertions can be verified by simulating $2N$ patterns, two for each memory location. Starting with all state variables set to X, each test writes a value to a location, and then checks that the value has been stored correctly. These patterns are called the "write" tests.

Second, a read operation should cause contents of the addressed memory bit to appear on *Dout* without altering the cell. For all $v \in \{0, 1\}$ and all $0 \leq i < N$:

$$Store(i, v) \left\{ A = i \wedge write = 0 \right\} Dout = v \wedge Store(i, v). \tag{2}$$

These assertions can be verified by simulating a total of $2N$ patterns, two for each memory location. Each test involves initializing one memory cell to a value, all other locations to X, and then reading from the cell's address. The test passes if the stored bit appears on *Dout*, and the cell contents remain unchanged. These simulation patterns are called the "read" tests.

Finally, any memory operation on one cell should not affect the value stored in any other memory cell. For all $v \in \{0, 1\}$, and all $0 \le i, j < N$, such that $i \ne j$:

$$Store(i, v) \left\{ A = j \right\} Store(i, v). \tag{3}$$

This set of assertions represents $2N^2$ combinations of address and data values. However, we can obtain the same effect with just $2N \log N$ combinations. For an address i with bit representation $\langle i_{n-1}, \ldots, i_0 \rangle$, all addresses j such that $j \ne i$ are covered by the n patterns of the form $\langle X, \ldots, X, \neg i_k, X, \ldots, X \rangle$ for $0 \le k < n$. Thus, the assertions can be replaced by the following assertions for $v \in \{0, 1\}$, $0 \le i < N$, and $0 \le k < n$:

$$Store(i, v) \left\{ A_k = \neg i_k \right\} Store(i, v). \tag{4}$$

These assertions can be verified by patterns in which a memory cell is initialized to some value, one of the address inputs is set to the complement of the corresponding bit in the cell's address, and all other input and state variables are set to X. Following the simulation of one cycle, the cell value is compared to its original value. These simulation patterns are termed the "address" tests.

4. Circuit Dependent Refinements

Equations 1, 2, and 4 translate directly into a total of $4N + 2N \log N$ simulation patterns. However, on our example circuit, the simulator gives false negative responses for all of the read and address tests. By adding one new assertion and refining the existing ones, we can devise an equally rigorous test that the circuit passes.

Refining the specification into a set of simulation patterns requires a more detailed consideration of the control sequencing and of the row and column addressing structure. Even with these details, we can ignore many aspects of the design, letting the simulator capture their behavior by its simulation model. Assume that the circuit is organized as a $\sqrt{N} \times \sqrt{N}$ array of

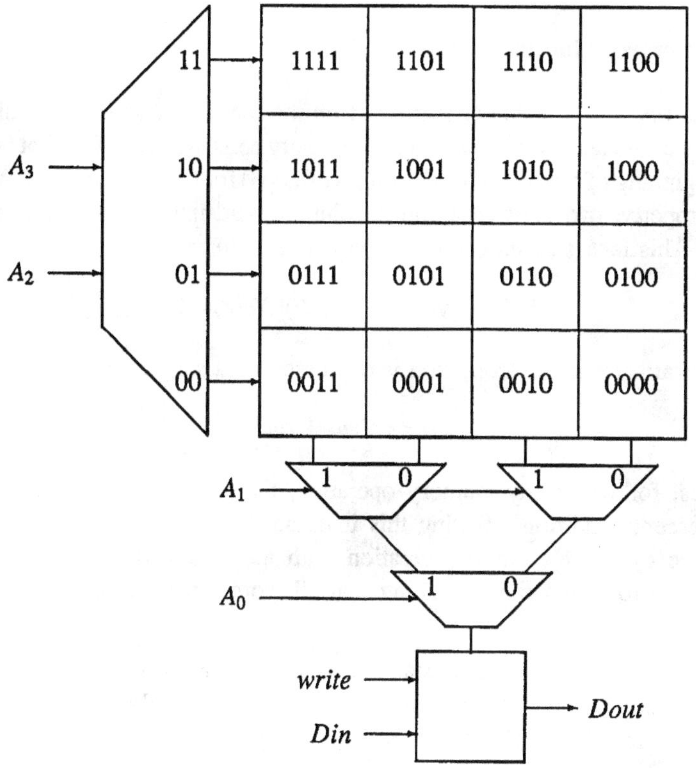

Figure 2: *Detailed Addressing Structure of a 16-Bit RAM.* Each cell is labeled with the binary representation of its address.

memory cells, where address bits $Arow = A_{n-1}, \ldots, A_{n/2}$ select the row, and address bits $Acol = A_{n/2-1}, \ldots, A_0$ select the column.

As an example, Figure 2 shows the addressing structure for a 16-bit RAM. Address inputs A_3 and A_2 are decoded to generate the signals on the 4 word lines. Address inputs A_1 and A_0 control a tree of bidirectional multiplexors to create a path between the selected column and the data input or output.

4.1. Control Line Initialization

Correct operation of this circuit relies on the fact that when the circuit is quiescent, the access transistors to all memory cells are shut off. That is, at the beginning of every memory cycle, $Word_r = 0$ for $0 \leq r < \sqrt{N}$. Without this property, two cells in a single column could interact in undesirable ways. This fact is formulated as a *system invariant*

$$ Inv \equiv \forall(0 \leq r < \sqrt{N})[Word_r = 0]. $$

The invariance of this condition is expressed by a single assertion:

$$ True \left\{ \ True \ \right\} Inv \tag{5} $$

That is, following any memory operation, the word lines will return to a quiescent condition. Testing this invariant involves simply simulating a single cycle of memory operation with all state and input variables initialized to X and then checking that all word lines are set to 0 at the end.

Once the assertion has been established, the invariant Inv can be assumed as a precondition in all other assertions, giving a revised assertion for the read tests for all $v \in \{0, 1\}$, and all $0 \leq i < N$:

$$ Inv \land Store(i, v) \left\{ A = i \land write = 0 \right\} Dout = v \land Store(i, v). \tag{6} $$

That is, we can begin all simulation read cycles with the word lines initialized to 0. With this refinement, the circuit passes the read tests.

Most circuits require some form of system invariant expressing conditions about the control logic that can be assumed true at the beginning of every input cycle. Devising the invariant requires a combination of analysis and experimentation. An insufficient system invariant will become immediately apparent during subsequent simulations, because output or state variables that should have Boolean values will equal X.

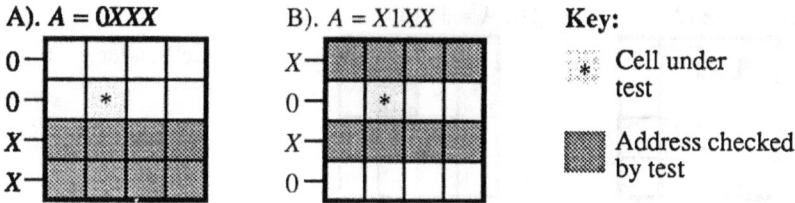

A). A = 0XXX   **B). A = X1XX**   **Key:**

Figure 3: *Row Address Tests for Memory Location 9.* Signals on the left indicate the values on the word lines. The word line controlling cell 9 remains at 0.

4.2. Row and Column Decoding

Even with the invariant our circuit still passes only half of the the address tests, namely those corresponding to the following equations for $v \in \{0, 1\}$, $0 \le i < N$, and $n/2 \le k < n$:

$$Inv \wedge Store(i, v) \left\{ A_k = \neg i_k \right\} Store(i, v). \tag{7}$$

For these tests, some bit k of the row address is set to $\neg i_k$, a controlling value for the NOR gate of word line decoder for memory cell i. The word line stays at 0 and the bit stored in cell i remains unchanged. These tests are called the "row address" tests. They prove that no memory cell is affected by an operation on a cell in a different row.

As an example, Figure 3 illustrates the addressing patterns for the row address tests for memory location 9 (1001 binary) in the 16-bit RAM of Figure 2. The two address settings: 0XXX and X1XX cause the word lines to have the values shown on the left. In both cases, cell 9 remains isolated from all others. The dark shaded areas indicate the cell addresses covered by these two tests. The union of these areas includes all addresses in other rows of the memory.

For the cases that fail, the NOR gates of the word line decoders have all X's on their inputs, causing sneak paths to form between the cell under test and other cells in the column. Figure 4A shows an example of such a pattern for memory location 9 in the 16-bit RAM of Figure 2. Although no connection is formed between this cell and the data input or output, (indicated by the tree structure at the bottom), the stored bit is corrupted by the other cells in the column (indicated by the lightly shaded area.)

Fortunately, we can overcome this problem by removing some of the

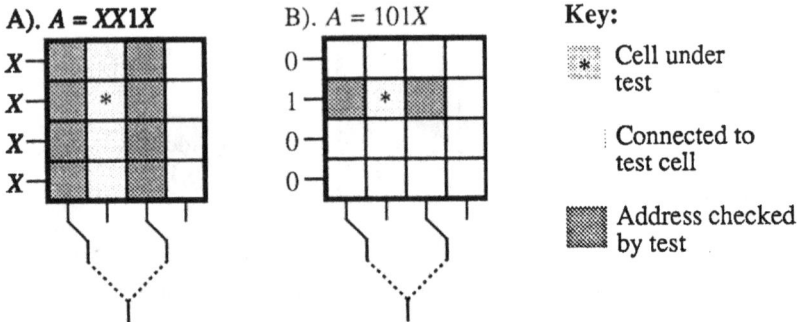

Figure 4: *Example of Initial (A) and Refined (B) Column Address Test for Memory Location 9.* The tree on the bottom indicates the connections formed by the column multiplexors, with dotted lines representing pass transistors with gate value X. In A), the word line value of X causes cell 9 to be corrupted. This is avoided in B).

redundancy from the tests. Once a circuit passes the row address tests, we need only show that no memory cell is affected by an operation on a cell in a different column of the same row. This can be expressed by the following equations for $v \in \{0, 1\}$, $0 \leq i, j < \sqrt{N}$, and $0 \leq k < n/2$:

$$Inv \wedge Store(j + i\sqrt{N}, v) \left\{ Arow = i \wedge A_k = \neg j_k \right\} Store(j + i\sqrt{N}, v).$$

These assertions define a series of tests in which the memory cell at row i, column j is initialized to a value v, the row address is set to i, and some bit of the column address is set to the complement of the corresponding bit in j. Figure 4B shows an example of such a pattern for memory location 9. The word lines are set so that only cells in a single row are accessed. Furthermore, the column addresses are set so that the column containing cell 9 remains isolated. This test covers the two cell addresses indicated by the darkly shaded area.

Even with this refinement, our circuit encounters a new problem due to the tree structure of the column selector. Under normal operation of the circuit, all cells in the selected row are read, and the pass transistors of the column multiplexors form a path between the selected column and the data input and output. When some of the column address lines equal X, however, the simulator finds false sneak paths throught the column multiplexor, causing a connection between the cell under test and one in another column. An example of this problem is shown in Figure 5A. Even

A). $A = 10X0$ B). $A = 1000$ Key:

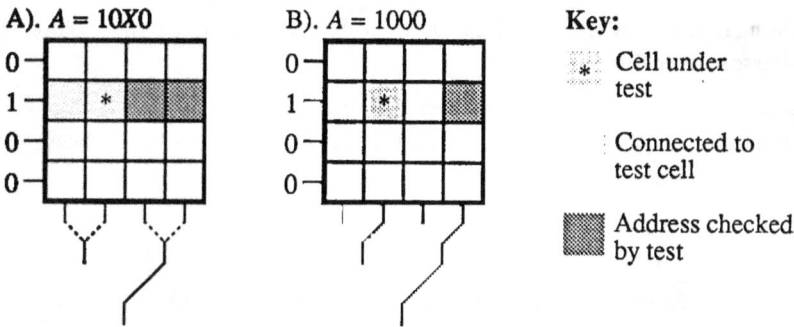

Figure 5: *Example of Refined (A) and More Refined (B) Column Address Test for Memory Location 9.* In A), a sneak path forms through the column multiplexor between cell 9 and an adjacent cell. This is avoided in B).

though only a single row of cells is accessed, a sneak path forms between the column containing cell 9 and an adjacent column (indicated by the lightly shaded area).

Again, this problem can be overcome by removing some of the redundancy from the tests. For a column address j having bit representation $\langle j_{n/2-1}, \ldots, j_0 \rangle$, all column addresses not equal to j are covered by patterns of the form $\langle \neg j_{n/2-1}, X, \ldots, X \rangle$, $\langle j_{n/2-1}, \neg j_{n/2-2}, X, \ldots, X \rangle$, and so on up to $\langle j_{n/2-1}, \ldots, j_1, \neg j_0 \rangle$. Each of these patterns has the property that the simulator will never find a path of potentially conducting transistors (i.e., with gate value 1 or X) between the bit lines of column j, and those of any other column. These tests can be expressed by a revised set of equations for $v \in \{0, 1\}$, $0 \le i, j < \sqrt{N}$, and $0 \le k < n/2$:

$$Inv \wedge Store(j + i\sqrt{N}, v) \tag{8}$$
$$\left\{ Arow = i \wedge A_k = \neg j_k \wedge \forall (k < t < n/2)[A_t = j_t] \right\}$$
$$Store(j + i\sqrt{N}, v).$$

These tests are called the "column address" tests.

The pattern of Figure 4B shows one of the column address tests for location 9 in the 16-bit RAM of Figure 2. The other is shown in Figure 5B. Observe that in both cases, the column containing cell 9 remains isolated, avoiding any corruption of the value stored there. The darkly shaded areas indicate the cell addresses tested by these patterns. The union of these areas includes all other cells in the row containing cell 9. These, combined with the two row address tests of Figure 3 cover all possible addresses other than location 9.

Equations 1, 5, 6, 7, and 8 together define a total of $1 + 4N + 2N \log N$ simulation patterns that our circuit passes and that prove its correctness.

5. Simulator Performance

The simulation operations called for by our memory verification tests differ markedly from those used in more traditional simulation methodologies. Each involves resetting the simulator to a condition where all input and state variables equal X, setting a small number of inputs and state variables to Boolean values, and then simulating a single cycle. In contrast, most simulators are designed to simulate long sequences of Boolean patterns. The differences between these two styles of simulator usage place differing demands on simulator functionality and performance. In developing the switch-level simulator COSMOS, we attempted to satisfy the needs of both forms of simulation.

Most simulators employ very pessimistic or inefficient algorithms for computing the behavior of a circuit in the presence of X's. With conventional usage, there is no need to do better, because most X's are eliminated at the start of simulation and never arise again. For our verification patterns, however, X's are the rule rather than the exception, and hence the algorithms must be as accurate and efficient as possible. The algorithms used by COSMOS satisfy these goals reasonably well, although, as the static RAM example shows, developing a set of verification patterns requires some understanding of both the circuit design and the simulation algorithm.

In the design of the switch-level simulator COSMOS, we were also able to optimize the efficiency when simulating many short sequences. Most of these optimizations involved simply tuning the performance of code that is normally considered non-critical, such as the code to reset all state variables of a circuit to X. More significantly, however, we were able to exploit the bit-level parallelism available with computer logic operations to simulate up to 32 sequences in parallel on a machine with a 32 bit word size. The COSMOS preprocessor transforms a transistor network into a set of evaluation procedures that utilize only memory references and logical operations. Hence, bit-level parallelism adds little extra cost. Experiments indicate that it increases simulation performance by a factor of 10–30. Although this would appear to be an obvious source of speed-up,

N	Transistors	Marching Test	Serial Verification	Parallel Verification
4	113	1.0s.	2.0s.	0.6s.
16	235	8.4s.	22.6s.	2.0s.
64	611	117s.	385s.	19.3s.
256	1931	30.8m.	122m.	4.4m.
1024	6875	10.4h.	47.9h.	1.5h.

Table 1: COSMOS CPU Times on DEC MicroVax-II

most simulators make no use of bit-level parallelism. Many simulation algorithms cannot exploit it. Furthermore, with conventional simulator usage, the simulation patterns are not formulated as a set of independent tests that can be run in parallel.

6. Experimental Results

The verification methodology has been applied to memory sizes ranging from 4 to 1024 bits. The performance of the program is shown in Table 6.. The last 3 columns of this table show simulation CPU times, measured on a Digital Equipment Corporation MicroVax-II. The first of these columns shows the time to simulate a marching test, giving a minimal test that all locations can be written and read, but not proving the circuit's correctness. The second shows the time to simulate the verification patterns without using bit-level parallelism. The final column shows the time to simulate the verification patterns using 32 way bit-level parallelism.

As can be seen, the parallel verification is faster than a simple marching test! Although a marching test requires simulating only $O(N)$ cycles, the extra $\log N$ factor of the verification patterns is more than compensated for by the speed-up provided by bit-level parallelism. Observe, however, that the overall simulation time in all 3 cases grows roughly quadratically with the memory size. As the memory size grows, both the number of patterns and the time to simulate a single pattern grow at least linearly. This complexity becomes noticeable for larger memory sizes. We estimate that the verification of a 4096-bit memory will require between 1 and 2 days of CPU time even in parallel mode. Clearly, this is approaching the limit of practicality.

7. Observations and Conclusions

This paper has shown that a typical CMOS static RAM design can be formally verified easily and efficiently by three-valued, switch-level logic simulation. Although these patterns were developed specifically for this design, similar techniques can develop patterns for almost all RAM designs. Other classes of memory designs can also be verified by simulating a linear, or nearly-linear number of patterns. Included among these are shift registers, FIFO's, and stacks. On the other hand, content-addressable memories do not seem to fit into this class, since it is not as easy to identify where a particular datum will be stored.

Other classes of circuits cannot be verified by simulating a polynomial number of patterns. Many functions computed by logic circuits, such as addition and parity, depend on a large number of input or state variables. For these circuits, we propose *symbolic* simulation [3] as a feasible and straightforward approach to design verification. A symbolic simulator resembles a conventional logic simulator, except that the user may introduce symbolic Boolean variables to represent input and initial state values, and the simulator computes the behavior of the circuit as a function of these Boolean variables. Symbolic simulation can utilize a methodology similar to that shown in this paper, by allowing the formulas in an assertion to be predicates containing universally-quantified Boolean variables. For example, we could view Equations 1, 2, and 3 as each representing a single assertion, and verify the RAM by simulating just 3 symbolic patterns. Although the additional overhead required by symbolic simulation would probably cause it to require longer for the RAM verification, there would be less need for circuit-dependent refinements. In addition, symbolic simulation would yield polynomial-time performance for a much wider range of circuits.

Although we have set a new standard for the size and class of circuit that can be verified formally, it is clear that some other technique is required to verify very large memories. Ideally, a verifier should be able to prove the correctness of an entire family of circuits given a parameterized description of the family [6]. Families of RAM circuits have very concise descriptions and hence seem ideal for this style of verification. However, developing such a verifier that can handle a sufficiently detailed MOS circuit model is no easy task.

Acknowledgements

This research was supported by the Defense Advanced Research Projects Agency, ARPA Order Number 4976.

References

[1] Barrow, H. G. VERIFY: a program for proving correctness of digital hardware designs. *Artificial Intelligence 24* (1984), 437–491.

[2] Browne, M. C., Clarke, E. M., Dill, D. L., and Mishra, B. Automatic verification of sequential circuits using temporal logic. *IEEE Transactions on Computers C-35*, 12 (Dec. 1986), 1035–1044.

[3] Bryant, R. E. Symbolic verification of MOS circuits. *1985 Chapel Hill Conference on VLSI*, Fuchs, H., Ed. Computer Science Press, Rockville, MD, 1985, 419–438.

[4] Bryant, R. E., Beatty, D., Brace, K., Cho, K., and Sheffler, T. COSMOS: a compiled simulator for MOS circuits. *24th Design Automation Conference*, 1987, 9–16.

[5] Bryant, R. E. A methodology for hardware verification based on logic simulation. Technical Report CMU-CS-87-128, Carnegie Mellon University, June, 1987.

[6] German, S. M., and Wang, Y. Formal verification of parameterized hardware designs. *Int. Conf. on Computer Design*, IEEE, 1985, 549–552.

[7] Glasser, L. A., and Dobberpuhl, D. W. *The Design and Analysis of VLSI Circuits*, Addison-Wesley, Reading, MA, 1985.

[8] Hoare, C. A. R. An axiomatic basis for computer programming. *Comm. ACM 12* (1969), 576–580.

[9] Moore, E. F. Gedanken-experiments on sequential machines. *Automata Studies*, Shannon C. E., and McCarthy, J., Eds. Princeton University Press, Princeton, NJ, 1956, 129–153.

[10] Weise, D. Functional verification of MOS circuits, *24th Design Automation Conference*, ACM and IEEE, 1987, 265–270.

Integrating Schematics and Programs for Design Capture

Richard Barth, Bertrand Serlet, and Pradeep Sindhu
Xerox PARC Computer Science Laboratory
3333 Coyote Hill Road
Palo Alto, CA 94304

We present a design capture system that allows parameterized schematics and code to be intermixed freely to produce annotated net lists. Users can easily add new abstractions to the small base set supplied by the system. The system allows convenient graphical specification of layout generators and has been used to produce several large VLSI chips.

1. Introduction

Traditional design capture systems are either based on schematics or texts. Most of the schematic-based systems represent designs as static entities in which decisions such as the width of buses and the size of memories are fixed, thereby limiting the potential expressive power of the graphical description. The text-based systems use a conventional programming language or a specialized hardware description language [1, 4, 8, 9, 10]. These systems offer great flexibility, but a textual description of structure is often harder to understand and manipulate than a graphical one. More recent work overcomes these drawbacks by defining graphical languages that permit flexible specification via schematics [5, 6]. Typically, the graphical language provides iterators and conditionals that are translated into a textual description. While this approach provides parameterized schematics, it fails to achieve a synthesis of graphical and textual descriptions in which either may be used with equal facility.

This paper describes an integrated text and graphics design capture system that has been used to produce several large (>50,000 transistor) VLSI chips. The central thesis of this system is that permitting a designer to freely intermix graphical and textual specifications offers significant advantages. First, the description tends to be more compact and comprehensible because the designer can choose the most appropriate way to express each piece of his design. Consequently, the design is easier to create, modify, and maintain. Second, the intermixing allows schematics to be parameterized, thereby offering all the benefits of better abstraction. Finally, it permits convenient graphical description of layout generators via schematics [3].

The particular way in which we have integrated text and graphics was influenced strongly by our programming environment. Rather than design a new graphical programming language, we chose to provide tight links between graphical objects and programs written in an existing strongly typed, block-structured programming language. Not only has this had the benefit of avoiding a proliferation of concepts, but it has also had two other effects which have been even more important. First, it has resulted in an *extensible* system, one in which there is no distinction between built-in abstractions and user-defined ones and where new abstractions may be added easily. Second, it has made graphical and textual descriptions symmetrical and interchangeable, permitting the designer to freely intermix the two in a given design.

The next section begins with a conceptual model for generating net lists from intermixed specifications. Subsequent sections provide implementation detail, describe our experience with using the system to design a number of large chips, and point out directions for future work.

2. Net List Generation Model

It is convenient to think of an intermixed specification as a *net list generator* that produces an annotated net list when evaluated. A net list generator is like a layout generator, except that it operates at a higher level of abstraction (in fact, the output of a net list generator typically forms the input to a layout generator). We believe that net list generators represent a more appropriate level of abstraction for design systems than layout generators because they simplify description while retaining the tight control provided by layout generators.

2.1. Computational Model

A *net list* is a hierarchy of instantiated *devices* connected by *nets*. Some devices are primitive, while others are compositions eventually built using primitive devices. Net lists may be *annotated* with arbitrary properties such as a name, a transistor size, or a net capacitance.

A net list generator is a function that takes arbitrary parameters and returns a net list. The function is represented concretely either by code that will be executed or a schematic that will be *extracted* (interpreted). When a net list generator is evaluated, it either returns a primitive device or it merges the results of subsidiary net list generators into a net list and returns it. Thus, the evaluation of a single net list generator may entail many levels of schematic extraction and/or code execution, possibly interleaved. Code and schematics are tightly linked, and the linkage works both ways.

2.2. Schematic Extraction

A *schematic* is a hierarchy of instantiated graphical objects composed of rectangles, icons, satellites, and compositions. Each instance of these graphical objects has a procedure, called the *extract method*, that knows how to convert the instance into a net list. It is this extract method that provides the link from geometry to code. The reverse link is provided by allowing user code to call the extractor directly.

In a schematic, *rectangles* usually become nets, with touching rectangles inside a composition forming a single net. An *icon* is a graphical abstraction of a net list. Typically, it has some named rectangles to which connections are made and some arbitrary pictorial geometry that denotes its function. *Satellites* are textual expressions attached to a non-text object or its instance; the entity to which a satellite is attached is called its *master*. They are evaluated during schematic extraction and serve principally to pass parameters to their master's extract method. Figure 1a shows an icon whose net list happens to be represented by a simple schematic. The schematic is a composition that contains transistor icons, an inverter icon, the rectangles that connect them, and satellites that serve to name nets and specify transistor sizes. Figure 1b shows a more complicated example.

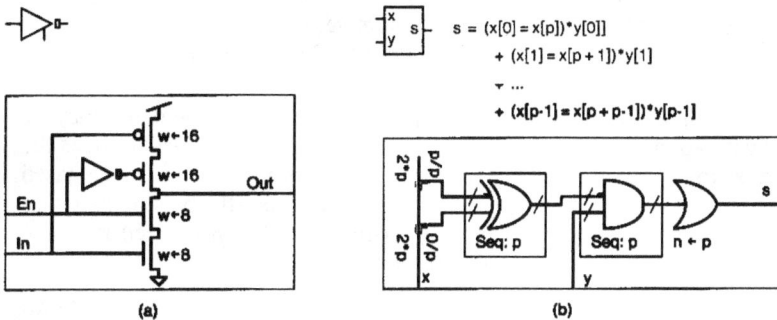

Figure 1. Two Icons and their Schematics

Our model of schematic extraction is best understood by capitalizing on the close analogy between a schematic and a program in a dynamically-scoped, block-structured programming language. Icon instances are analogous to procedure calls, compositions to blocks, and satellites to variable declarations and variable assignments. Variable declarations and assignments in satellites have scopes determined by the geometric and icon hierarchies, analogous to scopes determined by the dynamic nesting of blocks and procedures. Schematic extraction, then, is analogous to interpreting a program in the above language. It follows that extraction is a naturally recursive process in which each step computes the net list for some icon or composition instance.

2.3. Attributes of the Model

The combination of code and schematics embodied in this model can be thought of as a more powerful language for describing designs. This language has three desirable attributes: *extensibility*, *orthogonality*, and *uniformity*.

The language is extensible in the sense that there is a small set of abstractions for design description that can be easily augmented by users for widely different styles of schematics. This extensibility has been facilitated primarily by the fact that we concentrated on providing close links from graphical objects to *code*. Icons, for example, provide a direct graphical link to arbitrarily complex user procedures. Users can exploit this connection to define simple constructors such as arrays of cells and subrange selectors from a bus, or more complicated ones such as data-path and finite-state machine generators.

The language is orthogonal because we avoided duplicating graphically concepts that are better handled by code. We concentrated on the declarative power of schematics by introducing the equivalents of variables, blocks, and functions, but left constructs such as general iterators and conditionals to code. This works well because net list generators that capture structural decompositions are often declarative, so they are better expressed graphically, while generators that reflect algorithms are more easily expressed by code.

The language exhibits uniformity in that code and schematics may be used with equal ease and may be intermixed easily. This attribute frees the designer to choose the best way to express a given piece of his design based on whether it is *inherently* easier to express algorithmically, by schematics, or by an appropriate mixture of the two. Homogeneity has derived from the bidirectional linkage between geometry and code.

3. Implementation

At the time implementation of this system began, our laboratory already had an environment for the design of integrated circuits. Like most of the applications written locally, this environment had been built using the Cedar programming environment [11, 12] running on the Dorado personal computer [7]. A high quality layout editor comprised the major preexisting design tool. Each of these components influenced the implementation of our design capture system in both substance and style.

3.1. Net List Representation

Net lists are represented using four data types: *property, wire, cell type,* and *cell class* [2]. These basic types permit the definition of an extensible set of abstractions that can represent a design at any desired level of detail. This extensibility derives from the ability to define new cell classes without affecting the code for those that already exist.

The *property* data type represents an annotation consisting of a ⟨key, value⟩ pair. A property may be attached to a wire, a cell type, a cell class, or another property.

A *wire* represents a collection of nets. An *atomic wire* represents a single net, while a *structured wire* represents a collection that is organized hierarchically for convenience. A structured wire and its subwires form a directed acyclic graph whose leaves correspond to atomic wires. Only the atomic wires represent actual nets; the others simply provide structure.

A *cell type* represents a device. Cell types consist of an interface and some implementation data. The interface contains properties along with a *public wire* that specifies how an instance of this cell type is to be connected to other cell types. The implementation data contains the cell type's cell class and class-specific data. A *cell instance* represents an instantiation of a cell type within a net list.

A *cell class* groups the implementations of different cell types into sets with common characteristics. Some cell classes are primitive, while others are composite. The *primitive* cell classes are used to encode the structural leaves of a net list; in MOS design, these leaves are transistors. There is also a special primitive cell class, *unspecified*, that allows a cell type to have its interface defined while leaving its implementation unspecified. *Composite* cell classes describe how cell types may be combined to form more complex cell types. One composite cell class is well known. All others must be able to expand into this well known class.

3.2. Creating Net Lists by Program

The system provides a complete set of utilities to help create and manipulate annotated net lists by program. These utilities consist of modules for creating cell types and wires layered on top of more primitive creation functions. Figure 2 illustrates the use of these utilities to construct a decoder with n inputs.

```
Decoder: PROC[n: NAT] RETURNS[cell: CellType]
    size: NAT ← 2**n; array: TileArray ← NEW[TileArrayRec[size]];
    FOR i: NAT IN [0..size) DO {
        array[i] ← NEW[TileRowRec[n]];
        FOR j: NAT IN [0..n) DO {
            array[i][j] ← NEW[TilingClass.TileRec];
            array[i][j].type ← IF XthBitOfN[j, i] THEN One[] ELSE Zero[];
            array[i][j].renaming ← ["In", Index["In", j]], ["nIn", Index["nIn", j]];
            };
        };
    cell ← CreateTiling[Wires[Seq[n, "In"], Seq[n, "nIn"],
        Seq[size, "Out"], "Gnd", "Vdd"], array];
    }
```

Figure 2. Program to Construct a Decoder with n Inputs

Although this approach of providing creation utilities for textual specification is more verbose than a specification based on a hardware description language, it has two major advantages. It avoids a separate language for hardware design, so it is easier to implement. It is also better suited to an evolutionary system development style since utilities may be added and improved piecemeal. This is much harder to do in a language-based system once the language is fixed.

3.3. Schematic Extraction

The schematic extractor converts a combination of pictures and code into a net list. It proceeds top-down in a recursive fashion starting at the instance to be extracted. Each recursion computes the net list for some icon or composition instance in two steps. The first evaluates satellites associated with the instance, while the second invokes the instance's extract method. The parameters passed to the method include the set of variable definitions visible in the current scope, the *context*, in effect immediately after satellite evaluation. To speed up extraction for incremental changes, the extractor caches results on objects. Before extracting an instance, it checks whether the result (if any) cached on that instance's object is still valid. If it is, this result is returned, otherwise the instance is reextracted.

The context is central to the first step. Included in it are variables defined at higher levels in the recursion, as well as certain procedures and global variables accessible in the extractor's run time environment. To facilitate evaluation, the extractor maintains contexts in a stack, one frame per pending recursion level. A new context is created by copying variable definitions from the existing stack top and then evaluating satellites in the new context. These copy semantics ensure that simple use of the context during extraction is side-effect-free.

Satellite evaluation itself proceeds in two phases. Satellites bound to the instance's object are evaluated first, while those bound to the instance are evaluated second. This order allows defaults to be specified by object satellites and overridden by instance satellites. Table 1 provides a summary of satellite usage. Note that net list annotations may be specified directly via satellites.

Syntax	Semantics	Examples
⟨variable⟩ ~ ⟨value⟩	variable definition and initialization	lines ~ 42 lines ← maximumWidth/lineWidth
⟨variable⟩ ← ⟨value⟩	variable assignment	w ← 2 l ← 4 ratio ← IF lines<3 THEN 2.5 ELSE Ratio[Lines]
⟨key⟩: ⟨value⟩	net list annotation	Layout: Abut Simulation:"CombinatorialEval" width: DesignRules.MetalWidth ["CMos"] + 2*lambda
other	names satellite's master	arbitraryName

Table 1. Summary of Satellite Usage

The actual net list computation for an instance is performed by its extract method. For icons, the extract method proceeds differently depending on whether the icon is associated with a schematic or a user-defined procedure. In the former case, the method invokes the extractor on the schematic; in the latter, it simply calls the user-defined procedure. In either case, the method subsequently checks that the interface of the result net list conforms with the interface of the icon. The result of extracting an icon is either a cell type that is instantiated within the parent composition or a wire that is used to make one or more connections.

The extract method for compositions is considerably more complex, because it is here that most of the geometric work of extraction takes place. For each icon or subcomposition it encounters, the method recursively invokes the extractor and then instantiates the result within the cell type it is constructing. For icons, this result is a cell type or a wire, but for compositions it is always a cell type. When instantiating a cell type, the method determines how the cell instance is to be connected to other nets, based on intersecting rectangles that represent connection points on the instance with rectangles in the parent composition.

This brings us to the other major task of the method: rectangle intersection to compute connectivity. When the method encounters a rectangle, it creates a new wire and merges it with an existing wire if rectangle intersection indicates the two wires are connected. The geometrical engine used for this purpose is the same as for layout, since much of the code can be shared. In schematics, it is convenient to specify connectivity based on name equality as well as geometrical intersection, so the method also merges wires with the same name.

3.4. Annotations

In addition to processing annotations placed via satellites, the extractor annotates its result net list with all the graphical objects that were part of the net list's schematic. These graphical annotations establish a link between a wire and its rectangles, and a cell and its corresponding graphical object. Tools that need a wire as input can ask the user to designate a rectangle by pointing to it with the mouse. For example, signal waveforms are displayed this way during simulation. Similarly, tools that specify a wire or a cell as output can highlight the corresponding cells or rectangles. Analogous to symbolic debugging of source code, this approach permits real time interaction and has proven extremely effective in reducing the time traversing the extract-simulate-debug loop.

Some of the net list annotations are interpreted by a silicon assembler that does a walk of the net list hierarchy. At each level, the assembler uses the value of a distinguished annotation to select a layout method. The method recovers its parameters from further annotations, including the positions of graphical objects in the schematics. It then generates layout and adds annotations that link the layout to the original net list. For example, the annotation ⟨Layout:AbutX⟩ on a cell type selects the AbutX method, which computes the layout of the cell type by abutting the layouts of the cell type's instances in the X direction.

4. Results

Since its initial implementation in early 1986, the system has been used to describe approximately a dozen VLSI chips of varying sizes and complexities. Some of these chips have been fabricated while others are in the final stages of design.

This section provides data that characterize our experience with the system to date. To put this data in perspective, we first describe how the system fits within the larger context of a chip's overall design cycle and then provide the numbers that form the basis of the discussion in the retrospective section.

4.1. Chip Design Cycle

Figure 3 shows a highly idealized view of the design cycle of a VLSI chip. The bold boxes indicate stages that use the design capture system. The detailed specification stage involves paper design and/or high-level simulation. Upon its completion, the designer has a specification detailed enough to enter pieces of his design into the system. The design of a real chip rarely follows the clean-cut path shown. More frequently, there is overlap between stages, the design involves more than just a two-level description hierarchy, and the order of stages is hard to determine. Nevertheless, this figure provides us with a useful basis for discussion by identifying significant tasks involving the design capture system.

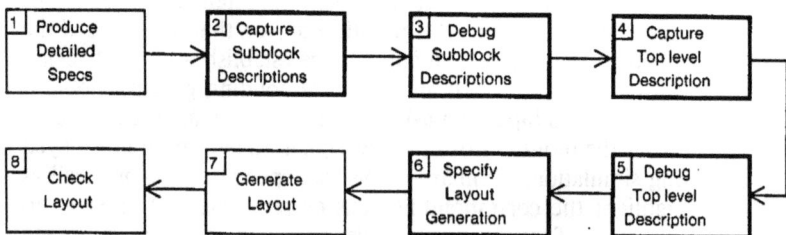

Figure 3. Idealized View of Chip Design Cycle

Often, a major portion of the overall design time is spent in debugging a description once it has been entered into the system (stages 3 and 5 in figure 3), so it is useful to examine debugging more closely as shown in figure 4. This process closely resembles the load-run-think-modify-compile cycle so familiar to programmers. And, as with programs, ideally one would like a system in which most of the time traversing the loop is spent in the thinking needed to find a problem (stage C).

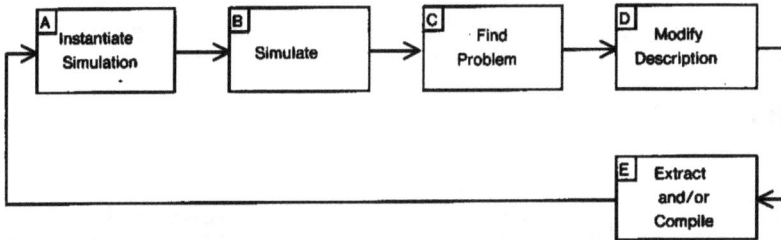

Figure 4. The Debug Loop

4.2. Chip Designs and Statistics

Table 2 shows the layout methods and development times for chips that have been designed using the system (the parenthesized numbers in column headers refer to steps in the design cycle). The figures in the table are subjective measures obtained by interviewing designers and, as such, are approximate. The layout methodology varies all the way from program-generated to standard cell to full custom (FC), while the design times range from a day to over two years. About half the chips have been fabricated and tested, and the remainder are just entering the layout production and verification stage. A figure that helps puts the design capture system in proper perspective is the fraction of time spent in design capture/debugging

Chip	Methodology	Current Status	(1) Detailed Specs	(2, 4) Design Capture	(3, 5) Debugging	(6, 7, 8) Layout
C1	Std Cell	At Layout	2 months	1.5 months	1 month	1 month
C2	Std Cell + FC	Fabricated	6 months	1 month	1 month	1 month
C3	FC	Being Debugged	4 months	3 months	5 months	3 months
C4	Pgm Gen	Fabricated	2 hours	3 hours	.	.
C5	FC	Fabricated	3 days	15 days	15 days	2 months
C6	Std Cell	Being Debugged	2 months	2 months	4 months	.
C7	FC	Fabricated	3 months	4 months	1 month	6 months
C8	FC	Fabricated	9 months	2.5 months	2.5 months	1 year
C9	Std Cell	At Layout	3 months	3 months	6 months	.
C10	Std Cell + FC	At Layout	1 month	1.5 months	3 months	15 days
C11	FC	Being Debugged	9 months	5 months	6 months	7 months
C12	FC	Fabricated	15 days	1 month	1.5 months	2 months

Table 2. Layout Methods and Development Times

over the entire design cycle. For our sample, this figure ranges from ~20% to ~75%, with the average being close to 50%. This indicates that improvements to the design capture system would have a significant impact on the overall design time.

Table 3 shows various size statistics for the same chips. The area and transistor counts indicate that several of the chips are moderate to large by today's standards, and so constitute a serious test of the design system. The schematic and code sizes show that, while most of the descriptions are dominated by schematics, sizable fractions of a few descriptions are in code, and most utilize some code.

Chip	# Transistors	Area (sq. mm)	Schematic Size (bytes)	Code Size (bytes)
C1	30K	70	100K	9K
C2	11K	45	86K	0
C3	340K	150	385K	27K
C4	1.5K	9	0	9K
C5	150K	42	135K	8K
C6	188K	156	115K	0
C7	67K	100	124K	45K
C8	61K	110	304K	150K
C9	36K	120	300K	8K
C10	73K	108	194K	67K
C11	130K	80	367K	92K
C12	13K	20	332K	0

Table 3. Chip Size Statistics

Finally, table 4 provides data that helps give a feel for the time around the debug loop (the parenthesized letters in column headers refer to stages in the debug loop). The first column gives the time to produce a net list from scratch, while the second gives the same time after a "typical" change made during debugging. The next two columns indicate the time to set up a simulation and the number of clock cycles of the chip simulated per minute once the simulation is under way. We should emphasize that these numbers are for complete chips rather than for chip subblocks.

5. Retrospective

This section uses the above data to discuss some of the system's basic design decisions and points out which decisions turned out well, which ones didn't, and where we encountered unexpected difficulty.

5.1. Parameterized Schematics

Our provision of parameterized schematics has worked well. Parameterization allowed our designers to delay decisions such as the

Chip	(E) Extraction	(E) Incr Extraction	(A) Simulation Set-up	(B) Simulation Speed
C1	20 min	5 min	10 min	60 cycles/min
C2	8 min	6 min	3 min	120 cycles/min
C3	25 min	5 min	12 min	4 cycles/min
C4	1 min	1 min	·	·
C5	15 min	5 min	45 min	60 cycles/min
C6	15 min	5 min	10 min	4 cycles/min
C7	15 min	4 min	30 min	5 cycles/min
C8	45 min	30 min	60 min	5 cycles/min
C9	20 min	4 min	6 min	12 cycles/min
C10	20 min	3 min	5 min	24 cycles/min
C11	40 min	30 min	20 min	6 cycles/min
C12	10 min	2 min	·	·

Table 4. Interaction Times

number of lines in a processor cache, the number of bits per line, and even the number of bits in a data path. Designers can therefore debug their designs with the smallest feasible sizes and expect, with high confidence, the full-size version to work. For chip C3, for example, this technique changed the time around the debug loop by an order of magnitude (the time shown is after reduction). Delaying size decisions had the additional benefit of allowing analysis to proceed in parallel with description, potentially speeding up the schedule.

Parameterized schematics also permit us to annotate schematics to describe how layout is to be produced by a layout generator. Typical industry practice consists of describing the layout via a floor plan that is independent of the schematic (the floor plan describes the placement while the schematic specifies the connectivity). We have successfully linked net list and layout generators so that designers' schematics drive layout synthesis.

Parameterization did, however, complicate implementation. Generating net lists incrementally is considerably harder because determining what changed and what didn't is more difficult. Parameterization also made it more difficult to implement commands that use connectivity (for example, the command to highlight a whole net) because this information is only available after a parameter-dependent extraction and cannot be derived from the geometry alone.

5.2. Extensible System

Our decision to build an extensible system has also worked out well. The object oriented model used in the extractor's implementation is an advantage for the CAD programmer who maintains the system since it results in a simple structure. It is also beneficial to the VLSI designer because it keeps his mental model of the extractor simple.

The extensibility of this model has made the addition of, and experimentation with, new graphical modes of expression very easy. It freed us from having to write a large number of graphical primitives before the system became useful. It allowed these primitives to be added on demand and provided freedom to experiment with several alternatives before picking one. Because of this extensibility, we have been able to build several powerful operators such as one and two dimensional sequencers, a data path generator, a finite automaton generator, and a rich set of wire structure manipulators. Thus far a unified sequence operator that is general enough to handle all the applications at hand has eluded us. We encountered surprising difficulty both in defining semantics and in implementation.

5.3. Extractor Implementation

The schematics extractor and the layout extractor use a common geometry engine. This may seem strange because layout extraction is context independent while schematic extraction is strongly context dependent. As it turns out, it is useful to do layout in a context dependent manner because it enables neighbor overlaps to either be considered or not and it allows net list construction to depend on the names of wires rather than just on geometric features. On the other hand, layout extractors have fairly sophisticated rectangle intersection algorithms which may appear to be overkill for schematic extraction. In our experience, sophistication is required since some schematics are quite large. The use of a common geometry engine is also valuable from a system structuring standpoint because it avoids redoing a substantial amount of intricate code.

The connection of graphics to code utilizes an interpreter for the local programming language. This provides great flexibility, since it allows general expressions to be used directly within schematics. Variables in these expressions can be either variables in the context or in the load state of the machine, thereby providing direct graphical access to any program running on the machine. A speed liability came with this flexibility. About half the time for extraction is spent in the interpreter because every time a schematic is reextracted expressions are reparsed and reevaluated.

In our object oriented model, the order in which instances are extracted cannot be controlled. This causes the same information to be specified more often than is strictly necessary. For example, when we select a component wire from a bus, the size of the bus must be specified within each selector because there is no way to guarantee that the bus size is known when the extractor encounters a particular selector.

From table 4 it is clear that the system's speed is not really adequate to debug large designs in an interactive manner. Extraction and simulation set up should be an order of magnitude faster to be compatible with the time spent thinking. Caching already buys a factor of around 4 over extraction from scratch, but as the figures for C2, C8 and C11 show, the gains are sometimes considerably less. The current implementation, which caches results on geometric objects, is inadequate in two respects. First, if a design description is relatively flat (the case with C2), then the system has to redo most of the work. Second, any piece of the design described by code is always recomputed (the case with C8 and C11), even if doesn't need to be. Improving incremental extraction is only part of the problem, however, since simulation setup also accounts for a significant fraction of the time. Moreover, our system currently does not have a way to do the set up incrementally, so the gains to be had are large.

In our first implementation, we used syntax to divide satellites into parameter expressions and result expressions. Parameter expressions computed values for parameters, while result expressions called the extract method for the instance being extracted. This approach was unsatisfactory because it required satellite syntax to be embellished each time we added a new object class. It was also inefficient because the interpreter was invoked unnecessarily in calling the extract method. Our second attempt eliminates result expressions and has hidden properties from which the extract method can be inferred. This solution is not satisfactory, either, because it does not allow the extract method information to be easily seen. At least part of our trouble here is the use of an existing editor.

5.4. Existing Editor

When we started building the system, we had a high-quality layout editor available to us. In one respect, this was a big plus because it left us free to implement higher levels of the design capture system. In other respects, however, the editor had drawbacks, particularly with regard to displaying annotations. For example, we found it useful to provide designers the option of making information visible for emphasis or invisible to avoid clutter. However, our editor required *all* information to be hidden, forcing us to implement satellites, or properties that could be made visible or not and whose position and looks could be controlled by the designer.

The editor does not impose any semantics on geometry and so the same editor can be used for both layout and schematics. This is, at the same time, a hindrance because we have to reintersect rectangles each time an object is extracted. The slowdown caused by reintersection is ameliorated somewhat by result caching but, as we noted earlier, caching does not work as well as we would like. The editor's lack of knowledge about schematic syntax makes syntactic errors difficult to detect interactively. The extractor finds some of these errors and a separate checker weeds out the rest.

6. Future Work

6.1. Operator Extension

The design of operators that succinctly express the structure of VLSI is a difficult task, especially within our simple bottom up construction model. We do not yet have a common set of operators for all our designs because extending the operator set is a tedious trial and error process that is best done in the context of many real designs.

The cell sequence operator has received a great deal of attention but we have yet to find an expression of sequence that covers the frequent tilings of a two dimensional plane with a single cell type and also combines power with conceptual simplicity. Decoders are obvious examples of such tilings. The wiring between cells, aggregation of wires into buses, and index dependence are just a few of the issues that must be considered.

6.2. Refinements to the Existing Model

When we designed the computational model, we decided not to consider the frequent updates that occur during debugging. For instance, the model allows arbitrary side effects to occur within procedures called during extraction. We have added manual specification of procedure arguments and a cache mechanism so that a procedure need not be reexecuted every time it is called. Really fixing this defect requires fundamental changes to the system. Most importantly, the underlying programming language needs to be modified to allow only side-effect-free procedures.

The greatest limitation to extraction speed is the interpreter, closely followed by rectangle intersection. The rectangles that make up the picture always intersect the same way, even when the parameters of a cell are changed. Maintaining an intermediate representation to avoid rectangle reintersection and enhancing the interpreter to decouple parsing from evaluation could improve the speed by a factor of two to ten.

Even with the existence of satellites, there is information that guides the extraction process that is always hidden. This is because the concept of satellites appeared after the editor was written and, as such, is not well integrated. A better way to allow users to control the display of useful information would be to add properties that have user-definable display procedures.

6.3. A New Computational Model

Our current computation model is batch oriented. Whenever a new net list is built, the top-level net list generation function is called to do so. There is caching that makes this process partially incremental, but the model is still one of complete reexecution of the generation function. One alternative is to change the contents of the net list incrementally when the net list generator is changed. In our model, every cell can be an arbitrary function of the subcells. Thus, every cell type, starting from the one that is changed all the way to the root cell type, must be rebuilt. Frequently, a change in a low-level cell type does not really require changes all the way to the root. Differentiating the portions of the net list that need to be changed from those that do not requires a fundamentally different computational model.

A much more restricted functional model of evaluation would facilitate the definition of such a model and make it easier to construct an incremental system. The chief benefit of such a system would be its interactive nature.

6.4. Editor Framework

A schematic specific editor, rather than a general purpose editor followed by an extractor, would allow more effective capture and feedback mechanisms. The decision to use an existing editor and translate afterwards was sound because we have sorted out many thorny semantic issues, but we now need to revisit this decision.

We would like to integrate representation specific editors into a single editor framework. Currently, the granularity at which we can intermix information is whole documents. We need to reduce this granularity so that, in a single document, we can freely mix commentary, code, schematics, layout, timing diagrams, and other information.

7. Summary

We have described an extensible design capture system based on the notion of net list generators. Specifying generators with a combination of schematics and programs in an existing language leverages the familiarity designers have with these modes of expression. We have discussed some of the issues that arose during the construction and use of this system. We have also sketched both incremental and radical changes that need to be made to the system to improve it.

Acknowledgments

Many people have contributed to the evolution of graphical design capture systems within the Computer Science Laboratory. The members of CSL provided the Cedar system that enabled rapid development and provided a solid foundation. Many PARC hardware designers supplied feedback and code that enriched the system. Finally, none of this work would have been possible without the research environment that Xerox has so graciously provided.

References

[1] M.R. Barbacci, "Instruction Set Processor Specification (ISPS): The Notation and its Applications," *IEEE Transactions on Computers*, Vol C-30, No 1, pp. 24-39, January 1981.

[2] R. Barth and B. Serlet, "A Structural Representation for VLSI Design," submitted for publication.

[3] R. Barth, L. Monier and B. Serlet, "PatchWork," submitted for publication.

[4] S. German and K. Lieberherr, "Zeus: A Language for Expressing Algorithms in Hardware," *IEEE Computer*, Vol 18, No 2, pp. 55-65, February 1985.

[5] K. Lieberherr, "A Two-Dimensional Hardware Design Language for VLSI," North Holland, *Proc. EUROMICRO Symposium on Microprocessing and Microprogramming*, pp. 131-142, March 1984.

[6] J. Nash and S. Smith, "A Front End Graphic Interface to the First Silicon Compiler," *Proc. EDA*, pp. 120-124, March 1984.

[7] K. Pier, "A Retrospective on the Dorado, a High-Performance Personal Computer," *Proc. 10th Annual International Symposium on Computer Architecture*, pp. 252-269, Dec. 1983.

[8] R. Piloty and D. Borrione, "The Conlan Project: Concepts, Implementations, and Applications," *IEEE Computer*, Vol 18, No 2, pp. 81-92, February 1985.

[9] M. Shahdad, R. Lipsett, E. Marschner, K. Sheehan, and Howard Cohen, "VHSIC Hardware Description Language," *IEEE Computer*, Vol 18, No 2, pp. 94-102, February 1985

[10] M. Shahdad, "An Overview of VHDL Language and Technology," *Proc. DAC* 1986, pp. 320-326, July 1986.

[11] D. Swinehart, P. Zellweger, R. Beach, and R. Hagmann, "A Structural View of the Cedar Programming Environment," *ACM Transactions on Programming Languages and Systems* 8, 4, October 1986.

[12] W. Teitelman, "A Tour Through Cedar," *IEEE Software*, Vol 1, No 2, pp. 44-73, April 1984.

Invited Talk

Design Representation and Design Management for Electronic Systems

A. Richard Newton

Dept of Electrical Engineering and Computer Sciences
University of California
Berkeley, CA

During the design of a complex electronic system, information from a wide variety of sources, and in a variety of forms, must be used. The design of a practical data management system for such information is very much an engineering, rather than scientific, task and involves a number of complex tradeoffs. In addition, if the goal of a fully-automated design system from behavioral description to implementation can only be achieved if the behavioral description is precise and unambiguous.

In this presentation, the state-of-the-art for design information management will be presented, along with a number of important issues and topics for further research. The presentation will touch on integrated database approaches, the role of interchange formats and design languages such as EDIF and VHDL, as well as the need for a well-defined semantic base for behavioral information. In particular, a number of approaches to the representation of time will be compared and a new approach will be presented.

Author Index

The MIT Press, with Peter Denning, general consulting editor, and Brian Randell, European consulting editor, publishes computer science books in the following series:

ACM Doctoral Dissertation Award and Distinguished Dissertation Series

Artificial Intelligence, Patrick Winston and Michael Brady, editors

Charles Babbage Institute Reprint Series for the History of Computing, Martin Campbell-Kelly, editor

Computer Systems, Herb Schwetman, editor

Exploring with Logo, E. Paul Goldenberg, editor

Foundations of Computing, Michael Garey and Albert Meyer, editors

History of Computing, I. Bernard Cohen and William Aspray, editors

Information Systems, Michael Lesk, editor

Logic Programming, Ehud Shapiro, editor; Fernando Pereira, Koichi Furukawa, and D. H. D. Warren, associate editors

The MIT Electrical Engineering and Computer Science Series

Scientific Computation, Dennis Gannon, editor

www.ingramcontent.com/pod-product-compliance
Lightning Source LLC
Chambersburg PA
CBHW060756220326
41598CB00022B/2448